Surface Science: Polymer Thin Films, Coatings and Adhesives

Surface Science: Polymer Thin Films, Coatings and Adhesives

Guest Editor

Maurice Brogly

MDPI

Basel • Beijing • Wuhan • Barcelona • Belgrade • Novi Sad • Cluj • Manchester

Guest Editor
Maurice Brogly
Laboratoire de Photochimie
et d'Ingénierie Macromoléculaires
Université de Haute Alsace
Mulhouse
France

Editorial Office
MDPI AG
Grosspeteranlage 5
4052 Basel, Switzerland

This is a reprint of the Special Issue, published open access by the journal *Surfaces* (ISSN 2571-9637), freely accessible at: https://www.mdpi.com/journal/surfaces/special_issues/68Z1Z57950.

For citation purposes, cite each article independently as indicated on the article page online and as indicated below:

Lastname, A.A.; Lastname, B.B. Article Title. *Journal Name* **Year**, *Volume Number*, Page Range.

ISBN 978-3-7258-6017-3 (Hbk)
ISBN 978-3-7258-6018-0 (PDF)
https://doi.org/10.3390/books978-3-7258-6018-0

Contents

About the Editor

Maurice Brogly

Maurice Brogly is a Full Professor at the Université de Haute Alsace (UHA) (Mulhouse, France) and a member of the Laboratory of Photochemistry and Macromolecular Engineering (LPIM). He teaches material chemistry (bachelor's degree) and polymer science (master's degree) at the Faculty of Science and Technology (FST). His research is focused on surface and interface phenomena in polymers, coatings and composites and on their impact on adhesion and durability. He received the Rolf Schubert Award from the Adhesion Society, is a member of the editorial board of the *Journal of Adhesion*, and is the author or co-author of 81 articles or book chapters, 64 invited conferences and 241 communications (H-factor 21). He has developed international academic collaborations and numerous research partnerships with world-leading industries in the field of polymers, thin films, composites, coatings and adhesion.

Preface

The aim of this Reprint of the Special Issue of the journal *Surfaces*, entitled "Surface Science: Polymer Thin Films, Coatings and Adhesives", is to highlight original research related to the modification, control, tuning, and characterisation of polymer surfaces and their properties at interfaces, by gathering cutting edge original research articles discussing how mastering surface and interface properties and phenomena, allowing the development of new polymer-based thin films, coatings and adhesives. This Reprint seeks to capture the latest breakthroughs in synthesis, characterisation and applications of advanced polymer thin films and coatings, particularly targeting surface treatments, wettability challenges, and the development and performance of new coatings. Indeed, dedicated articles address topics related to new chemical (nanoimprinting coupled to controlled graft polymerization) and physical (plasma activation) surface treatments to prepare surface-functionalized polymer-based films and fibres or to enhance corrosion protection. Wettability challenges are also highlighted in this Reprint, demonstrating how Marangoni flow drives the self-assembly of hydrogels. The core of the Reprint concerns the development of advanced coatings with original articles on new polyurethane preventive coatings, functional polyacrylate textile coatings and waterborne acrylic coatings. Finally, the chosen articles discuss the enhancement of ultimate properties (fracture toughness, wear) and performances of polymer films and coatings. This Reprint also explores how recent advances in polymer surface science address key challenges to enhance the efficiency, durability, and environmental sustainability of new functional thin films and coatings.

Maurice Brogly

Guest Editor

surfaces

MDPI

Editorial

On the Crucial Role of Surfaces and Interfaces in Polymer Thin Films, Coatings and Adhesives

Maurice Brogly

Laboratoire de Photochimie et d'Ingénierie Macromoléculaires, Université de Haute Alsace, 3b Rue Alfred, Werner, 68093 Mulhouse, France; maurice.brogly@uha.fr

Polymeric materials are increasingly used as thin films or coatings with end-use dimensions approaching those of individual polymer molecules. As they become smaller, the surface-to-volume ratio increases, emphasizing the crucial role of polymer surfaces and interfaces [1]. Research in this field investigates phenomena such as wettability, surface energy, surface structuration, chain dynamics in confined conditions, adhesion, and friction, which are important for applications in coatings; paints; adhesives; polymer hybrids and composites; multilayer polymer films for electronics; and biomaterials [2].

To successfully incorporate polymer thin films and coatings into functional devices, several prerequisites must be met, including dedicated surface chemical functions, controlled surface energy, micro- and/or nano-scale topography, porosity, adapted stiffness or flexibility, and improved durability and sustainability [3]. Tuning these properties necessitates knowing and controlling the polymer chains characteristics at interfaces [4]. However, the complexity of interfaces, as well as the difficulties of studying hidden structures, make it difficult to unravel the relationships between the interface properties, enhanced performance and final applications of polymer thin films and coatings. Therefore, fundamental knowledge of surface science associated with multi-scale advanced microscopic and spectroscopic surface characterization techniques [5] is necessary to investigate these pathways.

Thus, even though broad generalization is challenging, important concepts have to be addressed regarding the crucial role of surfaces and interfaces in polymer thin films, coatings and adhesives. These are based on the control of wetting phenomena at surfaces and/or interfaces by amending the surface chemistry and/or roughness and porosity—considering that interfacial forces act per the "actual" contact area and require close molecular contact—and on the gradient of properties at surfaces and interfaces such as polymer chain-selective adsorption, diffusion across interfaces and/or formation of an interphase/weak boundary layer that will control interfacial strength and energy dissipation mechanisms during the separation process. Considering that high-dissipation mechanisms depend on bonding chemistry, mechanical properties, temperature and speed [6], these factors must be targeted, as they mitigate crack propagation and increase resistance to failure and, thus, final properties, durability and sustainability.

Understanding these concepts substantially aids in determining the relationship between surfaces and interface characteristics, and in improving the properties of polymer thin films, coatings or adhesives. This is because these concepts are interrelated and constitute practical and efficient guidelines to promote interdisciplinary, multi-technique and multi-scale approaches in the field of polymer surface science.

Considering these guidelines, the Reprint of the Special Issue of *Surfaces* "Surface science: polymer thin films, coatings and adhesive" presents cutting-edge original research articles discussing surface and interface properties and phenomena, contributing to the development of new polymer-based thin films, coatings and adhesives. The presented articles address new chemical (nanoimprinting coupled with controlled graft polymerization) and physical (plasma activation) surface treatments to prepare surface-functionalized polymer-based films and fibers or to enhance corrosion protection. Wettability challenges are also highlighted in this Reprint, demonstrating how Marangoni flow drives self-assembly of hydrogels. The core focus of the Reprint is on the development of advanced coatings, with original articles discussing new polyurethane preventive coatings, functional polyacrylate textile coatings and waterborne acrylic coatings. Finally, some of the published articles discuss enhancement of the properties (fracture toughness, wear), performance and sustainability of polymer films and coatings.

Funding: This research received no external funding.

Conflicts of Interest: The author declares no conflicts of interest.

References

1. Stamm, M. *Polymer Surfaces and Interfaces: Characterization, Modifications and Applications;* Springer: Berlin/Heidelberg, Germany, 2008.
2. da Silva, L.F.M.; Öchsner, A.; Adams, R.D. *Handbook of Adhesion Technology*, 2nd ed.; Springer: Cham, Switzerland, 2018.
3. Kausar, A. Polymer coating technology for high performance applications: Fundamentals and advances. *J. Macromol. Sci. Part A* **2018**, *55*, 440–448. [CrossRef]
4. Fabbri, P.; Messori, M. Surface Modification of Polymers: Chemical, Physical, and Biological Routes. In *Modification of Polymer Properties;* Jasso-Gastinel, C.F., Kenny, J.M., Eds.; William Andrew Publishing: Norwich, NY, USA, 2017; pp. 109–130.
5. Brogly, M.; Bistac, S.; Bindel, D. Advanced surface FTIR spectroscopy analysis of poly(ethylene)-block-poly(ethylene oxide) thin film adsorbed on gold substrate. *Appl. Surf. Sci.* **2022**, *603*, 154428–154437. [CrossRef]
6. Awada, H.; Noel, O.; Hamieh, T.; Kazzi, Y.; Brogly, M. Contributions of chemical and mechanical surface properties and temperature effect on the adhesion at the nanoscale. *Thin Solid Film.* **2011**, *519*, 3690–3694. [CrossRef]

surfaces

MDPI

Article

Preparing Surface-Functionalized Polymer Films with Hierarchically Ordered Structure by a Combination of Nanoimprinting and Controlled Graft Polymerization

Masahiko Minoda *, Daichi Shimizu, Tatsuya Nohara and Jin Motoyanagi *

Faculty of Molecular Chemistry and Engineering, Graduate School of Science and Technology, Kyoto Institute of Technology, Matsugasaki, Sakyo-ku, Kyoto 606-8585, Japan; daichishimizu78@gmail.com (D.S.); tatsuyanohara78@gmail.com (T.N.)

* Correspondence: minoda@kit.ac.jp (M.M.); jinmoto@kit.ac.jp (J.M.); Tel.: +81-75-724-7513 (M.M.); +81-75-724-7537 (J.M.)

Abstract

It is widely recognized that fine surface structures found in nature contribute to surface functionality, and studies on the design of functional materials based on biomimetics have been actively conducted. In this study, polymer thin films with hierarchically ordered surface structure were prepared by combining both nanoimprinting using anodically oxidized porous alumina (AAO) as a template and surface-initiated atom transfer radical polymerization (SI-ATRP). To prepare such polymer films, we designed a new copolymer (poly{[2-(4-methyl-2-oxo-2H-chromen-7-yloxy)ethyl methacrylate]-*co*-[2-(2-bromo-2-methylpropionyloxy)ethyl methacrylate]}; poly(MCMA-*co*-HEMABr)) with coumarin moieties and α-haloester moieties in the pendants. The MCMA repeating units function to fix the pillar structure by photodimerization, and the HEMABr ones act as the polymerization initiation sites for SI-ATRP on the pillar surfaces. Surface structures consisting of vertically oriented multiple pillars were fabricated on the spin-coated poly(MCMA-*co*-HEMABr) thin films by nanoimprinting using an AAO template. Then, the coumarin moieties inside each pillar were crosslinked by UV light irradiation to fix the pillar structure. SEM observation confirmed that the internally crosslinked pillar structures were maintained even when immersed in organic solvents such as 1,2-dichloroethane and anisole, which are employed as solvents under SI-ATRP conditions. Finally, poly(2,2,2-trifluoroethyl methacrylate) and poly(*N*-isopropylacrylamide) chains were grafted onto the thin film by SI-ATRP, respectively, to prepare the hierarchically ordered surface structure. Furthermore, in this study, the surface properties as well as the thermoresponsive hydrophilic/hydrophobic switching of the obtained polymer films were investigated. The surface morphology and chemistry of the films with and without pillar structures were compared, especially the interfacial properties expressed as wettability. Grafting poly(TFEMA) increased the static contact angle for both flat and pillar films, and the con-tact angle of the pillar film surface increased from 104° for the flat film sample to 112°, suggesting the contribution of the pillar structure. Meanwhile, the pillar film surface grafted with poly(NIPAM) brought about a significant change in wettability when changing the temperature between 22 °C and 38 °C.

Keywords: nanoimprinting; graft polymerization; hierarchical surface structure; surface-initiated ATRP; anodically oxidized porous alumina; wettability

1. Introduction

The concept of biomimetics has great significance in creating excellent functional materials. In particular, various surface functional materials have been designed, inspired by the unique shapes, geometric patterns, and hierarchical structures of living organisms and plants [1–4]. Some of them have realized unique surface functions based on the synergistic effect between surface morphology and chemical properties. For example, lotus leaves with fractal surface structures containing hydrophobic waxes exhibit superhydrophobic surfaces, which is the so-called lotus effect [5–8]. The shell of a snail is made of a composite of minerals, mainly calcium carbonate and proteins, and has a superhydrophilic surface due to a thin layer of water trapped in the submicron-sized micro-grooves that cover its surface, and it exhibits oil-repellent and antifouling properties [9,10]. Biomimetic materials with surface microstructures are being developed based on nanotechnology. Essentially, nanotechnology manufacturing techniques can be divided into two broad categories: "top–down" and "bottom–up" approaches [11,12]. The "top–down" approach includes microfabrication techniques such as photolithography, dip-pen nanolithography, electron beam patterning, nanoimprint lithography [13–20]. On the other hand, the "bottom–up" approaches involve self-assembly methods that allow for fine patterning at the nanoscale level, providing well-defined nanoscale structures via molecular self-assembly [21,22]. Recently, the combination of nanoimprinting (a "top–down" approach) and SI-ATRP (a "bottom–up" approach) has been developed to prepare hierarchical surface structures [23]; however, there are few reports of surface functional materials produced using this method, and they have a lower surface roughness that can be more appropriately described as a disk-like morphology rather than a surface pillar morphology. The objective of this study was to develop new surface functional materials by combining the features of these two techniques. The fabrication process is outlined as follows (Figure 1): The first step in the fabrication process was to prepare a polymer thin film with vertically aligned pillar structures as the primary structure, which was achieved by using a hot-press nanoimprinting technique. A through-hole anodic alumina oxide (AAO) mold was used as the template for this process. In order to carry out graft modification on the pillar surfaces in organic solvents in the subsequent process, the pillar structure needed to be fixed. This was achieved by crosslinking formation inside each pillar by photo-irradiating the reactive groups previously introduced in the base polymer material. As a result, the pillar thin film was structurally fixed and insoluble in organic solvents. Next, we performed surface-initiated graft polymerization using the controlled radical polymerization (ATRP) initiation site (α-haloester site) on the pillar surface that was incorporated into the base polymer along with the crosslinking sites, to form graft chains on the pillar surface as secondary structures. We hoped that by grafting functional monomers, the surface properties based on the vertically aligned multiple pillar structure (primary structure) and the chemical properties of the grafted polymer chains (secondary structure, corresponding to the sheath of each pillar) on the pillar surfaces would work cooperatively and synergistically to realize tunable and unique surface functional properties. In this study, we attempted to improve the hydrophobicity of the pillar film surfaces by grafting fluorine-containing polymer chains (poly(2,2,2-trifluoroethyl methacrylate), poly(TFEMA)) and also attempted to realize hydrophilic/hydrophobic switching by grafting a temperature-responsive poly(*N*-isopropylacrylamide) (poly(NIPAM)); this enabled us to explore the possibility of imparting hydrophilic/hydrophobic functions by designing the chemical structure of the grafted polymer chains (secondary structure) that covered each pillar.

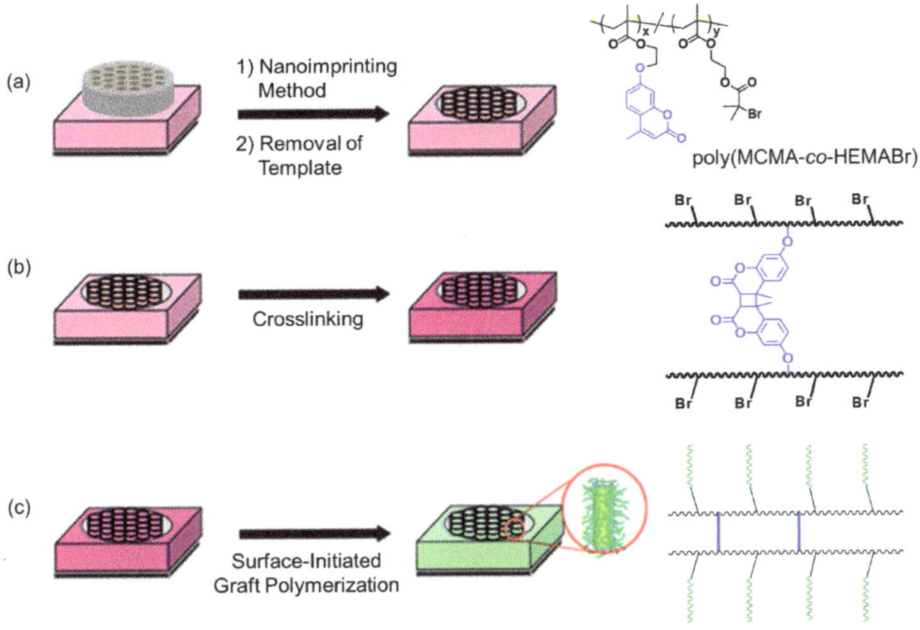

Figure 1. Preparation of a polymer thin film with designed hierarchically ordered surface structure by a combination of nanoimprinting and controlled polymerization. (**a**) Preparation of poly(MCMA-*co*-HEMABr) films with pillar structures on their surfaces. (**b**) Fixation of the pillar structure by photocrosslinking. (c) Surface-initiated graft polymerization from the crosslinked pillar films.

2. Materials and Methods

2.1. Preparation of Poly (MCMA-co-HEMABr) Films with Nano-Pillar Structure Surfaces (CL-Poly (MCMA-co-HEMABr))

The Si wafers underwent a 6 h treatment with heated 35% H_2O_2 aqueous solution, and Si wafers with amino groups on their surfaces were produced by subjecting them to 3 wt% 3-Aminopropyltrimethoxysilane aqueous solution for 30 min. Thin films were prepared over the Si wafers by spin-coating (1200 rpm, 20 s) using 10 wt% poly(MCMA-*co*-HEMABr) toluene solution. Films with nano-pillar structure surfaces were prepared by thermal nanoimprinting using AAO with a pore diameter of 0.2 μm as a template. The nanoimprinting processes were accomplished by using an Eitre-3 Nano Imprint Lithography system. The thermal imprinting conditions were 130 °C for 3 min under 20 bar. After the imprinting process, the temperature was decreased down to 60 °C under applied pressure. The AAO template (Al_2O_3) was dissolved in 1 molL^{-1} NaOH aqueous solution for 10 h. Thereafter, the resulting polymer film was irradiated with a 150 W high-pressure mercury lamp emitting at 365 nm for 5 h at room temperature (distance to sample = 5 cm).

2.2. SI-ATRP of TFEMA from Nano-Pillar Films of CL-Poly(MCMA-co-HEMABr) [24]

A Si wafer coated with CL-poly(MCMA-*co*-HEMABr) was added to a Schlenk flask containing CuCl (11 mg, 0.11 mmol) and $CuCl_2$ (1.0 mg, 7.4 μmol). TFEMA (1.84 g, 10.9 mmol), TPMA (32 mg, 0.11 mmol), EBI (21 mg, 0.11 mmol), and anisole (1.77 g) were added via a syringe. The flask was purged via three cycles of freezing, pumping, thawing, and backfilling with N_2. The reaction was sealed and heated at 85 °C for 5 h and then cooled. The reaction mixture was analyzed by SEC and ^{1}H NMR spectroscopy. The wafer was removed from the flask and rinsed liberally with anisole and MeOH.

2.3. SI-ATRP of NIPAM from Nano-Pillar Films of Poly(MCMA-co-HEMABr) [25]

A Si wafer coated with CL-poly(MCMA-*co*-HEMABr) was added to a Schlenk flask containing CuCl (6 mg, 0.06 mmol). NIPAM (0.67 g, 6.2 mmol), Me_6TREN (28 mg, 0.12 mmol), EBI (12 mg, 0.06 mmol), ascorbic acid (11 mg, 0.06 mmol), and 2-propanol/water (4/1, v/v) (0.67 g) were added via a syringe. The flask was purged via three cycles of freezing, pumping, thawing, and backfilling with N_2. The reaction was sealed and heated at 25 °C for 5 h and then cooled. The reaction mixture was analyzed by SEC and ^{1}H NMR spectroscopy. The wafer was removed from the flask and rinsed liberally with 2-propanol and water.

2.4. Contact Angle Measurements

Water contact angles on various films were measured using a static contact angle meter, CA-S150 (Kyowa Interface Science Co., Ltd., Niiza, Japan), by placing a drop of distilled water (10 μL) on a substrate equipped with a water circulation system at 20 to 45 °C. Samples for contact angle measurements (2 cm × 2 cm) were dried in a vacuum oven at 60 °C for 12 h before measurement. Water contact angles were measured in the temperature range of 20 to 45 °C using a plate temperature controller. Five measurements were taken at different positions on each substrate, and the average value was calculated.

3. Results and Discussion

3.1. Synthesis of Poly(MCMA-co-HEMABr)

For the preparation of polymer thin films with hierarchically ordered surface structure, we designed a new copolymer (poly{[2-(4-methyl-2-oxo-2H-chromen-7-yloxy)ethyl methacrylate]-*co*-[2-(2-bromo-2-methylpropionyloxy)ethyl methacrylate]}; poly(MCMA-*co*-HEMABr)) with coumarin moieties and α-haloester moieties. The synthesis of methacrylate monomers with coumarin moieties (MCMA) was carried out by the condensation of methacryloyl chloride with a coumarin derivative according to a previous report [26]. And the synthesis of methacrylate monomers with α-haloester moieties (HEMABr) was carried out by the condensation of 2-hydroxyethyl methacrylate with 2-bromoisobutyryl bromide according to a previous report [27]. MCMA and HEMABr were copolymerized using AIBN in 1,2-dichloroethane at 60 °C for 4 h ($[MCMA]_0/[HEMABr]_0/[AIBN]_0$ = 45/55/1, $[MCMA]_0 + [HEMABr]_0$ = 25 wt%). The obtained 1,2-dichloroethane solution of the reaction mixture was poured into a large amount of MeOH to remove the unreacted monomers and precipitate the polymer. The obtained polymer was characterized by ^{1}H NMR and size exclusion chromatography (SEC). Figure 2 shows the ^{1}H NMR spectrum of the isolated copolymer (M_n = 150,000; M_w/M_n = 2.89). ^{1}H NMR analysis of poly(MCMA-*co*-HEMABr) showed a series of characteristic resonances, including those of the MCMA moieties (peaks b at δ = 6.5–7.0 ppm and δ = 7.4–7.7 ppm, peak c at δ = 6.1 ppm, and peak d at δ = 2.4 ppm), the HEMABr moieties (peak e at δ = 1.9 ppm), and the MMA main chain protons (peaks f at δ = 0.8–1.4 ppm). The integration ratio of the characteristic signals (peaks b, c, d, and e in Figure 2) supported the formation of a copolymer of MCMA and HEMABr with a composition ratio of 40:60. The synthesized poly(MCMA-*co*-HEMABr) had both crosslinkable sites (coumarin moieties) and polymerization initiation sites (α-haloester moieties) and was used to produce a thin film with vertically aligned pillars on its surface.

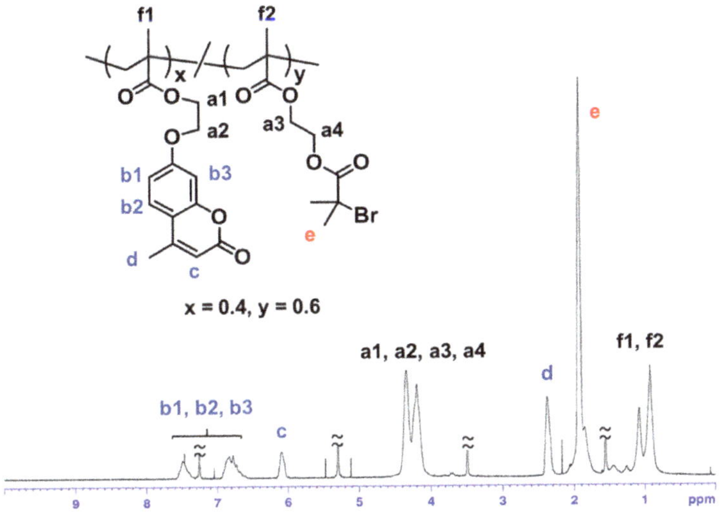

Figure 2. ^{1}H NMR spectrum of poly(MCMA-*co*-HEMABr) in $CDCl_3$.

3.2. Preparation of Poly(MCMA-co-HEMABr) Films With Pillar Structures on Their Surfaces

Since temperature setting is important in hot-press nanoimprinting, the glass transition temperature of the obtained poly(MCMA-*co*-HEMABr) was examined by differential scanning calorimetry (DSC) before nanoimprinting and found to be 75 °C (heating rate; 10 °C/min). (Figure S1). Therefore, the set temperature for hot-press nanoimprinting was 130 °C, which is sufficiently higher than the glass transition temperature of the same bath copolymer. A 1,2-dichloroethane solution of the copolymer was then spin-cast (1200 rpm, 20 s) onto a Si wafer pretreated with 3-aminopropyltrimethoxysilane to produce a thin film. This pretreatment was performed to improve the interaction between the copolymer and the Si wafer, thereby preventing the copolymer thin film from peeling off under SI-ATRP conditions. The spin-coated film showed uniform coloring due to structural coloring, suggesting a constant film thickness. Furthermore, the thickness of the resulting film was measured at five points using contact probe analysis, which also indicated a uniform film thickness at approximately 1400 nm. The obtained polymer film was nanoimprinted using an AAO as a template at 130 °C for 3 min under 20 bar. The formation of pillar structures is determined by the combination of temperature and pressure. In this experiment, the pressure was fixed at 20 bar based on preliminary considerations, and the formation behavior of pillar structures was examined by changing the temperature. When the temperature was increased to 140 °C, which is above 130 °C, it was observed that the length of the formed pillars increased, but some defects existed in the pillar structure. On the other hand, when the temperature was reduced to 120 °C and 110 °C, the length of the formed pillars decreased. Furthermore, at 100 °C, the template was not transferred to the polymer film. The AAO template was removed by etching with 1 $molL^{-1}$ NaOH aqueous solution. After washing with water and drying, the surface morphology of the polymer film was observed by SEM (Figure 3a). Vertically aligned pillar structures with a diameter of 260 ± 30 nm and a length of 1000 ± 200 nm were observed, confirming that the fine pore structure of the AAO used as a template had been transferred to the polymer film surface. As shown in Figure 4a, no Al peaks were observed in the XPS analysis of the pillar film surface after the AAO was removed by dissolving, confirming the complete removal of the AAO. Furthermore, no traces of undissolved AAO were observed in the SEM observation of the resulting pillar film.

Figure 3. SEM images of the surface of the (**a**) poly(MCMA-*co*-HEMABr) pillar film and (**b**) CL-poly(MCMA-*co*-HEMABr) pillar film.

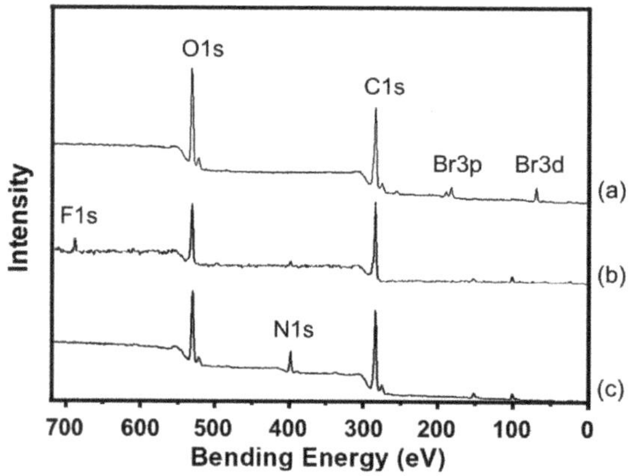

Figure 4. XPS spectra of the surfaces of the (a) CL-poly(MCMA-*co*-HEMA-Br) pillar film, (b) CL-poly((MCMA-*co*-HEMABr)-*g*-poly(TFEMA)) pillar film and (c) CL-poly((MCMA-*co*-HEMABr)-*g*-poly(NIPAM)) pillar film.

Next, the pillar structure was fixed by photocrosslinking inside each pillar. As shown in Figure 5, UV absorption spectroscopy confirmed that the photochemical dimerization of the pendant coumarin moieties in the copolymer film proceeded in the solid state. Upon irradiation with UV light (150 W high-pressure mercury lamp; λ = 365 nm), the intensity of the absorption band originating from the coumarin groups at 280–350 nm gradually decreased [24,28]. The reaction rate of the coumarin moiety was calculated from the decrease in the peak intensity (absorbance at 320 nm) of the UV-vis spectra in Figure 4. Consequently, the reaction rate of the coumarin moiety was estimated to be approximately 80%. The photodimerization of the coumarin moieties occurred both intramolecularly and intermolecularly, the latter contributing to the fixation of the pillar structure. When the surface morphology of a film sample (crosslinked poly(MCMA-*co*-HEMABr); CL-poly(MCMA-*co*-HEMABr)) obtained by the photodimerization of the coumarin pendants was observed by SEM, it was confirmed that the aligned individual pillar structures were maintained and their size was almost unchanged (Figure 3b) compared to those in the SEM image before photoirradiation (Figure 3a). When the light irradiation time was 1 to 3 h, the polymer pillar structure could not be maintained and partially dissolved. Furthermore, the crosslinked pillar film was insoluble even when immersed in 1,2-dichloroethane, a good solvent for the starting copolymer, suggesting that the fixation of the pillar structure was achieved by the formation of crosslinks inside the pillars.

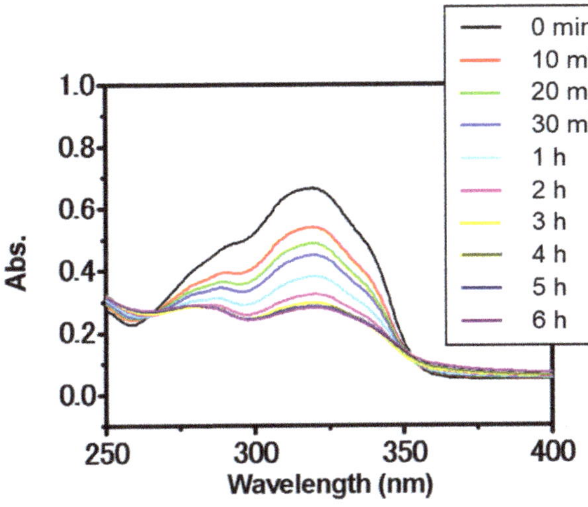

Figure 5. UV spectral change in poly(MCMA-*co*-HEMA-Br) film upon irradiation with UV light (λ = 365 nm).

3.3. SI-ATRP from the Pillar Films of CL-Poly(MCMA-co-HEMABr)

For smooth, flat surfaces, the hydrophilicity or hydrophobicity is derived from the chemical properties of the material. On the other hand, for uneven surfaces, the hydrophilicity or hydrophobicity of the surface is determined not only by the size and hierarchical structure of the fine uneven morphology but also by the contribution of the chemical properties of the material covering the surface. The polymer film used in this study had a vertically aligned pillar structure fabricated by nanoimprinting as primary structure, which resulted in an uneven surface structure. Furthermore, surface-initiated graft polymerization was performed on the surface of this pillar film to form polymer graft chains as secondary structures. In other words, the obtained polymer film had a unique hierarchical structure in which, in addition to the pillar structure with submicron-sized unevenness, a nanoscale surface structure based on the grafted polymer chains was formed. Therefore, we envisaged that by selecting the type of polymer chain to be grafted, it would be possible to widely tune the properties of the pillar film surface through the synergistic effect of the surface morphology of the pillar structure and the chemical properties of the grafted polymer chain. Specifically, we expected that the hydrophobicity based on the uneven structure would be further improved by grafting a fluorine-containing monomer, and furthermore, we expected that the hydrophobic/hydrophilic switching ability of the surface due to an external stimulus (temperature change) would be amplified by grafting polymer chains whose hydrophilic/hydrophilic properties could be changed depending on the temperature.

To perform functionalization with grafted polymer chains, the surface-initiated ATRP of 2,2,2-trifluoroethyl methacrylate (TFEMA) and *N*-isopropylacrylamide (NIPAM) was carried out using the polymerization initiation sites on the pillar surfaces. For the graft polymerization of TFEMA, the polymer thin films were heated at 85 °C for 5 h in a Schlenk flask with the monomer anisole as a solvent, Cu catalysts, a ligand, and the initiator after degassing by three freeze–pump–thaw cycles ($[TFEMA]_0/[CuCl]_0/[CuCl_2]_0/[TPMA]_0/[EBI]_0$ = 100/1/0.1/1/1, TPMA: tris(2-pyridylmethyl) amine, EBI; ethyl 2-bromoisobutyrate) [26]. To ensure surface-initiated ATRP, EBI was added to the polymerization system as a free initiator. After graft polymerization, the polymer film (CL-poly(MCMA-*co*-HEMABr)-*g*-poly(TFEMA)) was thoroughly rinsed and carefully dried to remove the physisorbed polymers and solvents from the film sample. Since it was practically difficult to directly analyze the molecular characteristics of the polymer grafted onto the pillar film, the polymer derived from the free initiator (free polymer) generated in the solution was analyzed as an indicator of the grafted polymer. When the polymerization behavior in solution is controlled, the molecular weight and molecular weight distribution of surface grafted chains are expected to be very close to those of the polymers derived from the free initiators [29]. The molecular weight (M_n) and molecular weight distribution (M_w/M_n) of the free poly(TFEMA) were 5600 and 1.26, respectively, according to SEC analysis.

Next, we investigated the SI-ATRP of NIPAM from the polymer thin films. Graft polymerization was conducted at 25 °C in a Schlenk flask with the monomer 2-propanol/water (4/1, v/v) as a mixed solvent, a Cu catalyst, a ligand, a reducing agent, and the initiator after degassing by three freeze–pump–thaw cycles ($[NIPAM]_0/[CuCl]_0/[Me_6TREN]_0/$[ascorbic acid]$_0/[EBI]_0$ = 100/1/2/1/1, Me_6TREN: tris [2-(dimethylamino)ethyl]amine) [27]. After graft polymerization, the polymer films (CL-poly(MCMA-*co*-HEMABr)-*g*-poly(NIPAM)) were thoroughly rinsed with 2-propanol to remove the physisorbed polymers from the film sample. The M_n and M_w/M_n of poly(NIPAM) were 7200 and 1.55, respectively, which were determined by SEC based on the free poly(NIPAM). SEM observation of the film surfaces after graft modification confirmed that the pillar structures were maintained even after the graft polymerization for both TFEMA and NIPAM (Figure 6). The growth of graft polymers on the pillar surfaces was confirmed by X-ray photoelectron spectroscopy (XPS) (Figure 4). All the pillar films exhibited C1s (285 eV) and O1s (531 eV) peaks corresponding to the methacrylate polymer chains. The poly(TFEMA)-grafted sample showed an F1s (685 eV) peak corresponding to the TFEMA component, while the poly(NIPAM)-grafted sample showed an N1s (398 eV) peak corresponding to the NIPAM component. To further confirm the progress of graft polymerization, Fourier transform infrared (FTIR) spectroscopy was used to investigate the structural changes in the polymer film before and after the graft modification (Figure 7). The FTIR spectrum of the crosslinked pillar film measured in transmission mode showed a group of peaks originating from the methacrylate polymer, with the main peaks assignable to C=O stretching (1730 cm^{-1}) and C-H stretching (2950 cm^{-1}). In the grafted pillar thin films, new peaks due to the chemical structure of the polymer brush were observed. For example, when TFEMA was polymerized, the absorption due to C=O stretching vibration (1730 cm^{-1}) increased relatively compared to the C=C stretching (1620 cm^{-1}), and when poly(NIPAM) was grafted, a new absorption due to the amide N-H bending vibration (1550 cm^{-1}) was observed. These results confirmed that the grafted polymers were successfully introduced onto the pillar film surfaces by SI-ATRP and the expected secondary structure was constructed.

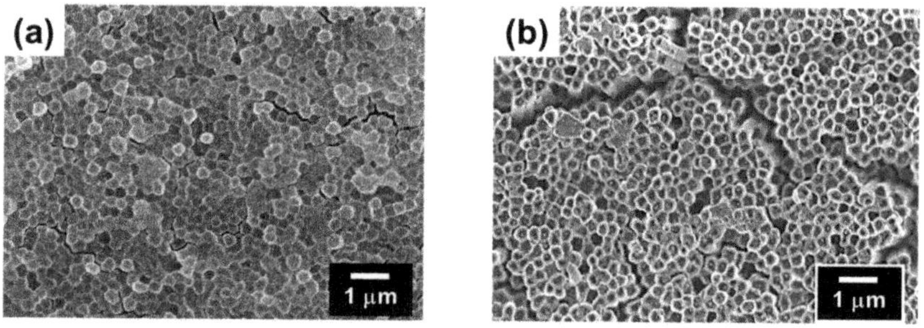

Figure 6. SEM images of the surface of the (**a**) CL-poly((MCMA-*co*-HEMABr)-*g*-poly(TFEMA)) pillar film and (**b**) CL-poly((MCMA-*co*-HEMABr)-*g*-poly(NIPAM)) pillar film.

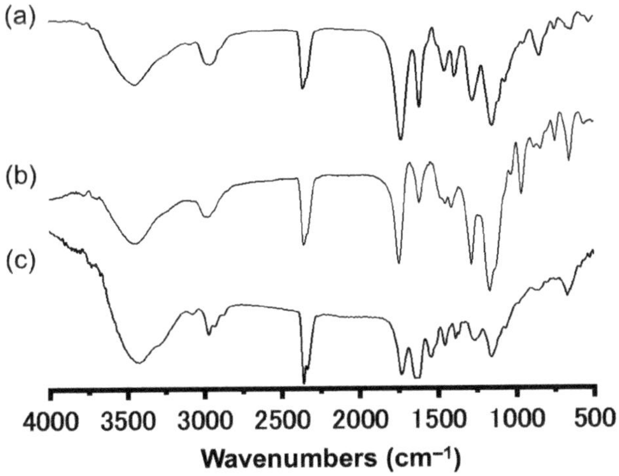

Figure 7. IR spectra obtained for the (a) CL-poly(MCMA-*co*-HEMA-Br) pillar film, (b) CL-poly((MCMA-*co*-HEMABr)-*g*-poly(TFEMA)) pillar film, and (c) CL-poly((MCMA-*co*-HEMABr)-*g*-poly(NIPAM)) pillar film.

3.4. Wattability of Polymer Film Surfaces

In order to ascertain the effect of the pillar structure on the interfacial properties of the film surface (especially the gas-phase interface), a flat film prepared by spin-coating was used as a control sample, and a comparative study was conducted focusing on wettability, which is one of the evaluation indices for interfacial activity. First, to clarify the change in surface properties caused by the formation of the pillar structure, a comparative analysis of the static water contact angles of the flat film and the pillar film was conducted. The contact angles were measured to be 80° for the flat film and 16° for the columnar film (Figure 8a,b), which was probably due to the capillary action induced by the pillared structure of the surface, as shown in Figure 3b, rather than an increase in hydrophilicity due to increased surface roughness based on the pillar structure (Wenzel's theory). Since poly(TFEMA) is hydrophobic [30], the wettability of the surface was reduced by grafting a fluoropolymer onto the film surface. The static contact angle of a water droplet on the flat film surface grafted with poly(TFEMA) was 104° (Figure 8c), whereas the value was 112° on the pillar film surface grafted with poly(TFEMA) (Figure 8d), both of which were significantly more hydrophobic than the sample before grafting. It is noteworthy here that the difference in wettability of the pillar film before and after grafting with poly(TFEMA) was significantly large, and the sample grafted with poly(TFEMA) retained its pillared structure (Figure 6b) but no longer showed the capillary effect and became highly water-repellent.

Poly(NIPAM), a representative thermoresponsive polymer, changes its conformation and hydrophilicity/hydrophobicity in response to temperature changes in an aqueous environment. That is, below the phase transition temperature (T_{pt}) of about 32 °C, it forms an extended conformation due to hydration and becomes hydrophilic, and when the temperature rises above T_{pt}, it dehydrates and assumes a collapsed conformation and becomes hydrophobic. Furthermore, this hydrophilic/hydrophobic switching due to the phase transition is reversible with temperature changes [31,32]. When poly(NIPAM) is fixed to a material surface, the change in the properties of the polymer chain determines the surface properties, and the interfacial properties (hydrophilicity/hydrophobicity) of the material surface can be controlled by changing the temperature [33]. A limited number of reports have demonstrated that the introduction of nanoscale roughness onto a surface by grafting can have a significant effect on the wettability of the surface [23,34]. Therefore, it is interesting to investigate the effect of poly(NIPAM)-grafted pillar surfaces on interfacial properties, such as wettability. Figure 9a shows the static contact angle of a water droplet on the surface of the poly(NIPAM)-grafted flat film and pillar film as a function of temperature. For both sample surfaces, an increase in the water contact angle was observed between 31 °C and 35 °C during the heating process, and it was found that this temperature-responsive change in wettability was significantly amplified by the presence of the pillar structure. As shown in Figure 9b, when the temperature was increased from 22 °C to 38 °C, the water contact angle of the poly(NIPAM)-grafted flat film increased by only 13°, whereas the water contact angle of the poly(NIPAM)-grafted pillar film increased significantly to 68°. This improvement in wettability due to the temperature response of poly(NIPAM) chains is consistent with another report [23,34]. Furthermore, when the temperature was repeatedly increased and decreased in the range from 22 °C to 38 °C, the water contact angles of both sample surfaces changed reversibly, with the amplitude of change being significantly larger for the grafted pillar film. In addition, the amplitude of the water contact angle was maintained within the range of the number of trials in this experiment, indicating that the hydrophilic/hydrophobic switching of the sample surface could be reversibly controlled without attenuation (Figure 9c). These results indicate that the functionality of the film surface was amplified by the synergistic effect between the surface morphology of the pillar structures (primary structure) and the chemical properties of the grafted polymer chains (secondary structure) introduced onto the surface.

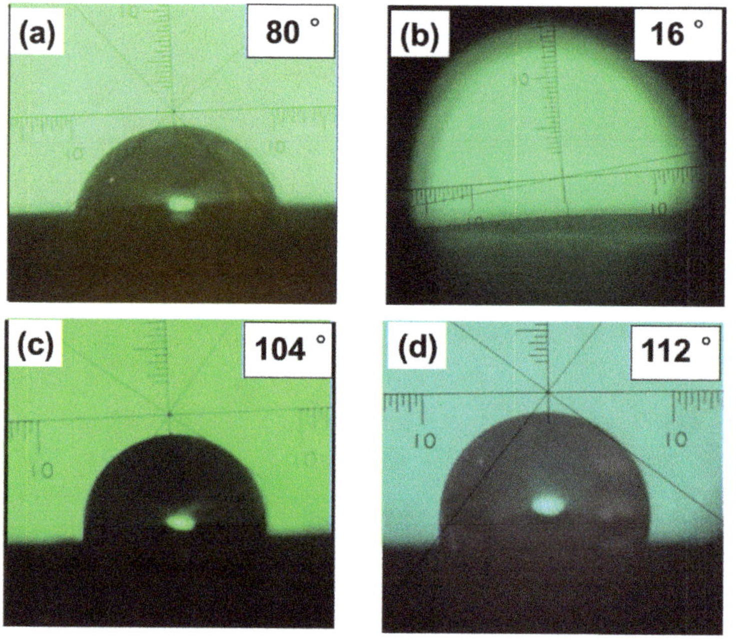

Figure 8. Images of water droplets in static contact angle measurements on surfaces of several specimens: (**a**) CL-poly(MCMA-*co*-HEMABr) flat film; (**b**) CL-poly(MCMA-*co*-HEMABr) pillar film; (**c**) CL-poly(MCMA-*co*-HEMABr) flat film grafted with poly(TFEMA); (**d**) CL-poly(MCMA-*co*-HEMABr) pillar film grafted with poly(TFEMA).

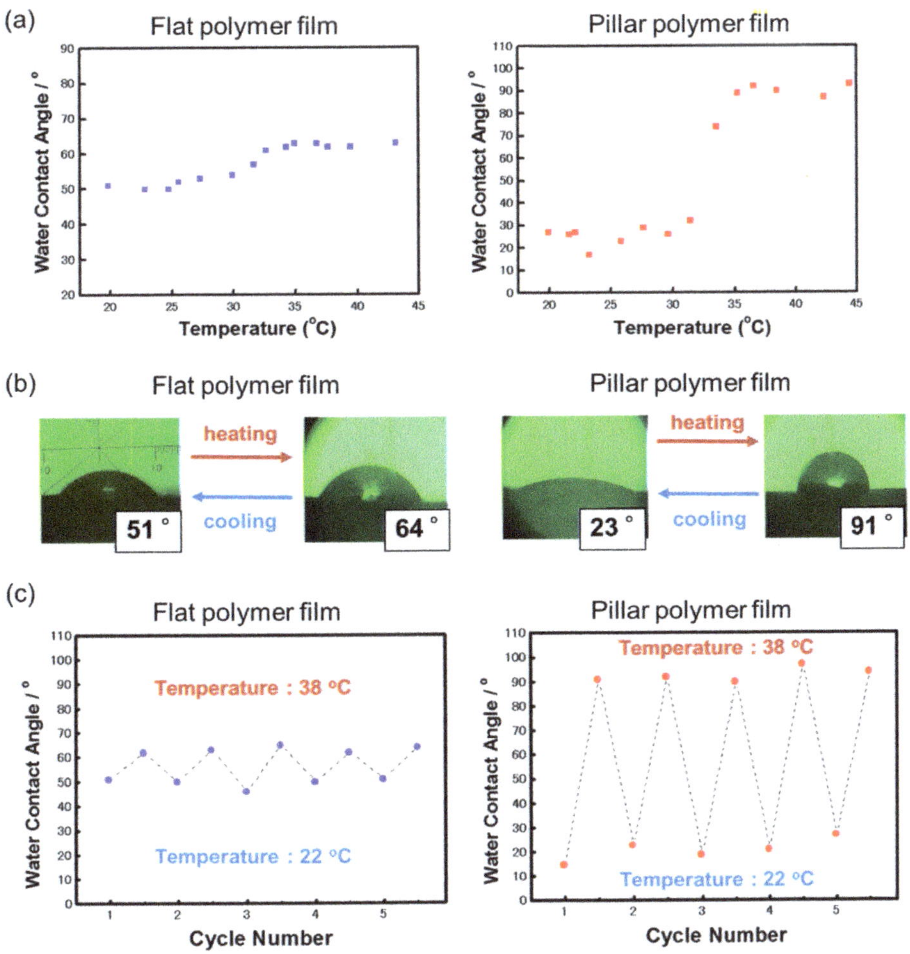

Figure 9. (**a**) Change in water contact angle upon the heating of the poly(NIPAM)-grafted flat and pillar films. (**b**) Images of water droplets in static contact angle measurements on these sample surfaces at 22 °C and 38 °C. (**c**) Changes in water contact angle on the sample surfaces after five heating and cooling cycles in the temperature range of 22 °C and 38 °C.

4. Conclusions

A new method to fabricate polymer thin films with hierarchically ordered surface structures was achieved by combining nanoimprinting using an AAO template with the SI-ATRP of TFEMA and NIPAM. In this study, the surface morphology and chemistry of the films with and without pillar structures were compared, especially the interfacial properties expressed as wettability. Grafting poly(TFEMA) increased the static contact angle for both flat and pillar films, and the contact angle of the pillar film surface increased from 104° for the flat film sample to 112°, suggesting the contribution of the pillar structure. Meanwhile, the pillar film surface grafted with poly(NIPAM) brought about a significant change in wettability by changing the temperature between 22 °C and 38 °C. Furthermore, it was shown that the interfacial properties of the film surface (hydrophilic/hydrophobic switching) could be reversibly and reproducibly controlled by temperature. This experiment also revealed that the presence of pillar structures significantly affected the interfacial properties of the film surfaces. These findings indicate that the functionality of the film surfaces is expressed through a synergistic effect between the surface morphology of the pillar structure (primary structure) and the chemical properties of the polymer chains introduced to the pillar surface by graft modification (secondary structure). The results presented in this study are expected to be applied to new nanomaterials research, such as the creation of hierarchical surface functional materials by combining top–down and bottom–up techniques.

Supplementary Materials: The following supporting information can be downloaded at https://www.mdpi.com/article/10.3390/surfaces8030048/s1. Methods, Chemical and Reagents, poly(MCMA-*co*-HEMABr), and Figure S1: DCS trace of poly(MCMA-*co*-HEMABr).

Author Contributions: Conceptualization, M.M.; methodology, J.M. and M.M.; validation, J.M.; formal analysis, D.S. and T.N.; investigation, J.M.; resources, J.M. and M.M.; data curation, D.S., T.N., and J.M.; writing—original draft preparation, J.M.; writing—review and editing, M.M. visualization, J.M.; supervision, M.M.; project administration, M.M.; funding acquisition, J.M. and M.M. All authors have read and agreed to the published version of the manuscript.

Funding: This research was funded by the JSPS KAKENHI, Grant-in-Aid for Scientific Research (C), grant numbers 20K05605 and 22K05216.

Institutional Review Board Statement: Not applicable.

Informed Consent Statement: Not applicable.

Data Availability Statement: Data are contained within the article.

Conflicts of Interest: The authors declare no conflicts of interest.

References

1. Wang, Y.; Zheng, G.; Jiang, N.; Ying, G.; Li, Y.; Cai, X.; Meng, J.; Mai, L.; Guo, M.; Zhang, Y.S.; et al. Nature-inspired micropatterns. *Nat. Rev. Methods Primers* **2023**, *3*, 68. [CrossRef]
2. Himel, M.H.; Sikder, B.; Ahmed, T.; Muhaimin, S. Biomimicry in nanotechnology: A comprehensive review. *Nanoscale Adv.* **2023**, *5*, 596–614. [CrossRef]
3. Li, J.; Li, M.; Koh, J.J.; Wang, J.; Lyu, Z. 3D-printed biomimetic structures for energy and environmental applications. *DeCarbon* **2024**, *3*, 10026. [CrossRef]
4. Yang, M.; Kotov, N.A. Quantitative biomimetics of high-performance materials. *Nat. Rev. Mater.* **2025**, *10*, 382–395. [CrossRef]
5. Yamamoto, M.; Nishikawa, N.; Mayama, H.; Nonomura, Y.; Yokojima, S.; Nakamura, S.; Uchida, K. Theoretical explanation of the lotus effect: Superhydrophobic property changes by removal of nanostructures from the surface of a lotus leaf. *Langmuir* **2015**, *31*, 7355–7363. [CrossRef]
6. Zhang, M.; Feng, S.; Wang, L.; Zheng, Y. Lotus effect in wetting and self-cleaning. *Biotribology* **2016**, *5*, 31–43. [CrossRef]
7. Nine, M.J.; Tung, T.T.; Alotaibi, F.; Tran, D.N.H.; Losic, D. Facile Adhesion-Tuning of Superhydrophobic Surfaces between "Lotus" and "Petal" Effect and their Influence on Icing and Deicing Properties. *ACS Appl. Mater. Interfaces* **2017**, *9*, 8393–8402. [CrossRef]
8. Wang, L. A critical review on robust self-cleaning properties of lotus leaf. *Soft Matter* **2023**, *19*, 1058–1075. [CrossRef]
9. Liu, X.; Zhou, J.; Xue, Z.; Gao, J.; Meng, J.; Wang, S.; Jiang, L. Clam's Shell Inspired High-Energy Inorganic Coatings with Underwater Low Adhesive Superoleophobicity. *Adv. Mater.* **2012**, *24*, 3401–3405. [CrossRef]
10. Liu, M.; Wang, S.; Jiang, L. Nature-inspired superwettability systems. *Nat. Rev. Mater.* **2017**, *2*, 17036. [CrossRef]
11. Cook, A.B.; Clemons, T.D. Bottom-Up versus Top-Down Strategies for Morphology Control in Polymer-Based Biomedical Materials. *Adv. Nanobiomed. Res.* **2022**, *2*, 2100087. [CrossRef]
12. Hou, J.; Zhao, C.; Zhang, H. Bio-Inspired Subnanofluidics: Advanced Fabrication and Functionalization. *Small Methods* **2023**, *8*, 2300278. [CrossRef]
13. Wang, Y.; Hu, H.; Shao, J.; Ding, Y. Fabrication of Well-Defined Mushroom-Shaped Structures for Biomimetic Dry Adhesive by Conventional Photolithography and Molding. *ACS Appl. Mater. Interfaces* **2014**, *6*, 2213–2218. [CrossRef]

14. Alameda, M.T.; Alameda, T.; Osorio, M.R.; Herna, J.J. Multilevel Hierarchical Topographies by Combined Photolithography and Nanoimprinting Processes To Create Surfaces with Controlled Wetting. *ACS Appl. Nano Mater.* **2019**, *2*, 4727–4733. [CrossRef]
15. Hirtz, M.; Oikonomou, A.; Georgiou, T.; Fuchs, H.; Vijayaraghavan, A. Multiplexed biomimetic lipid membranes on graphene by dip-pen nanolithography. *Nat. Commun.* **2013**, *4*, 2591. [CrossRef]
16. Liu, G.; Petrosko, S.H.; Zheng, Z.; Mirkin, C. Evolution of Dip-Pen Nanolithography (DPN): From Molecular Patterning to Materials Discovery. *Chem. Rev.* **2020**, *120*, 6009–6047. [CrossRef]
17. Kim, T.; Pang, C.; Suh, K.Y. Shape-Tunable Polymer Nanofibrillar Structures by Oblique Electron Beam Irradiation. *Langmuir* **2009**, *25*, 8879–8882. [CrossRef]
18. Qin, N.; Qian, Z.G.; Zhou, C.; Xia, X.X.; Tao, T.H. 3D electron-beam writing at sub-15 nm resolution using spider silk as a resist. *Nat. Commun.* **2021**, *12*, 5133. [CrossRef]
19. Yang, Y.; He, H.; Li, Y.; Qiu, J. Using Nanoimprint Lithography to Create Robust, Buoyant, Superhydrophobic PVB/SiO_2 Coatings on wood Surfaces Inspired by Red roses petal. *Sci. Rep.* **2019**, *9*, 9961. [CrossRef]
20. Nowduri, B.; Schulte, S.; Decker, D.; Schäfer, K.-H.; Saumer, M. Biomimetic Nanostructures Fabricated by Nanoimprint Lithography for Improved Cell-Coupling. *Adv. Funct. Mater.* **2020**, *30*, 2004227. [CrossRef]
21. Tu, R.S.; Tirrell, M. Bottom-up design of biomimetic assemblies. *Adv. Drug Deliv. Rev.* **2004**, *56*, 1537–1563. [CrossRef]
22. Song, J.; Malathong, V.; Bertozzi, C.R. Mineralization of synthetic polymer scaffolds: A bottom-up approach for the development of artificial bone. *J. Am. Chem. Soc.* **2005**, *127*, 3366–3372. [CrossRef]
23. Nagase, K.; Onuma, T.; Yamato, M.; Takeda, N.; Okano, T. Enhanced Wettability Changes by Synergistic Effect of Micro/Nanoimprinted Substrates and Grafted Thermoresponsive Polymer Brushes. *Macromol. Rapid Commun.* **2015**, *36*, 1965–1970. [CrossRef]
24. Flejszar, M.; Ślusarczyk, K.; Hochół, A.; Chmielarz, P.; Wytrwal, M.; Wolski, K.; Spilarewicz, K.; Awsiuk, K.; Raczkowska, J. Sequential SI-ATRP in μL-scale for surface nanoengineering: A new concept for designing polyelectrolyte nanolayers formed by complex architecture polymers. *Eur. Polym. J.* **2023**, *194*, 112142. [CrossRef]
25. Gury, L.; Kamble, S.; Parisi, D.; Zhang, J.; Lee, J.; Abdullah, A.; Matyjaszewski, K.; Bockstaller, M.R.; Vlassopoulos, D.; Fytas, G. Internal Microstructure Dictates Interactions of Polymer-Grafted Nanoparticles in Solution. *Macromolecules* **2021**, *54*, 7234–7243. [CrossRef]
26. Onbulak, S.; Rzayev, J. Cylindrical Nanocapsules from Photo-Cross-Linkable Core-Shell Bottlebrush Copolymers. *Polym. Chem.* **2015**, *6*, 764–771. [CrossRef]
27. Matyjaszewski, K.; Gaynor, S.G.; Kulfan, A.; Podwika, M. Preparation of Hyperbranched Polyacrylates by Atom Transfer Radical Polymerization. 1. Acrylic AB* Monomers in "Living" Radical Polymerizations. *Macromolecules* **1997**, *30*, 5192–5194. [CrossRef]
28. Motoyanagi, J.; Nishimura, I.; Minoda, M. Living Cationic Polymerization of a Coumarin-Substituted Vinyl Ether and Reversible Photoinduced Crosslinking of the Resulting Homopolymers and Amphiphilic Block Copolymers. *J. Polym. Sci. Part A Polym. Chem.* **2011**, *49*, 4701–4707. [CrossRef]
29. Zoppe, J.O.; Ataman, N.C.; Mocny, P.; Wang, J.; Moraes, J.; Klok, H.-A. Surface-Initiated Controlled Radical Polymerization: State-of-the-Art, Opportunities, and Challenges in Surface and Interface Engineering with Polymer Brushes. *Chem. Rev.* **2017**, *117*, 1105–1318. [CrossRef]
30. Xia, Z.; Feng, Z.; Wu, R.; Niu, Z.; He, J.; Bai, C. Tough, hydrophobic, pressure-resistant, and self-cleaning underwater engineering materials based on copolymerization of butadiene and trifluoroethyl methacrylate. *ACS Appl. Polym. Mater.* **2023**, *5*, 8241–8249. [CrossRef]
31. Halperin, A.; Kröger, M.; Winnik, F.M. Poly(N-isopropylacrylamide) Phase Diagrams: Fifty Years of Research. *Angew. Chem. Int. Ed.* **2015**, *54*, 15342–15367. [CrossRef]
32. Osváth, Z.; Iván, B. The Dependence of the Cloud Point, Clearing Point, and Hysteresis of Poly(N-isopropylacrylamide) on Experimental Conditions: The Need for Standardization of Thermoresponsive Transition Determinations. *Macromol. Chem. Phys.* **2017**, *218*, 1600470. [CrossRef]
33. Thiele, S.; Andersson, J.; Dahlin, A.; Hailes, R.L.N. Tuning the Thermoresponsive Behavior of Surface-Attached PNIPAM Networks: Varying the Crosslinker Content in SI-ATRP. *Langmuir* **2021**, *37*, 3391–3398. [CrossRef]
34. Yu, Q.; Li, X.; Zhang, Y.; Yuan, L.; Zhao, T.; Chen, H. The synergistic effects of stimuli-responsive polymers with nano- structured surfaces: Wettability and protein adsorption. *RSC Adv.* **2011**, *1*, 262–269. [CrossRef]

surfaces

MDPI

Article

Wettability's Challenge to High-Voltage Insulators: Polyurethane as Preventive Coating

Touqeer Ahmad Raza [1,2,*], Muhammad Kamran [1,2], Syed Ahtisham Mehmood Shah [2] and Muhammad Mehran Bashir [2,*]

[1] Department of Electrical Engineering, University of Engineering and Technology, Lahore 54890, Pakistan; mkamran@uet.edu.pk

[2] Department of Electrical Engineering, Muhammad Nawaz Sharif University of Engineering and Technology, Multan 60600, Pakistan; ahtishamshah@yahoo.com

* Correspondence: touqeerahmadraza@mnsuet.edu.pk (T.A.R.); dr.mehran@mnsuet.edu.pk (M.M.B.); Tel.: +92-3327613923 (T.A.R.); +92-3027434327 (M.M.B.)

Abstract: The failure of a porcelain insulator on a transmission line is a crucial cause of power supply interruptions, leading to poor reliability and revenue loss. The insulator's performance is adversely affected by environmental contaminants, and wettability intensifies this adverse effect by developing a conductive path along the insulator's surface, leading to premature flashover and insulator failure. This work aims to analyze the response of the electric field distribution and current density using the finite element method (FEM) under different wettability conditions. Discrete water droplets were placed along the surface, and the contact angle was varied to represent different levels of surface hydrophobicity. Abrupt rises and spikes were observed on the plots for the electric field and current density distribution, indicating distortion; however, the distortion kept on decreasing with the increase in the contact angle. Overall, the average stress followed a declining pattern, where the values of the electric field were reduced from 2.588 to 2.412 kV/cm, and current the density was reduced from 0.187 to 0.068 nA/cm^2 for an increase in the contact angle from 60° to 140°. Simulation results advocate for hydrophobic insulator surfaces. Therefore, a proper coating is necessary to enrich hydrophobicity and mitigate the adversity of wettability. Polyurethane, due to its excellent hydrophobic and insulating properties, offers a potential coating. Flashover voltage tests have been performed for the coated insulator under dry and wet conditions, where the flashover voltage improved from 79.14 kV to 82.04 kV and 48.4 kV to 53.8 kV, respectively, which supports the simulations' outcomes.

Keywords: porcelain insulators; hydrophobicity; surface coatings; finite element method; polyurethane; high voltage

1. Introduction

For a reliable power system operation, consumers must receive an uninterrupted supply, and insulator failure is one of the key causes of interruption. Outdoor insulators are subjected to numerous factors, such as rain, dust, fog, and contamination, that affect their performance, making them vulnerable to repeated flashover [1,2]. The atmosphere contains diverse pollutants, including sea and road salts in coastal areas, as well as agricultural and road dust, bird droppings, fertilizers, and various industrial emissions depending on the location. Outdoor overhead transmission line insulators are exposed to these contaminants, with coastal regions being dominated by sodium chloride (NaCl) and areas near cement or coal plants experiencing higher levels of calcium oxide (CaO) salts [3,4].

The literature reveals that soluble contaminants have a more drastic influence on insulator performance [5]. This is due to the fact that the mobility of charged particles is enhanced with the increasing concentration of the soluble salts, which results in the formation of conductive path [6]. The path boosts the magnitude as well as the duration of the leakage current, producing heat dissipation. Then, dry bands are formed in highly stressed regions, which eventually bridge the insulator along the surface under certain applied voltages, and an arc discharge is developed, called a flashover [7,8]. Depending on the degree of wettability, the contamination distribution, and the shape of the insulator, the electric field distribution is differently affected [9].

Static contact angles offer a straightforward way to assess a surface's natural attraction for a liquid when the system is stable. This angle reflects the balance of forces between the liquid's surface tension (cohesive) and the adhesive forces between the liquid and the solid surface [10,11]. Consider a water droplet placed on a smooth surface; the angle the droplet's edge makes with the surface is the static contact angle, which is elaborated in Figure 1a, where γ_{SL}, γ_{SV}, and γ_{LV} are interfacial tensions, and θ is the tangent angle at the interface of the liquid and the surface. Dynamic contact angles, on the other hand, are measured when the three-phase boundary (contact line) is in motion, either during the wetting (advancing) or de-wetting (receding) of the surface. The static contact angle, due to its simplicity and ease of measurement, is widely used to understand a material's basic surface properties, such as desired wettability after treatments (coating or cleaning). A low static contact angle (usually less than 90°) indicates good wetting, meaning the liquid spreads easily on the surface and the surface is hydrophilic. A high static contact angle (usually greater than 90°) indicates poor wetting, and the liquid tends to take the form of a sphere and the surface is hydrophobic. The effect of the contact angle on the shape of the water droplet is explained in Figure 1b, where surface hydrophobicity decreases from top to bottom. It is obvious that the shape of the water droplet at the top surface, with a high contact angle, is close to a sphere, which helps it to roll from the surface, making it dry. Hydrophobic surfaces have also been witnessed in some plants and animal furs [12,13], and lotus leaf is the most prominent one [14,15].

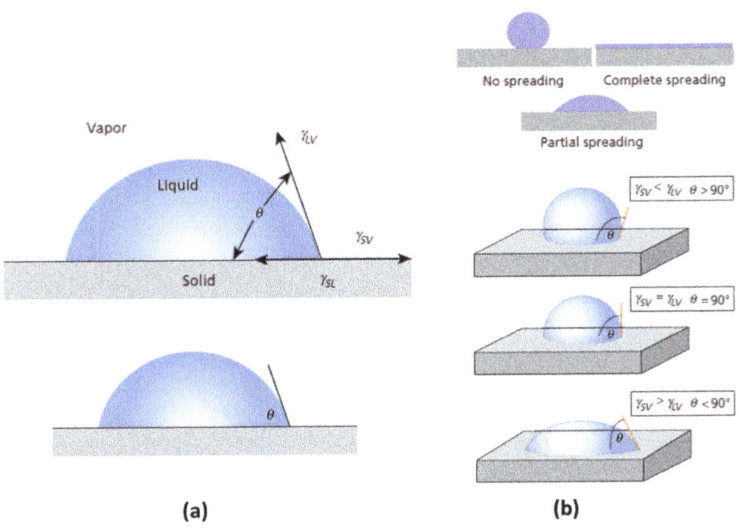

Figure 1. (**a**) Contact angle. (**b**) Impact of contact angle on the shape of the water droplet. Reprinted from [13].

Among insulators, polymers have displayed better hydrophobicity, but ceramic insulators are superior in terms of their insulating capabilities and mechanical strength [16–18]; even composite insulators cannot match ceramics in terms of electromechanical stress [19–23]. Porcelain insulators are usually preferred for their rigidity and excellent cantilever strength, which is important for supporting heavy busbars and apparatus. The surface of a porcelain insulator is glazed to cope with the external effects. However, the surface still has a reasonable degree of adhesivity and hydrophilicity, facilitating contaminants and wettability, and is prone to flashover more often. Thus, a proper coating is essential for a ceramic insulator to enrich the hydrophobicity of the surface and to minimize the influence of pollution, and polymer-based coatings are commonly used to protect surfaces from external effects [24–26].

The aim of this study is to examine the impact of the contact angle of water droplets on the electric field and current density distribution of the insulator surface as wettability distorts these field distributions. As software provides more flexibility in studying the wide range of circumstances and conditions, different wettability scenarios have been modeled in COMSOL Multiphysics software® version 6.1. To intuitively analyze field distribution for complex geometries, the finite element method (FEM) stands out as crucial numerical technique [27,28], and it has been applied here. In addition, the insulator has been tested for flashover voltage using polyurethane as a potential coating material under dry and wet conditions.

2. Materials and Methods

2.1. Coating Material

There are several coating materials that are used for different purposes; however, for insulator surfaces, a few of the essentially hydrophobic examples have been discussed below; these are poly meta-phenylene isophthalamide (PMIA) paper (with nano-SiO_2), silicon nitride (Si_3N_4), silicon carbide (SiC), and polyurethane (PU). Other important properties include relative permittivity, thermal conductivity, density, and insulating strength [29] (listed in Table 1).

Table 1. Characteristics of potential coating material.

Sr.	Material	Hydrophobicity	Thermal Conductivity ($Wm^{-1}K^{-1}$)	Relative Permittivity	Density (g/cm^3)	Breakdown Strength (kV/mm)
1	Polyurethane	73–140° [30]	0.16–0.2 [31]	5–6.5 [32]	1.125 [32]	15.75–19.7 [33]
2	PMIA Paper (with Nano-SiO_2)	~ 153° [34]	0.396 [35]	2.9–4.4 [36]	0.6 [35]	35 [34]
3	Silicon Nitride (Si_3N_4)	~ 143° [37]	86–120 [38]	7.5–8.3 [18]	3.17 [38]	18 [39]
4	Silicon Carbide (SiC)	100–135.3° [40]	~85 [38]	~6.5 [38]	3.21 [38]	8–14 [41]

In areas with high temperatures and the possibility of heat buildup, a material with higher thermal conductivity increases the risk of flashover more than the one with low thermal conductivity. This is due to the fact that low thermal conductivity means low heat transfer capability, which minimizes the heat transfer, preventing the insulator from overheating and potentially causing flashover. The lower density and viscosity of the varnishes and coatings offer better coating atomization, resulting in a finer and more even spraying application. It also improves leveling, producing a more uniform, smoother, and aesthetic surface [42]. Moreover, as high-voltage insulators are applied by means of strings, the high density of the material further adds to the weight of the string and creates a burden for the supporting tower. Therefore, polyurethane and PMIA paper (with nano-SiO_2) are of greater interest than other two materials.

Relative electrical permittivity is another crucial parameter to be discussed. The materials with lower permittivity are helpful for a more uniform distribution of the electric field along the surface; however, it must be kept in mind that when two materials with different permittivities are in contact, a significant portion of the electric field will be concentrated in the material with lower permittivity, potentially leading to overstressing and the failure of that component [43–45]. So, to achieve uniform electric field distribution, polyurethane, with comparable relative permittivity to that of ceramic materials, is better suited to being studied as a coating material.

2.1.1. Polyurethane

Polyurethane belongs to the class of block copolymers and possesses many useful features, such as biocompatibility, bio-stability, non-flammability, and nontoxicity [32]. It is also transparent, hard, and considered to be an anti-aging material with abrasion resistance. Its exceptional features are due to presence of hard–soft segments in its microstructure [46,47]. The general reaction for polyurethane is shown in Figure 2, and some other properties of polyurethane listed in Table 2.

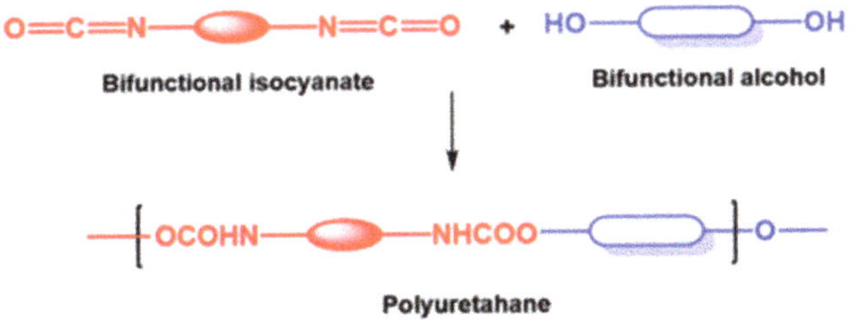

Figure 2. Addition reaction to synthesize polyurethane. Reprinted from [47].

Table 2. Polyurethane properties [28,29].

Sr.#	Properties	Values
1	Specific gravity	1.1–1.46
2	Tensile strength	18 MPa
3	Tensile modulus	0.8–1.1 GPa
4	Compressive strength	90–250 MPa
5	Dielectric strength	400–500 volt/mil (15.75–19.7) kV/mm

Thermoset polyurethane, owing to its rigidness and excellent hydrophobicity, is the preferred for coating on insulator surfaces. It is synthesized by a reaction between oxytetramethylene, which is a soft-segment polyol, and hexamethylene diisocyanate, providing the di-isocyanate functionality necessary for crosslinking. Dibutyltin dilaurate (DBTDL) was used as a catalyst to accelerate the reaction [48–50].

$$n(HO\text{-}O\text{-}(CH_2)_4\text{-}O\text{-}) + n(O{=}C{=}N\text{-}(CH_2)_6\text{-}N{=}C{=}O) \rightarrow \\ (HO\text{-}O\text{-}(CH_2)_4\text{-}O\text{-}U\text{-}O\text{-}(CH_2)_4\text{-}O\text{-})n \quad (1)$$

2.1.2. Application Process

Three methods—namely, the brushing method, spraying method, and dipping method—are commonly used to deposit a layer on the surface of the insulator [51]. The spraying method with air atomization is more effective than the other methods due to its small dispersion [51,52], and it was utilized to deposit the coating in this work. In this method, compressed air is mixed with the polymer and sprayed on the insulator. The process of atomization takes place through increased air speed, which scatters polyurethane into fine droplets, facilitating a fine and uniform coating [52]. The thickness of the coating for high-voltage insulators should neither be very low nor too high as low thickness does not provide adequate hydrophobicity, while high thickness leads to an increase in temperature due to surface leakage current. The thickness of polyurethane was kept around 0.4 mm. The curing condition for the polyurethane coating is 10–25 °C, with relative humidity of 45–50% [53].

2.2. *Finite Element Method (FEM)*

The electric field is computed by gradient of potential:

$$E = -\nabla V \tag{2}$$

Poisson's equation for electrostatics in terms of the potential distribution is given as follows:

$$\varepsilon\nabla(\nabla V) = -\rho \tag{3}$$

The finite element method (FEM) which was proposed by Courant in 1943 [54]. Later, it was applied in electromagnetics and microwave engineering to analyze wave propagation [55,56]. This method is employed in order to solve boundary value problems (BVPs), which are described by a partial differential equation along with a set of boundary conditions. The governing differential equation is converted into a set of linear algebraic equations to obtain the approximate solution of a given boundary value problem [57]. In this method, the region or domain of interest is subdivided into a finite number of subregions or subdomains known as "finite elements", whose shape is defined by nodes. The trial functions, in the form of polynomials, defined in each element are known as shape functions or interpolation functions since they are used to find the primary unknown quantity within the element via its value at the nodes [58]. A system of linear equations is obtained through the conversion of the governing differential equation, using either the variational method or the weighted residual method. The variational method requires a functional that depicts the energy related to the boundary value problem (BVP), which is further minimized by setting its partial derivatives to zero with respect to dependent variables. On the other hand, the weighted residual method does not require any functional and directly starts by finding the residual from the differential equation of the boundary value problem (BVP). The product of the weight function and the residual is integrated over the domain, resulting in weak form representation of the BVP.

Rectangular elements are mostly applied for simple geometries; to cover the complex geometry, triangular elements as shown in Figure 3 [59], are preferred over rectangular elements [60,61]. The numbers 1, 2, and 3 in Figure 3 indicate the nodes of triangular element covering the geometry for which electric field and current density are measured. COMSOL software uses a Galerkin approach to the weighted residual method as it is simple and begins directly with the differential equation of the associated boundary value problem. The obtained system of linear equations can be solved either by using the Gauss elimination method or the LU factorization method to find the potential at all nodes, and potential elsewhere can be interpolated [62–64].

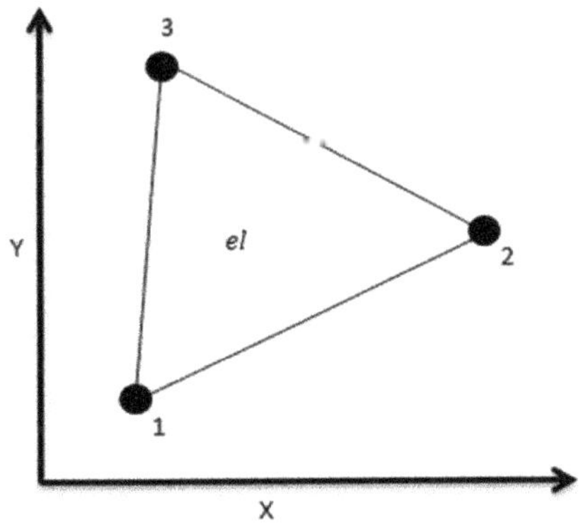

Figure 3. Triangular element [59].

2.3. Flashover Testing

The experimental setup for power frequency flashover testing included a testing transformer of 25 kVA as a regulated AC voltage source up to 150 kV. The applied voltage increased linearly (2% per s) up to the flashover voltage within 1 min, as per the IEC-60060-1standard [65]. The water pipe acts as a series impedance, with the value adjusted to ~100 kΩ, and a capacitive voltage divider was used for measurement purposes, as shown in Figure 4. A disc insulator from class 52-10 ANSI, having 160 kN strength with an AC flashover voltage of 80 kV under dry conditions and 50 kV under wet conditions, was taken as a test specimen.

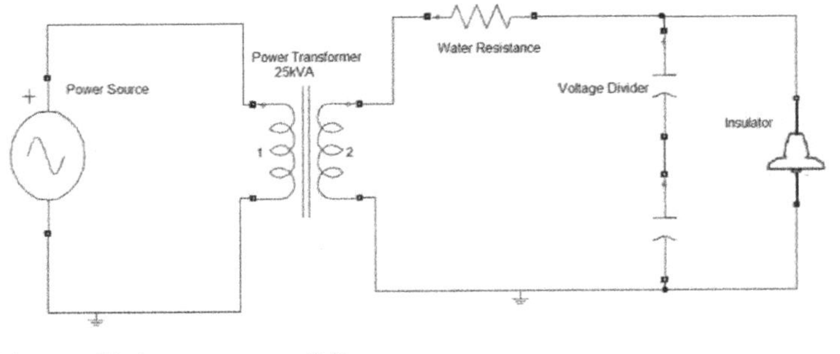

Figure 4. Flashover test setup [66].

3. Results and Discussion

3.1. Simulations for Electric Field Distribution

The insulator specimen considered has been modeled in COMSOL Multiphysics®, as shown in Figure 5a,b, to analyze the electric filed distribution and current density via the finite element method. The outdoor porcelain insulator comprises three main parts: (1) the main porcelain insulating material; (2) the cap and pin iron ends employed to connect the insulator to the power system structure; and (3) the intermediary cement that clenches the insulator to the metal fittings. The diameter of the insulator has been taken as 30 cm, and pin-to-cap clearance has been taken as 16.5 cm.

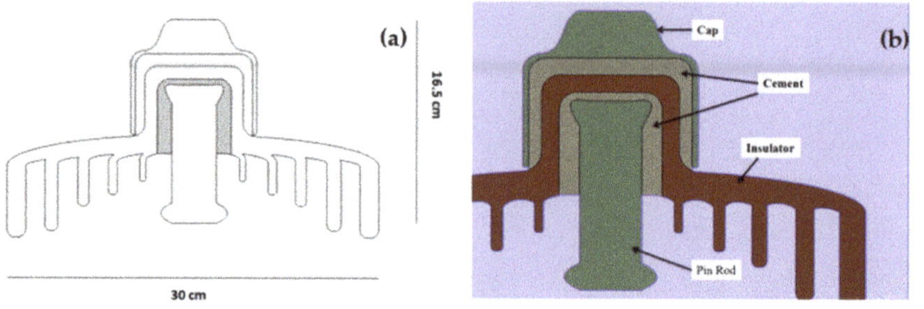

Figure 5. Porcelain disc insulator: (**a**) dimension; (**b**) labeling.

The relative permittivity of porcelain has been taken to be 6 [67], while this value is taken to be 1 for air and cast iron and 10 for moist concrete (cement) [68].

3.1.1. Dry Condition

The modeled insulator has been simulated under dry conditions, where 50 kV is applied to the insulator. Figure 6 shows the potential distribution contour, along with the electric field lines, where it can be observed that the lines are dense at the junction of the cap, air, and cement, indicating maximum electric field. However, the field lines are more distant towards the edges of the surface.

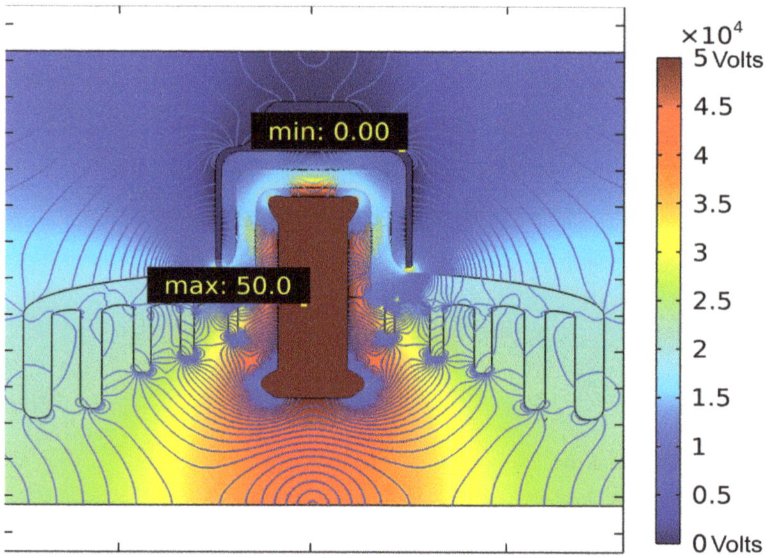

Figure 6. Electric potential contour and electric field lines (dry surface).

The electric field was plotted along the creepage path, as shown in Figure 7a,b. It can be observed that the electric field decreases and then starts increasing, with some sudden rises. To elaborate further, the plot has been divided into two portions: Figure 8a illustrates the electric field for the upper surface; and Figure 8b illustrates the electric field for the lower surface. Figure 8a shows a decrease in the electric field value, and the decrease is seen to be gradual as it is traverses from the center to the ends. This pattern is consistent with that in Figure 6, where field lines are more distant at the edges, far away from the potential source. Figure 8b demonstrates some peaks on the plot, and the amplitude rises progressively as we approach the center near, the potential source.

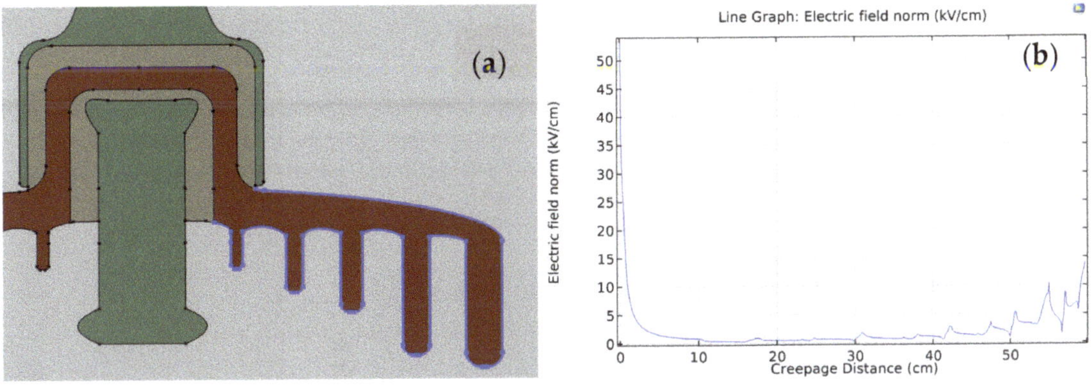

Figure 7. Electric field plot (dry case): (**a**) selection path; (**b**) electric field curve.

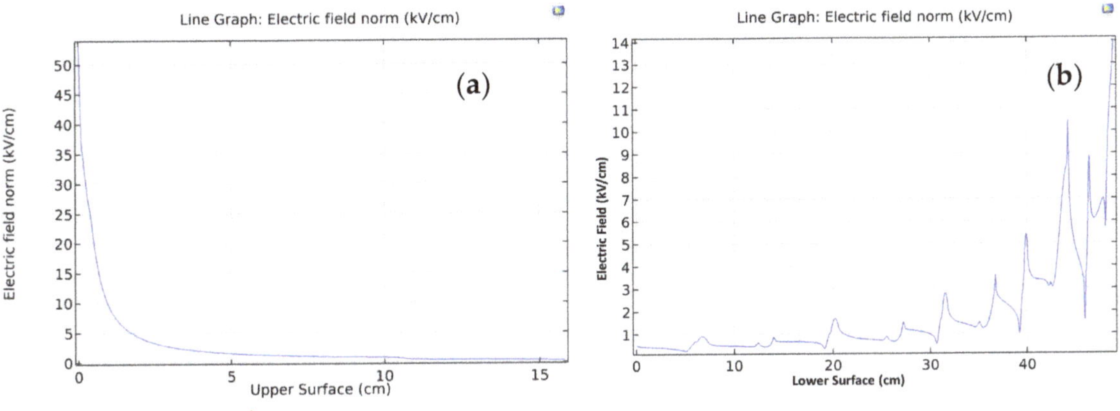

Figure 8. Electric field plot (dry case): (**a**) upper surface; (**b**) lower surface (from the outer edge to the inner curve).

3.1.2. Wet Condition

Hydrophilic case: Six water droplets with typical relative permittivity of 80 [69,70] and conductivity of 178 μS/cm/cm (0.0178 S/m) [71,72] were placed along the surface of the insulator with a contact angle of 60°, illustrating a hydrophilic situation, as shown in Figure 9.

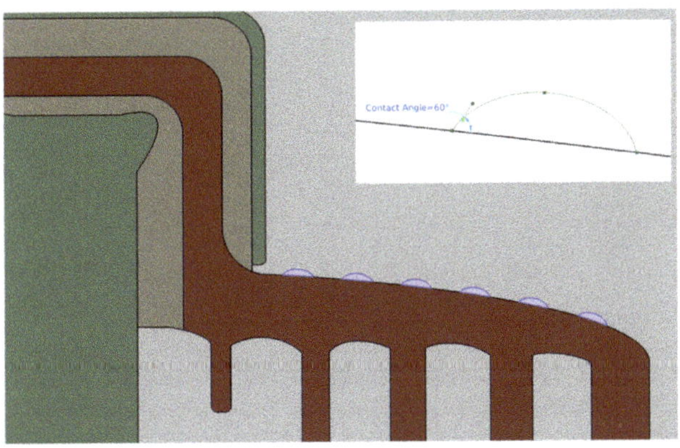

Figure 9. Water droplets for hydrophilic case (contact angle: 60°).

Water droplets with a radius of 2 mm were considered to have a volume of 33.5 mm^3 (33.5 μL). In 2D geometry, volume cannot be modeled, so the area of the circle was considered, which is 12.5664 mm^2. The droplets are considered at rest, and their contact line is stationary, so a static angle has been considered. The focus of this study is on the discrete droplets with a fair distance between them, reducing the likelihood of droplet interactions [73]. The voltage distribution, through colors, as well as the electric field patterns around the water droplet are shown in Figure 10. It can be seen that the field lines are denser around the water droplets, indicating high electric field stress.

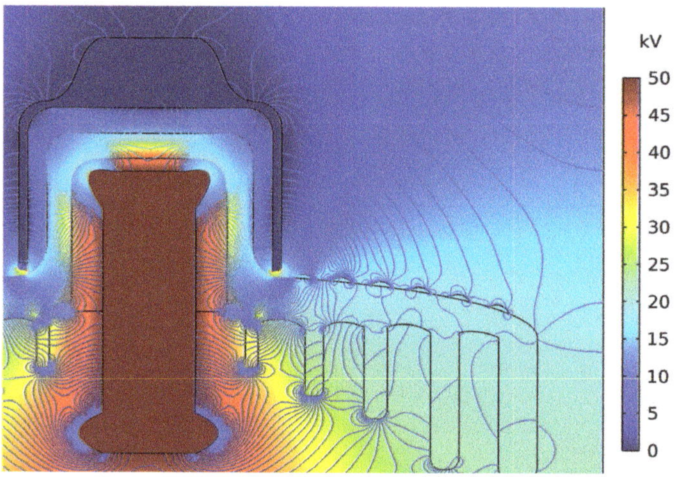

Figure 10. Electric potential contour and electric field lines (hydrophilic surface).

The selection path and plot for electric field distribution along the outer surface is shown in Figure 11. Although the pl3.2.ot follows a declining pattern towards outer edge, this is not gradual; rather, it has been distorted. The peaks appeared due to the fact that water droplets have different permittivity than the insulator, and electric field lines are thus denser around water droplets. The first peak, magnified at the top right corner, reveals its width to be 0.86 cm.

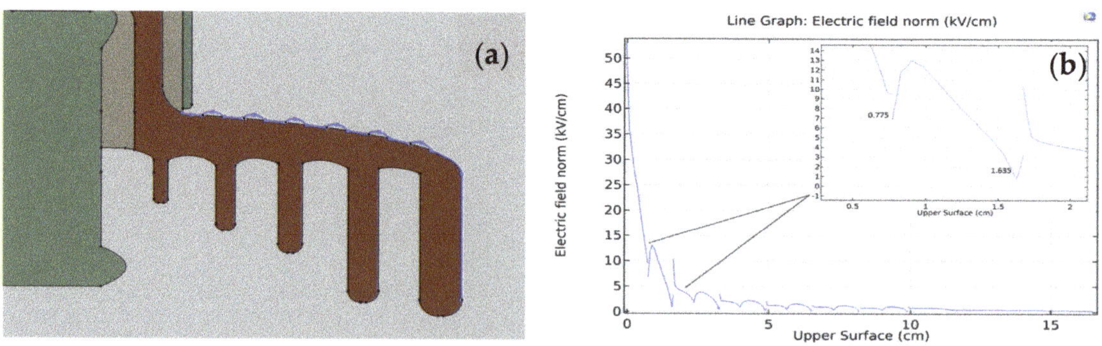

Figure 11. Electric field for hydrophilic case (60°): (**a**) selection path; (**b**) electric field curve.

(I) Hydrophobic case: The contact angle of the water droplets has now been changed to 120° , resembling a hydrophobic case, as shown in Figure 12, where voltage distribution is indicated through varying colors and electric field patterns are indicated through lines. As discussed in Section 1, referring to Figure 1b, the water droplet splits over the surface with a low contact angle, and it tends to be spherical when the contact angle is high (approaching to 180°); however, its volume remains the same. In 2D geometry, volume cannot be modeled; thus, the area of the shape has been kept the same.

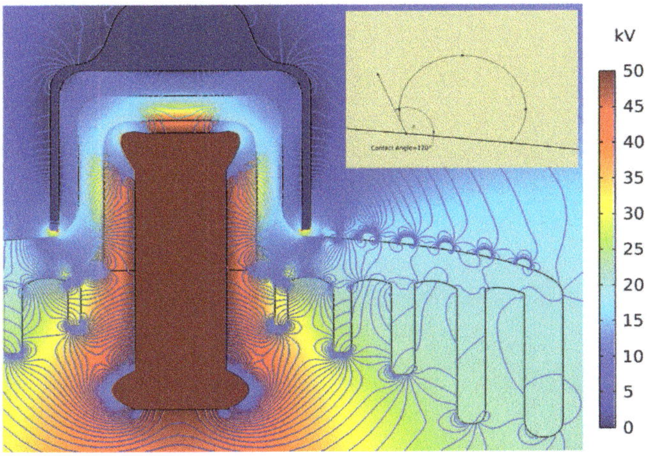

Figure 12. Electric potential contour and electric field lines (hydrophobic surface).

The shape of the water droplets changes, as does the electric field around them. Electric field distribution has been plotted along the outer surface, as shown in Figure 13, where it can be seen that the curve is still distorted; however, the peaks have been narrowed down. The magnified first peak is shown at the top right corner, where the reduction in width is evident, i.e., 0.835 cm. Though the amplitude of the first peak is bit higher, the overall stress (average) is reduced to 2.425 kV/cm, compared with 2.588 kV/cm for the hydrophilic surface (Table 3). The maximum value of the curve (at the start of the curve) also has a declining pattern, as can be seen in Table 3. Average values have been measured using a measuring probe in the software, which traverses the predefined path and integrates the data then dividesby the path length; thus, the unit remains same i.e: kV/cm. The maximum values presented in Table 3 refer to the highest point on entire curve that appearsat the beginning of the curve the region closer to center, in the vicinity of first droplet It was noted that the values decreased with the increase in contact angle.

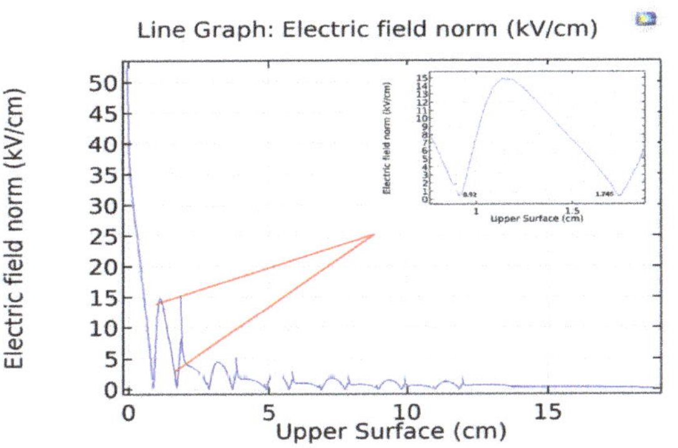

Figure 13. Electric field plot for the hydrophobic case (120°).

Table 3. Electric field measurement at the upper surface.

Sr.	Case	Wettability	Electric Field	
			Maximum (kV/cm)	Average (kV/cm)
1	Hydrophilic	Contact Angle: 60°	53.301	2.588
2	Hydrophobic	Contact Angle: 120°	52.708	2.425
3	Strongly Hydrophobic	Contact Angle: 140°	51.527	2.412

(II) Strongly hydrophobic case: The water droplets tend to be more sphere-shaped with a higher contact angle of the hydrophobic surface and have a tendency to roll off from the surface, making it dry before too long [10,11,74]. It has been noted elsewhere that a contact angle of polyurethane up to 140° is achievable [30]. Here, an example of strongly hydrophobicity, with a contact angle of 140°, where the voltage distribution is shown through colors and the lines represent electric field patterns, has been simulated for the worst-case scenario with the droplets still present on the surface, as illustrated in Figure 14.

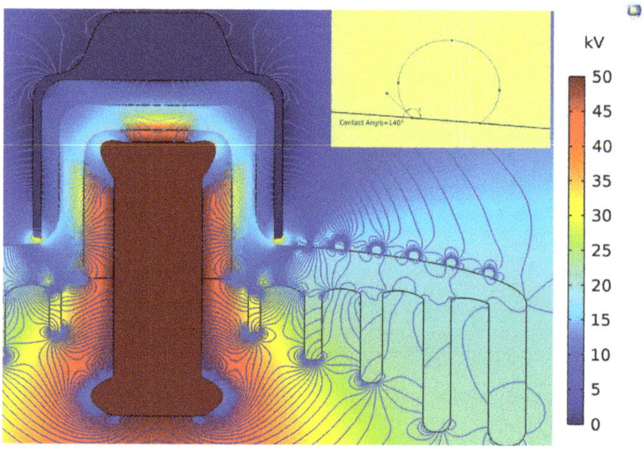

Figure 14. Electric potential contour and electric field lines (strong hydrophobic surface).

Referring to Figure 15, the electric field plot for the strongly hydrophobic case still shows distortion; however, peaks have been further narrowed down. The magnified first peak (the top right corner of Figure 15) shows a width reduction to 0.719 cm, and overall stress is noted to be 2.412 kV/cm (Table 3), making this case better than that of the 120° and 60° cases.

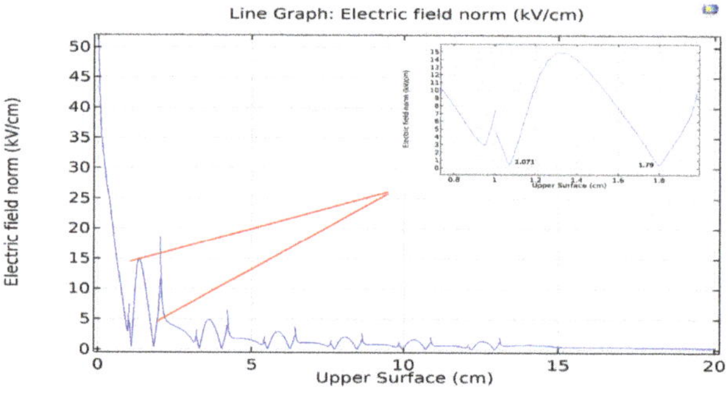

Figure 15. Electric field plot for strongly hydrophobic case (140°).

Maximum values are indeed observed to decline with higher contact angles, showing less accumulation of field lines in the vicinity of the water droplet for the higher contact angle. This is mainly due to the fact that water droplets spread more on hydrophilic surfaces, while in case of hydrophobic surfaces, water droplets tend to be spherical in shape, resulting in a smaller contact area.

3.2. Simulations for Current Density

Current density curves have also been plotted for the conditions under study, as illustrated in Figure 16. Conductivity values were set as follows: water (rain) at 0.0178 S/m [71,72]; porcelain at 10^{-12} S/m [75]; air at 1.8×10^{-14} S/m [76]; concrete at 10^{-4} S/m [77]; and the iron rod and cap at 10^{+7} S/m [78]. The curve for the dry condition (Figure 16a) depicts a gradual decrease towards the outer edges; however, the decrease is not gradual for wet conditions; rather, it exhibit distortion in the form of sudden elevations or spikes, as shown in Figure 16b–d. By focusing at the prominent spike, the amplitude has been significantly reduced from 0.187 nA/cm^2 for hydrophilic cases to 0.078 nA/cm^2 for hydrophobic cases. This decline continues to 0.054 nA/cm^2 for the strongly hydrophobic case. This declining trend strongly suggests that distortion or disruption is attenuated by surfaces that repel water.

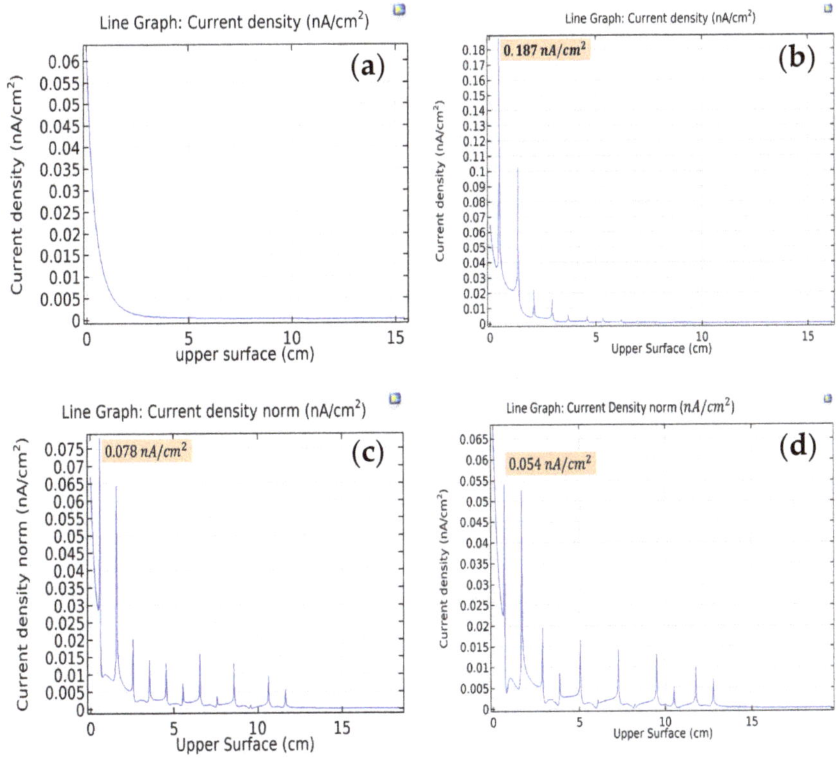

Figure 16. Current density plot: (**a**) dry case; (**b**) wet case with hydrophilic surface; (**c**) hydrophobic surface; (**d**) strong hydrophobic surface.

The maximum and average values of current density along the outer surface for three cases of wettability have also been noted and are listed in Table 4. Values support the earlier observation of a decreasing trend with an increase in the contact angle. The average value, showing stress, has been noted to see a tremendous decrease of almost 64.03% for an increase in the contact angle from 60° to 140°.

Table 4. Current density measurement at outer surface.

Sr.	Case	Wettability	Current Density	
			Average (nA/cm^2)	Maximum (nA/cm^2)
1	Hydrophilic	Contact Angle: 60°	0.004248	0.18743
2	Hydrophobic	Contact Angle:120°	0.003481	0.07807
3	Strongly Hydrophobic	Contact Angle: 140°	0.003174	0.06768

Hence, simulation results support the argument that a water droplet with a low contact angle intensifies distortion and stress. This is due to the fact that the water droplet spreads over the surface in low-contact-angle cases, boosting the surface conductivity. However, elevating the contact angle effectively lowers this stress. Therefore, a proper hydrophobic coating can significantly mitigate the distortion and adversity of wettability.

3.3. Polyurethane as a Remedial Coating

Figure 17 shows uncoated and polyurethane-coated disc insulators, which were tested to obtain AC flashover voltage under dry and wet conditions as per the experimental setup shown in Figure 4.

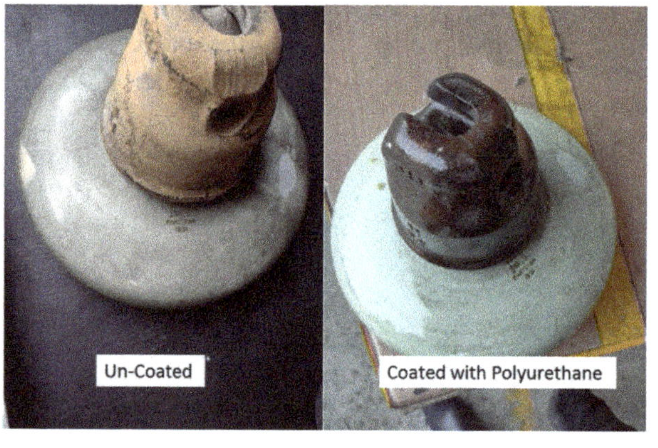

Figure 17. Uncoated and coated insulator.

The experiment was conducted five times to obtain the mean flashover voltage value, and standard deviation (σ) has also been calculated using the following formulas:

$$V_f = \frac{\sum_{k=1}^{N} V_k}{N} \tag{4}$$

$$\sigma = \frac{1}{V_f}\sqrt{\frac{\sum_{k=1}^{N}\left(V_k - V_f\right)^2}{N-1}} \tag{5}$$

where V_f represents the flashover voltage, V_k represents the flashover voltage for the kth time, and N represents the number of valid tests. The results are listed in Table 5.

Table 5. Flashover characteristics of the coated and the uncoated insulator.

Insulator	AC Flashover Voltage Dry (kV)	$\%\sigma$	AC Flashover Voltage Wet (kV)	$\%\sigma$
Uncoated	79.14 kV	1.2	48.4 kV	1.4
Coated	82.04 kV	1.3	53.8 kV	1.2

It is evident from Table 5 that the polyurethane coating has elevated the flashover voltage, making the insulator less vulnerable to premature flashover. However, extensive experimentation under various conditions of pollution and contamination is necessary before it can be used at the industrial level.

4. Conclusions

The disc insulator model has been evaluated for electric field and current density distribution under dry and wet conditions using the finite element method in COMSOL Multiphysics®. Three cases of wettability involving three surfaces—namely, hydrophilic, hydrophobic, and strongly hydrophobic surfaces—have been simulated by placing discrete water droplets at various contact angles. Abrupt increases have been witnessed on the plots for the electric field, showing distortion; however, this distortion was reduced from 2.588 kV/cm for the hydrophilic surface to 2.412 kV/cm for the strongly hydrophobic surface. Current density plots also showed spikes on the curve; however, they followed the same declining pattern, and overall stress was noticed, with an almost-64.03% reduction from 0.187 to 0.068 nA/cm^2 following an increase in the contact angle from 60° to 140°. This is mainly due to the fact that water droplets spread more on hydrophilic surfaces, while in case of hydrophobic surfaces, water droplets tend to be spherical in shape, thus lowering the effect of wettability. The polyurethane coating evidently improved the insulator's flashover voltage under both dry and wet conditions. The voltage improved from 79.14 kV to 82.04 kV for dry conditions, while that of wet conditions increased from 48.4 kV to 53.8 kV, surpassing the performance of the uncoated insulator. From the simulation and experimental results, it can be concluded that strongly hydrophobic surfaces make the insulator less vulnerable to flashover. This research work has not only introduced polyurethane as a potential coating but also opened new avenues for researchers working on improving the hydrophobicity of insulator surfaces. However, extensive experimental work under contaminated conditions is essential to fully characterize polyurethane-coated insulators' flashover behavior.

Author Contributions: Conceptualization, T.A.R. and M.K.; methodology, T.A.R. and S.A.M.S.; software, T.A.R. and S.A.M.S.; formal analysis, T.A.R. and M.M.B.; writing—original draft preparation, T.A.R. and S.A.M.S.; writing—review and editing, M.K. and M.M.B.; supervision, M.K. All authors have read and agreed to the published version of the manuscript.

Funding: This research was funded by office of research, innovation, and commercialization (ORIC), University of Engineering and Technology, Lahore, Pakistan (Funding number: ORIC/112ASRB/2075, date: 3 November 2021), and was carried in laboratory of High Voltage at the Department of Electrical Engineering, University of Engineering and Technology, Lahore, Pakistan.

Institutional Review Board Statement: Not applicable.

Informed Consent Statement: Not applicable.

Data Availability Statement: The original contributions presented in this study are included in the article. Further inquiries can be directed to the corresponding author.

Conflicts of Interest: The authors declare no conflicts of interest. The funders had no role in the design of the study; in the collection, analyses, or interpretation of data; in the writing of the manuscript; or in the decision to publish the results.

References

1. Lan, L.; Zhang, G.; Wang, Y.; Wen, X.; Wang, W.; Pei, H. The Influence of Natural Contamination on Pollution Flashover Voltage Waveform of Porcelain Insulators in Heavily Polluted Area. *IEEE Access* **2019**, *7*, 121395–121406. [CrossRef]
2. Salem, A.A.; Lau, K.Y.; Rahiman, W.; Abdul-Malek, Z.; Al-Gailani, S.A.; Mohammed, N.; Rahman, R.A.; Al-Ameri, S.M. Pollution Flashover Voltage of Transmission Line Insulators: Systematic Review of Experimental Works. *IEEE Access* **2022**, *10*, 10416–10444. [CrossRef]
3. Topalis, F.V.; Gonos, I.F.; Stathopulos, I.A. Dielectric Behavior of Polluted Porcelain Insulators. *IEE Proc. Gener. Transm. Distrib.* **2001**, *148*, 269. [CrossRef]
4. Jiang, X.; Yuan, J.; Zhang, Z.; Hu, J.; Sun, C. Study on AC Artificial-Contaminated Flashover Performance of Various Types of Insulators. *IEEE Trans. Power Deliv.* **2007**, *22*, 2567–2574. [CrossRef]
5. Yang, Z.; Jiang, X.; Han, X.; Zhang, Z.; Hu, J. Influence of Pollution Chemical Components on AC Flashover Performance of Various Types of Insulators. *High Voltage* **2019**, *4*, 105–112. [CrossRef]
6. Khatoon, S.; Khan, A.A.; Sharma, A. Study the effect of contaminants on flashover performance of porcelain disc insulator by artificial pollution testing. In Proceedings of the 2016 International Conference on Recent Advances and Innovations in Engineering (ICRAIE), Jaipur, India, 23–25 December 2016; IEEE: Piscataway, NJ, USA, 2016. [CrossRef]
7. Gencoglu, M.T.; Cebeci, M. The Pollution Flashover on High Voltage Insulators. *Electr. Power Syst. Res.* **2020**, *181*, 106180.
8. Salem, A.A.; Rahman, R.A.; Al-Ameri, S. Pollution Flashover Characteristics of High-Voltage Outdoor Insulators: Analytical Study. *Arabian J. Sci. Eng.* **2021**, *47*, 2711–2729. [CrossRef]
9. Cui, L.; Ramesh, M. Prediction of flashover voltage using electric field measurement on clean and polluted insulators. *Int. J. Electr. Power Energy Syst.* **2020**, *116*, 105574. [CrossRef]
10. Shirtcliffe, N.J.; McHale, G.; Atherton, S.; Newton, M.I. An introduction to superhydrophobicity. *Adv. Colloid Interface Sci.* **2010**, *161*, 124–138. [CrossRef]
11. Bhushan, B.; Jung, Y.C. Wetting, adhesion and friction of superhydrophobic and hydrophilic leaves and fabricated micro/nanopatterned surfaces. *J. Phys. Condens. Matter* **2008**, *20*, 225010. [CrossRef]
12. Koch, K.; Bhushan, B.; Barthlott, W. Diversity of Structure, Morphology and Wetting of Plant Surfaces. *Soft Matter* **2008**, *4*, 1943–1963. [CrossRef]
13. Drelich, J.; Marmur, A. Physics and Applications of Superhydrophobic and Superhydrophilic Surfaces and Coatings. *Surf. Innov.* **2014**, *2*, 211–227. [CrossRef]
14. Farzane, M.; Chisholm, W.A. Protective Coatings for Overhead Lines in Winter Conditions. In *Techniques for Protecting Overhead Lines in Winter Conditions*; Springer: Berlin/Heidelberg, Germany, 2022; pp. 195–309. [CrossRef]
15. Nguyen, T.V.; Nguyen, T.A.; Nguyen, T.H. The Synergistic Effects of SiO_2 Nanoparticles and Organic Photostabilizers for Enhanced Weathering Resistance of Acrylic Polyurethane Coating. *J. Compos. Sci.* **2020**, *4*, 23. [CrossRef]
16. Aman, A.; Abdullah, A.R.; Yaacob, M.M. Dielectric Property of Waste Tire Dust Polypropylene (WTD-PP) Composite for High Voltage Outdoor Insulation Application. In Proceedings of the IEEE International Power Engineering and Optimization Conference (PEOCO 2012), Melaka, Malaysia, 6–7 June 2012.
17. Sawane, Y.B.; Ogale, S.B.; Banpurkar, A.G. Low Voltage Electrowetting on Ferroelectric PVDF-HFP Insulator with Highly Tunable Contact Angle Range. *ACS Appl. Mater. Interfaces* **2016**, *8*, 24049–24056. [CrossRef]
18. Xia, Y.; Chen, J.; Zhu, Z.; Zhang, Q.; Yang, H.; Wang, Q. Significantly Enhanced Dielectric and Hydrophobic Properties of SiO2@MgO/PMMA Composite Films. *RSC Adv.* **2018**, *8*, 4032. [CrossRef]
19. Ahmadi-Joneidi, I.; Majzoobi, A.; Shayegani-akmal, A.A.; Mohseni, H.; Jadidian, J. Aging Evaluation of Silicone Rubber Insulators Using Leakage Current and Flashover Voltage Analysis. *IEEE Trans. Dielectr. Electr. Insul.* **2013**, *20*, 212–220. [CrossRef]
20. Cardoso, R.; Balestro, A.C.; Dellallibera, A.; Costa, E.C.M.; Angelini, J.M.G.; Mei, L.H.I. Silicone Insulators of Power Transmission Lines with a Variable Inorganic Load Concentration: Electrical and Physiochemical Analyses. *Measurement* **2014**, *50*, 63–73. [CrossRef]
21. Othman, N.A.; Piah, M.A.M.; Adzis, Z. Contamination Effects on Charge Distribution Measurement of High Voltage Glass Insulator String. *Measurement* **2017**, *105*, 34–40. [CrossRef]
22. Verma, A.R.; Subba Reddy, B. Interpretation of Surface Degradation on Polymeric Insulators. *Eng. Failure Anal.* **2019**, *95*, 214–225. [CrossRef]
23. Mahmoodi, J.; Mirzaie, M.; Shayegani-Akmal, A.A. Surface Charge Distribution Analysis of Polymeric Insulator Under AC and DC Voltage Based on Numerical and Experimental Tests. *Int. J. Electr. Power Energy Syst.* **2019**, *105*, 283–296. [CrossRef]
24. Al-Karawi, S.; Al-Taie, A. Insulators Used in Outdoor Power Applications: Materials and Coatings. In Proceedings of the 2023 6th International Conference on Engineering Technology and its Applications (IICETA), Al-Najaf, Iraq, 15–16 July 2023; IEEE: Piscataway, NJ, USA, 2023. [CrossRef]
25. Küchler, A. *High Voltage Engineering: Fundamentals—Technology—Applications*; Springer: Berlin/Heidelberg, Germany, 2017.

26. Zhang, Z.; Du, J.; Li, J.; Huang, X.; Kang, T.; Zhang, C.; Wang, S.; Ajao, O.O.; Wang, W.-J.; Liu, P. Polymer Nanocomposites with Aligned Two-Dimensional Materials. *Prog. Polym. Sci.* **2021**, *114*, 101360. [CrossRef]
27. Lan, L.; Gorur, R. Computation of AC Wet Flashover Voltage of Ceramic and Composite Insulators. *IEEE Trans. Dielectr. Electr. Insul.* **2008**, *15*, 1346–1352. [CrossRef]
28. Bala, P.; Bose, R.; Chatterjee, S. Electric Stress Analysis of a 11kV RTV Silicone Rubber Coated Porcelain Insulator. In Proceedings of the 2016 Biennial International Conference on Power and Energy Systems: Towards Sustainable Energy (PESTSE), Bangalore, India, 21–23 January 2016; IEEE: Piscataway, NJ, USA, 2016. [CrossRef]
29. Allahdini, A.; Momen, G.; Munger, F.; Brettschneider, S.; Fofana, I.; Jafari, R. Performance of a Nanotextured Superhydrophobic Coating Developed for High-Voltage Outdoor Porcelain Insulators. *Colloids Surf. A Physicochem. Eng. Asp.* **2022**, *649*, 129461. [CrossRef]
30. Gu, X.; Gao, T.; Meng, X.; Zhu, Y.; Wang, G. Enhanced Hydrophobicity of Polyurethane with the Self-Assembly of Perfluoropolyether-Based Triblock Copolymers. *Prog. Org. Coat.* **2022**, *163*, 106561. [CrossRef]
31. De Souza, F.M.; Kahol, P.K.; Gupta, R.K. Introduction to Polyurethane Chemistry. In *Polyurethane Chemistry: Renewable Polyols and Isocyanates*; American Chemical Society: Washington, DC, USA, 2021; pp. 1–24.
32. Kausar, A. Polyurethane Nanocomposite Coatings: State of the Art and Perspectives. *Polym. Int.* **2018**, *67*, 1470–1477. [CrossRef]
33. John, K.; Naidu, S.V. Sisal Fiber/Glass Fiber Hybrid Composites: The Impact and Compressive Properties. *J. Reinf. Plast. Compos.* **2004**, *23*, 1253–1258. [CrossRef]
34. Yu, X.; Wang, Q.; Fu, L.; Shen, G.; Zhang, W.; Yang, R.; Sun, K.; Lv, F.; Fan, S. Superhydrophobic Meta-Aramid Papers Prepared by the Surface-Embedded Spray Coating Strategy. *Appl. Surf. Sci.* **2023**, *648*, 159044. [CrossRef]
35. Liu, B.; Lv, F.; Fan, X.; Li, Y.; Jiang, B. Molecular Dynamics Study of the Influence of Nano SiO_2 on the Thermodynamic Properties of PMIA Composites. *Polymers* **2022**, *14*, 3134. [CrossRef]
36. Liu, B.; Lv, F.; Fan, X.; Xiao, H.; Bi, H. Insulation Performance and Simulation Analysis of SiO_2-Aramid Paper under High-Voltage Bushing. *Nanomaterials* **2022**, *12*, 748. [CrossRef]
37. Biesuz, M.; Tomasi, M.; Santhosh, B.; Sglavo, V.M.; Soraru, G.D. Polymer-Derived Si_3N_4 Nanofelts as a Novel Oil Spills Clean-Up Architecture. *J. Environ. Chem. Eng.* **2020**, *8*, 104134. [CrossRef]
38. Huang, X.; Jiang, P.; Tanaka, T. A Review of Dielectric Polymer Composites with High Thermal Conductivity. *IEEE Electr. Insul. Mag.* **2011**, *27*, 8–16. [CrossRef]
39. Matsunaga, C.; Zhou, Y.; Kusano, D.; Hyuga, H.; Hirao, K. Dielectric Breakdown of Silicon Nitride Substrates with Various Thicknesses. *J. Ceram. Soc. Jpn.* **2018**, *126*, 693–698. [CrossRef]
40. Feoktistov, D.V.; Kuznetsov, G.V.; Sivkov, A.A.; Ivashutenko, A.S.; Nikitin, D.S.; Shanenkov, I.I.; Abdelmagid, A.M.; Orlova, E.G. Expanding the Scope of SiC Ceramics through Its Surface Modification by Different Methods. *Surf. Coat. Technol.* **2022**, *435*, 128263. [CrossRef]
41. Han, Y.; Wang, Z.; Sun, J.; Li, Z. Enhanced Switching Electric Field and Breakdown Strength of Epoxy Composites with Core-Shell Silicon Carbide Nanoparticles. *High Voltage* **2023**, *8*, 583–589. [CrossRef]
42. Jayasinghe, S.N.; Edirisinghe, M.J. Effect of Viscosity on the Size of Relics Produced by Electrostatic Atomization. *J. Aerosol Sci.* **2002**, *33*, 1379–1388. [CrossRef]
43. Qasim, S.A.; Gupta, N. Comparison of the Dielectric Properties of Functionally Graded Material Dielectrics and Layered Dielectrics Used for Electric Stress Control. *IET Sci. Meas. Technol.* **2020**, *14*, 471–477. [CrossRef]
44. Rachmawati, N.; Kojima, H.; Hayakawa, N.; Kato, K.; Zebouchi, N. Electric Field Simulation of Permittivity and Conductivity Graded Materials (ε/σ-FGM) for HVDC GIS Spacers. *IEEE Trans. Dielectr. Electr. Insul.* **2021**, *28*, 736–744. [CrossRef]
45. Zaputra, S.; Fabio, M.R.; Hidayat, S. Application of Non-Linear Discrete Spatial Permittivity Distributions in Reducing Electric Field Concentration in GIS Spacer. In Proceedings of the 2023 4th International Conference on High Voltage Engineering and Power Systems (ICHVEPS), Denpasar Bali, Indonesia, 6–10 August 2023; pp. 717–721. [CrossRef]
46. Peng, S.; Iroh, J.O. Dependence of the Dynamic Mechanical Properties and Structure of Polyurethane-Clay Nanocomposites on the Weight Fraction of Clay. *J. Compos. Sci.* **2022**, *6*, 173. [CrossRef]
47. Zafar, F.; Sharmin, E. *Polyurethane/Monograph*; Intech: Rijeka, Croatia, 2012.
48. Moody, V.; Needles, H.L. Polyurethane Coating. In *Tufted Carpet*; Elsevier: Amsterdam, The Netherlands, 2004; pp. 105–108. [CrossRef]
49. Sonnenschein, M.F. Introduction to Polyurethane Chemistry. In *Polyurethanes*; John Wiley & Sons, Inc.: Hoboken, NJ, USA, 2014; pp. 105–126. [CrossRef]
50. Barksby, N.; Dormish, J.F.; Haider, K.W. Polyurethane Synthesis. In *Encyclopedia of Polymeric Nanomaterials*; Kobayashi, S., Müllen, K., Eds.; Springer: Berlin/Heidelberg, Germany, 2014. [CrossRef]
51. Dong, B.; Jiang, X.; Hu, J.; Shu, L.; Sun, C. Effects of Artificial Polluting Methods on AC Flashover Voltage of Composite Insulators. *IEEE Trans. Dielectr. Electr. Insul.* **2012**, *19*, 714–722. [CrossRef]

52. Balakrishnan, A.; Saha, M.C. Influence of Air-Draft on Fabrication of Polyurethane Thin Films via Ultrasonic Atomization. *At. Sprays* **2012**, *22*, 23–35. [CrossRef]
53. High Voltage Insulator Coatings: State of the Art & Future Development. INMR. Available online: https://www.inmr.com/high-voltage-insulator-coatings-state-of-the-art-future-development/ (accessed on 13 January 2025).
54. Courant, R. Variational Methods for the Solution of Problems of Equilibrium and Vibrations. *Bull. Am. Math. Soc.* **1943**, *49*, 1–24. [CrossRef]
55. Silvester, P. Finite Element Solution of Homogeneous Waveguide Problems. *Alta Freq.* **1969**, *38*, 313–317.
56. Inoue, K.; Hayata, K.; Koshiba, M. Finite-Element Solution of Three-Dimensional Periodic Waveguide Problems. *Electron. Commun. Jpn. Part II Electron.* **1989**, *72*, 68–77. [CrossRef]
57. Axelsson, O.; Barker, V.A. The Finite Element Method. In *Finite Element Solution of Boundary Value Problems*; SIAM: Philadelphia, PA, USA, 2001; pp. 163–267. [CrossRef]
58. Gokhale, N.S. *Practical Finite Element Analysis*; Finite To Infinite: Maharashtra, India, 2020.
59. Jin, J.-M. *Finite Element Method in Electromagnetics*; Wiley Sons, Inc., John: Hoboken, NJ, USA, 2014.
60. Sastry, S.S. *Introductory Methods of Numerical Analysis*; Prentice Hall India Pvt., Ltd.: New Delhi, India, 2012.
61. Rao, S.S. *Finite Element Method in Engineering*; Elsevier Science and Technology Books: Cambridge, MA, USA, 2017.
62. Zienkiewicz, O.C.; Taylor, R.L. *The Finite Element Method: Its Basis and Fundamentals*, 7th ed.; Elsevier: Amsterdam, The Netherlands, 2013. [CrossRef]
63. Arora, R.; Mosch, W. *High Voltage and Electrical Insulation Engineering*, 2nd ed.; IEEE Press: Piscataway, NJ, USA; Wiley: Hoboken, NJ, USA, 2022.
64. Polycarpou, A.C. *Introduction to the Finite Element Method in Electromagnetics*; Springer Nature: Berlin/Heidelberg, Germany, 2022.
65. *IEC 60060-1:2025*; High-Voltage Test Techniques—Part 1: General Terminology and Test Requirements, 4th ed. International Electrotechnical Commission: Geneva, Switzerland, 2025.
66. Pei, S.; Liu, Y.; Ji, X.; Geng, J.; Zhou, G.; Wang, S. UV-Flashover Evaluation of Porcelain Insulators Based on Deep Learning. *IET Sci. Meas. Technol.* **2018**, *12*, 770–776. [CrossRef]
67. Klüss, J.; Bichik, A.; Arshad, A.; Strandberg, J.; Kazmi, S.; Laiho, M.; Stark, T.; Panchal, J.; Ahmad, B.; Väkeväinen, K. Simulation Precision and the Human Factor. In Proceedings of the 24th Nordic Insulation Symposium on Materials, Components and Diagnostics, Copenhagen, Denmark, 15–17 June 2015. [CrossRef]
68. Meyer, K.; Erdogmus, E.; Morcous, G.; Naughtin, M. Use of Ground Penetrating Radar for Accurate Concrete Thickness Measurements. In Proceedings of the AEI 2008 Conference—AEI 2008: Building Integration Solutions, Denver, CO, USA, 24–27 September 2008; Volume 328, pp. 1–10. [CrossRef]
69. Archer, G.G.; Wang, P. The Dielectric Constant of Water and Debye-Hückel Limiting Law Slopes. *J. Phys. Chem. Ref. Data* **1990**, *19*, 371–411. [CrossRef]
70. Ellison, W.J. Permittivity of Pure Water, at Standard Atmospheric Pressure, over the Frequency Range 0–25THz and the Temperature Range 0–100 °C. *J. Phys. Chem. Ref. Data* **2007**, *36*, 1–18. [CrossRef]
71. Jones, R.G. Measurements of the Electrical Conductivity of Water. *IEE Proc. Sci. Meas. Technol.* **2002**, *149*, 320–322. [CrossRef]
72. Lekouch, I.; Mileta, M.; Muselli, M.; Milimouk-Melnytchouk, I.; Šojat, V.; Kabbachi, B.; Beysens, D. Comparative Chemical Analysis of Dew and Rain Water. *Atmos. Res.* **2010**, *95*, 224–234. [CrossRef]
73. Huhtamäki, T.; Tian, X.; Korhonen, J.; Ras, R.H.A. Surface-Wetting Characterization Using Contact-Angle Measurements. *Nat. Protoc.* **2018**, *13*, 1521–1538. [CrossRef]
74. Danish, M. Contact Angle Studies of Hydrophobic and Hydrophilic Surfaces. In *Handbook of Magnetic Hybrid Nanoalloys and their Nanocomposites*; Springer: Berlin/Heidelberg, Germany, 2022; pp. 761–782. [CrossRef]
75. Chaudhuri, S.P.; Sarkar, P.; Chakraborty, A.K. Electrical Resistivity of Porcelain in Relation to Constitution. *Ceram. Int.* **1999**, *25*, 91–99. [CrossRef]
76. Kamsali, N.; Prasad, B.S.N.; Datta, J.; Nejadkoorki, F. The Electrical Conductivity as an Index of Air Pollution in the Atmosphere. In *Advanced Air Pollution*; Springer: Berlin/Heidelberg, Germany, 2011; pp. 365–390.
77. Su, J.; Yang, C.; Wu, W.; Huang, R. Effect of Moisture Content on Concrete Resistivity Measurement. *J. Chin. Inst. Eng.* **2002**, *25*, 117–122. [CrossRef]
78. National Nondestructive Testing Education Center. Copper Properties. In NDT Techniques: Eddy Current. Available online: https://www.nde-ed.org/NDETechniques/EddyCurrent/ET_Tables/ET_matlprop_Copper.xhtml (accessed on 13 January 2025).

surfaces

MDPI

Article

Functional Polyacrylate Textile Coatings with *N,N*-Diethyl-3-methylbenzamide (DEET) Immobilized on Zirconia, Alumina and Silica Sorbents

Sergei Zverev [1,*], Sergei Andreev [1], Ekaterina Anosova [1], Varvara Morenova [1], Maria Rakitina [2] and Vladimir Vinokurov [2]

[1] Disinfectology Institute of F.F. Erisman FSCH of Rospotrebnadzor, 18 Nauchny Drive, Moscow 117246, Russia
[2] Department of Physical and Colloid Chemistry, Gubkin University, 65 Leninsky Prospekt, Moscow 119991, Russia
[*] Correspondence: zverev.s.chem01@gmail.com

Abstract: In this study, polymer films based on the inorganic sorbents Al_2O_3, ZrO_2 and SiO_2-phenyl with repellent *N,N*-diethyl-3-methylbenzamide were prepared and used as functional textile coatings. The high sorption activity of oxides with respect to *N,N*-diethyl-3-methylbenzamide (63–239 mg/g) allows for the use of these compounds as repellent carrier materials, and their mixture with polyacrylates allows for the formation of functional coatings–polymer films. Scanning electron microscopy and Fourier transform infrared spectroscopy results revealed that the inorganic sorbents Al_2O_3, ZrO_2 and SiO_2-phenyl were successfully anchored in the polyacrylate structure, and the FTIR spectra confirmed the presence of repellent molecules. The thermal diffusion parameters of *N,N*-diethyl-3-methylbenzamide were also calculated via thermogravimetric analysis and high-performance liquid chromatography with diode array detection. The highest thermal diffusion rates and concentrations were observed for the material with Al_2O_3 (up to $148.3 \cdot 10^{-9}$ mol at 200 °C), and lower values for ZrO_2 and SiO_2-phenyl (up to $15.2 \cdot 10^{-9}$ mol and $34.3 \cdot 10^{-9}$ mol at 200 °C, respectively). The heat flux parameter J_f was also calculated according to Onsager's theory and Fourier's law. The release of repellent from polymeric materials can be achieved by applying less heat than that required to reach the boiling point of *N,N*-diethyl-3-methylbenzamide.

Keywords: sorbents; repellent; DEET; polyacrylate; functional coatings; thermal diffusion; textile

1. Introduction

Functional coatings based on polyacrylate compounds are a class of polymeric materials that can be used on their own or can be applied to textiles [1,2]. Polyacrylates are widely used in the textile industry and other fields due to their properties such as high mechanical strength, chemical and biological inertness, and resistance to thermal and ultraviolet degradation [3–6]. The properties of polyacrylates can be varied by incorporating additives into the polymeric structure. Thus, the incorporation of inorganic oxides of silica, alumina and zirconia or zinc oxide and titanium dioxide leads to an increase in the mechanical strength of the polymer structure and in the photocatalytic activity of the materials and allows for the introduction of carbon nanotubes to improve the electrical conductivity of polymers [3]. In addition, polyacrylates and their copolymers are widely used in agriculture as pesticide carriers in the form of microcapsules or heterogeneous dispersions that can deliver the active ingredient in a targeted manner [5–7]. This approach

 https://doi.org/10.3390/surfaces8020033

not only reduces the toxicological impact of pesticides on the environment but also ensures their long-term effectiveness.

Functional polymer coatings are particularly important in the development of materials to protect against insect attack. Insects are a major source of dangerous infectious diseases [8]. Previous studies have shown that the polymeric binding of some pyrethroids to polyacrylates thermally fixed to textile materials results in prolonged activity against blood-sucking insects. Another advantage of polyacrylate coatings is their high resistance to environmental factors and to repeated washing [9].

However, the use of pyrethroids in textiles is limited by their high toxicity. Polymeric coatings based on repellents—compounds of synthetic and natural origin that can repel insects while maintaining the protective functions of textiles—can be an alternative approach [4,10–16]. These compounds were shown to be effective against blood-sucking insects as early as the mid-20th century [17] and are still widely used in various formulations, but there are a number of difficulties with their application and fixation to the surface of textiles, as these compounds are oily liquids that are poorly soluble in water. There are few technologies available to produce functional coatings with repellent properties. Therefore, to fix the repellent *N,N*-diethyl-3-methylbenzamide (DEET), electrospinning, which is based on the action of a high electric potential on a polymeric solution, is possible [11,12,15,18]. As a result of Coulomb forces, the electrostatically charged polymer is pulled towards the surface of the material, exceeding the critical speed of the liquid and resulting in the solidification of superstrong structures. This approach provides a long-lasting repellent effect of textiles against blood-sucking insects, but special laboratory equipment is required to obtain materials with a repellent coating [12]. In addition, the production of protective clothing from such textiles requires the scaling up of such technologies, and it is not clear whether the process can be replicated on an industrial scale.

Another approach to functional coatings with repellent properties is the sequential synthesis of repellent conjugates with dye molecules for subsequent application to textile fibers [19,20]. Owing to the -N=N- bonds, the modified dye molecule is formed, whereas the -N=C=O bond of the DEET molecule, which is responsible for the repellent activity, remains unchanged. However, the synthesis of derivatives requires the introduction of an amino group to form a conjugate. The effect of this modification on the repellent activity of the substance has also not been investigated. The production of safe functional repellent coatings for protective clothing against attacks by dangerous blood-sucking insects on humans therefore remains an urgent task.

The purpose of this study is to investigate the properties of functional textile coatings with repellent effects containing *N,N*-diethyl-3-methylbenzamide. First, polymeric materials containing inorganic sorbents such as zirconia, alumina, and silica with grafted phenyl groups were obtained. Previously, we studied and described the sorption properties of these oxides and demonstrated their high sorption capacity towards synthetic repellents, particularly DEET [21]. In this work, repellent-containing sorbents were fixed on a polyacrylate film. The structure of the films obtained was studied by scanning electron microscopy and FTIR spectroscopy, and the process of thermal diffusion of the repellent was also investigated. The resulting composition was applied to the textile, after which the material structure was examined by scanning electron microscopy, and the repellency was determined via high-performance liquid chromatography with diode array detection. The results obtained open new possibilities for the development of protective materials with reduced toxicity against attacks by blood-sucking insects with controlled release of repellents.

2. Materials and Methods

2.1. Materials

N,N-diethyl-3-methylbenzamide (DEET) was purchased from Merck (Darmstadt, Germany); acetonitrile was purchased from Honeywell (Charlotte, NC, USA); aluminum nitrate nonahydrate, zirconyl nitrate dihydrate, ammonium nitrate, citric acid hydrate, ethylene glycol and ethyl acetate were purchased from JSC LenReactiv (St. Petersburg, Russia); polyacrylic acid copolymer consisting of a mixture of 2-(perfluorodecyl)ethyl acrylate and stearyl acrylate was purchased from Zhangzhou F&S New Material Co., Ltd. (Zhangzhou, Fujian, China); and the Silasorb phenyl (SiO_2-phenyl) used was from Elsico (Moscow, Russia) without further purification. Water was purified via a Direct-Q system (Millipore, Darmstadt, Germany).

2.2. Methods

2.2.1. Synthesis of the Zirconia and Alumina Sorbents

The synthesis of zirconia and alumina, which are sorbents for the DEET repellent, has been described in detail previously [21]. The sol–gel method for the preparation of these oxides involves the interaction of zirconyl nitrate (or aluminum nitrate) with ammonium nitrate in the presence of citric acid and ethylene glycol. The mixture was then heated at 95 °C until evaporation occurred and subsequently ignited. The resulting product was purified from residual carbon at 800 °C for 2 h.

2.2.2. Static Sorption of DEET

A detailed description of the sorption of the repellent DEET on inorganic oxides has also been described in a previous paper [21]. When sorption was complete, the oxides were quantitatively removed from the solution and dried for 2 h using an IKA RV digital 10 V rotary evaporator (Staufen, Germany) to remove residual acetonitrile.

2.2.3. Preparation of Polymeric Films with DEET-Containing Inorganic Sorbents

In total, 2 g of sorbent (Al_2O_3, ZrO_2, SiO_2-phenyl) containing the DEET repellent was mixed with 7 g of polyacrylic acid copolymer in a 50 mL glass beaker and stirred for 10 min at room temperature. Then, 0.25 mL of the obtained mixture was placed in a mold (15 mm diameter, 0.5 mm height) and centrifuged using an EZ4 spin coater (Lebo Science, Beijing, China) for 5 min at 300 rpm. The polymeric opaque film (15 mm diameter, 0.5 mm thickness) was then removed from the mold and dried at 30 °C for 3 h.

2.2.4. Scanning Electron Microscopy (SEM)

The morphology of the samples was examined on a MIRA3 autoemission scanning electron microscope (TESCAN, Brno, Czech Republic) with an SE detector at an acceleration voltage of 1 kV.

2.2.5. FTIR Spectroscopy

The FTIR spectra were recorded on an InfraLUM FT-08 FTIR spectrometer in ATR FTIR mode (Lumex, St. Petersburg, Russia) via a KBr beam splitter in the spectral range of 600–3000 cm^{-1}. In OMNIC data acquisition software (OMNIC 8.3, Thermo Fisher Scientific, Waltham, MA, USA) of the FTIR spectrometer, the resolution was set at 4 cm^{-1}, and optical aperture as "open", number of points—3662 and 3527 for DEET in liquid film and polymer films, respectively.

2.2.6. Thermogravimetric Analysis of Polymer Films Containing DEET

Thermogravimetric analysis (TGA) (Thermal Analyser—STA 449 F5 Jupiter, Netzsch, Selb, Germany) was performed for samples (10 ± 2 mg weight) with different sorbents and for a sample of polyacrylate film (15 mm diameter, 0.5 mm thickness) without sorbents. The samples were heated to 800 °C under nitrogen at a heating rate of 10 °C/min.

2.2.7. Thermal Diffusion Parameters of DEET from Polymer Films

The polymeric film (15 mm diameter, 0.5 mm thickness) containing DEET immobilized on the sorbent was placed on a glass substrate and heated to 50, 100, 150 and 200 °C (Figure 1). After the material was held at a given temperature for 1, 5, 10, 15 and 30 min, an air sample was taken via a UOPV 4–40 aspirator (Niki-Mlt, St. Petersburg, Russia) with two Petri absorbers connected in series, each filled with 10 mL of acetonitrile. The air sample was taken at a rate of 2 $L \cdot min^{-1}$ for 5 min. The solvent was then evaporated, and the sample was diluted in 5 mL of acetonitrile. The DEET content of the samples was determined via high-performance liquid chromatography with diode array detection (DAD) on an Ultimate 3000 HPLC-system (Thermo Fisher Scientific, Waltham, MA, USA). The concentration of DEET in the samples was calculated via absolute calibration at a wavelength of 210 nm.

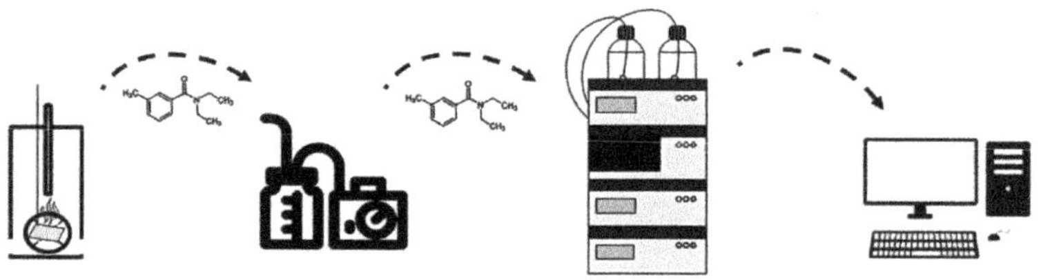

Figure 1. Scheme of the experiment for determining the diffusion flux of *N,N*-diethyl-3-methylbenzamide from a polymeric film material.

From the data obtained, DEET concentration values (mol, 10^{-9}) were calculated and plotted as a function of the flux J_f ($W \cdot m^{-1} \cdot K^{-1}$), which was calculated via the Fourier equation [22] (1)

$$J_f = \frac{Q}{\varepsilon \Delta x T_2} + K\frac{(T_1 - T_2)^2}{T_1 T_2}, \quad (1)$$

Q—heat supplied to the polymeric material, W;
Δx—thickness of the polymeric material layer, m;
ε—an arbitrary number smaller but close to 1;
T_1—material heating temperature, K;
T_2—material surface temperature, K;
K—thermal conductivity constant of the polymeric film material, $W \cdot m^{-1} \cdot s^{-1} \cdot K^{-1}$.
For each of the polymeric film formulations, 10 replicates were performed.

2.2.8. Coating of Textile Fibers with DEET-Containing Polymer Films

A 5 × 5 cm polyester textile sample (density of 160 g/m^2) was immersed in a solution of the copolymer of polyacrylic acid with the inorganic sorbents, obtained according to point 2.2.3 of the Methods Section, and kept for 3 min. The sample was then wiped out and dried at 170 °C for 3 min. The samples were then stored in heat-sealed polyester bags to prevent the loss of repellency.

2.2.9. Determination of DEET Concentration in Textile Samples

The textile samples obtained according to Section 2.2.8 in the Methods Section were cut into small pieces and placed in a 100 mL round-bottom flask to which 50 mL of ethyl acetate was added. The solution was refluxed for 5 h with stirring, the pieces of textile were removed, the contents of the flask were evaporated (IKA RV digital 10 V, Staufen, Germany), and the sample was diluted in 5 mL of acetonitrile. The DEET content was determined under the conditions described in the Methods Section 2.2.7. Five replicates were performed for each textile sample.

2.2.10. Data Analysis

Chromatographic data were collected and processed via Chromeleon 7 software (Thermo Fischer Scientific, Waltham, MA, USA). Excel 2019 (Microsoft Corporation, Redmond, WA, USA) and OriginPro 2018 version b9.5.1.195 (Origin Corp., Northampton, MA, USA) were used for detailed calculations and plotting. Correlation between the values was determined using Spearman's criteria [23]. Significance was set at $p = 0.05$ for all calculations.

3. Results and Discussion

3.1. Polymer Films Containing DEET

The inorganic oxides Al_2O_3, ZrO_2 and SiO_2-phenyl were selected as sorbents for the DEET repellent according to the results of previous studies [21], which showed pronounced sorption capacities around 63, 119 and 239 mg/g, respectively. To potentially exploit the sorption properties of the sorbents, inorganic substances were incorporated into a polyacrylate formulation based on 2-(perfluorodecyl)ethyl acrylate and stearyl acrylate. The choice of polyacrylates was due to their widespread use in the textile industry for fixing substances on the surface of fabrics and their ability to retain active substances on their surface when exposed to external influences. However, DEET is poorly soluble in water when mixed with selected polyacrylates; so, alternative solutions are needed to obtain functional coatings with repellent properties. It was possible to fix solid oxide particles in the polymeric layer when the sorbents were mixed with polyacrylates (Figure 2). Scanning electron microscopy results revealed that the sorbent particles were successfully incorporated into the polymer layer in all the cases. The particle size and shape of the inorganic sorbents are in agreement with the SEM data obtained in previous studies [21].

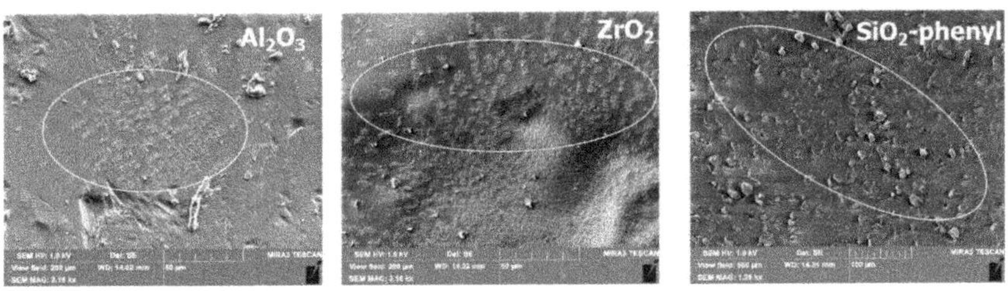

Figure 2. Scanning electron microscopy images of polymer films with embedded inorganic sorbents Al_2O_3, ZrO_2 and SiO_2-phenyl containing DEET.

The presence of the DEET repellent in the polymeric materials containing the Al_2O_3, ZrO_2 and SiO_2 phenyl sorbents was confirmed by FTIR spectroscopy, as signals characteristic of *N,N* diethyl 3 methylbenzamide (945, 1070, 1160, 1349, 1460, 1460, 1490, 1700, 2874, 2934, 2973 cm^{-1}) were detected (Figure 3), indicating that DEET was successfully incorporated into the polymeric materials in all the cases.

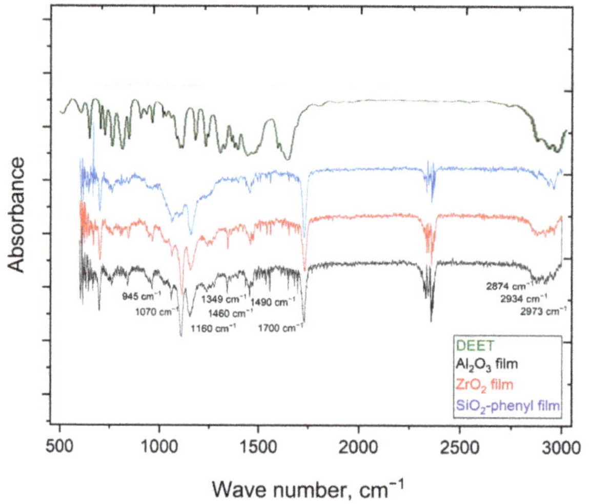

Figure 3. FTIR spectra of DEET in liquid film and polymer films with the incorporated inorganic sorbents such as Al_2O_3, ZrO_2 and SiO_2-phenyl.

3.2. Determination of the Thermal Diffusion Parameters of DEET from Polymer Films

To determine the thermal diffusion parameters of DEET from polymeric materials with embedded inorganic sorbents, TGA studies were performed, and their derivatives were plotted (Figure 4).

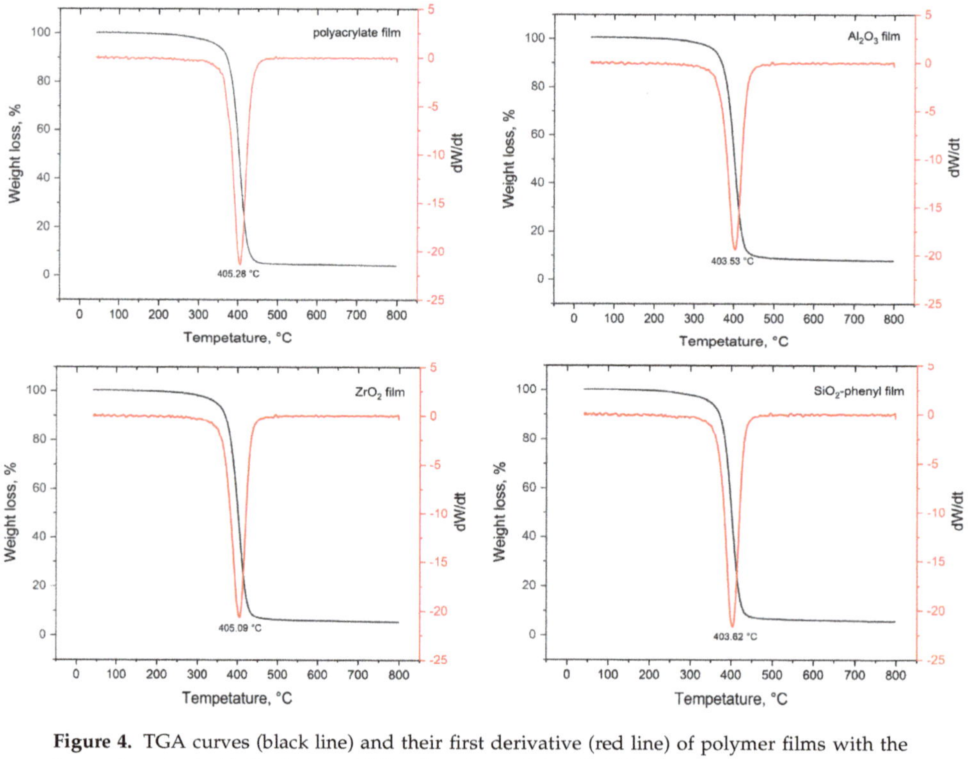

Figure 4. TGA curves (black line) and their first derivative (red line) of polymer films with the inorganic sorbents Al_2O_3, ZrO_2 and SiO_2-phenyl containing the DEET repellent.

The decomposition temperatures of all the polymeric materials ranged from 403.5 to 405.3 °C, which is the same as the decomposition temperature of the polyacrylate copolymer. In the 40–400 °C section of the TGA curves (Figure 4), other volatile components of the system are indicated, i.e., water, which is a component of the polyacrylate copolymer, and the DEET repellent, which has a boiling point of 288 °C. However, according to the TGA curves obtained, the study of the release rate of the repellent was complicated by the low amount of repellent fixed in the polymer material; so, it was not possible to determine the optimum energy for the controlled release of DEET from the polymer surface.

To calculate the thermal diffusion parameters of DEET, an experiment was carried out in which polymer films with the inorganic sorbents Al_2O_3, ZrO_2 and SiO_2-phenyl were heated, followed by air sampling over the surface of the materials (Figure 1). This approach made it possible to study the rate of release of the repellent during successive heating of the polymeric film material. DEET concentration versus temperature was plotted for all the polymeric materials (Figure 5).

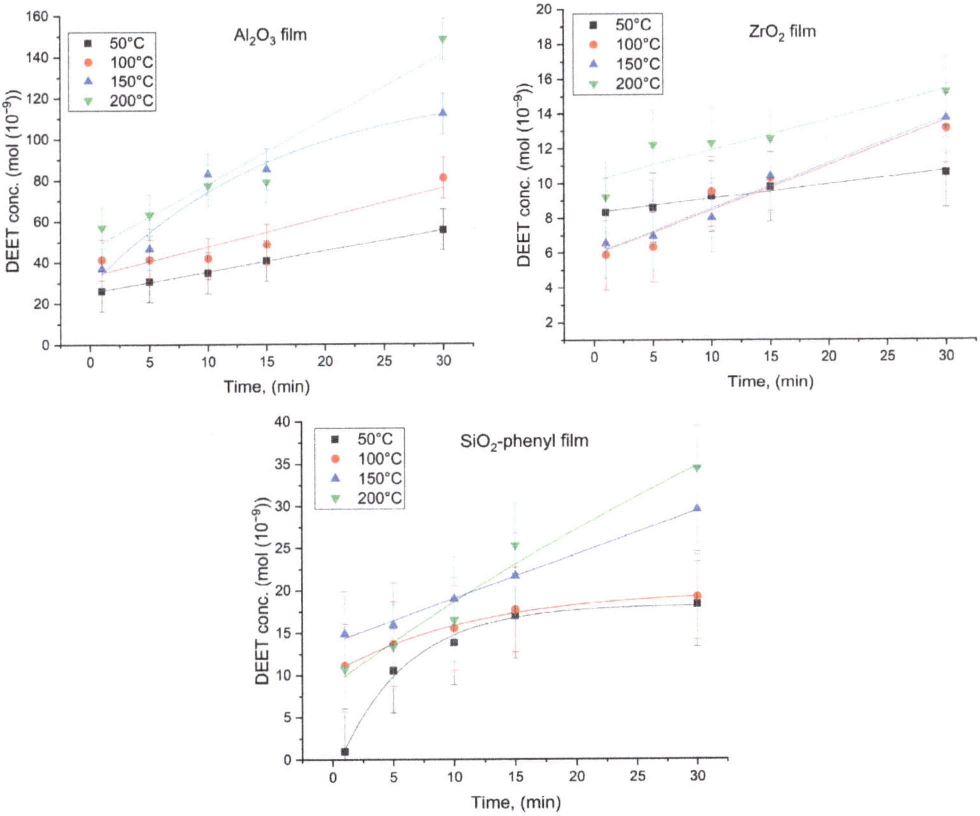

Figure 5. Concentration of DEET (mol) formed by thermal diffusion on the surface of polymer films containing the inorganic sorbents Al_2O_3, ZrO_2 and SiO_2-phenyl as a function of time (min) (n = 10).

The plots obtained show that the concentration of repellent gradually increased with increasing exposure time of the polymeric materials under the given conditions. Moreover, significantly higher DEET concentrations were observed for the Al_2O_3 sorbent-based materials (up to $148.3 \cdot 10^{-9}$ mol at 200 °C) and, to a lesser extent, for the ZrO_2 and SiO_2-phenyl materials (up to $15.2 \cdot 10^{-9}$ mol and $34.3 \cdot 10^{-9}$ mol at 200 °C, respectively). This could be due to the porosity of these inorganic sorbents, with the mean pore size of Al_2O_3

being the smallest (3.84 nm—Al_2O_3, 9.85 nm—SiO_2-phenyl, 46.0 nm—ZrO_2) [21]. Analysis of the 200 °C DEET concentration data and sorbent average pore size values revealed a rank correlation of these parameters with a coefficient of +1 by Spearman's test. It can be assumed that DEET thermal diffusion is significantly influenced by sorbent pore size. Moreover, we were not able to determine a single dependence that would describe the rate of release of the repellent for all the sorbents, because the shapes of the Al_2O_3, ZrO_2 and SiO_2-phenyl particles were different according to the SEM data. Notably, despite the greater sorption capacity of inorganic SiO_2-phenyl oxide (239 mg/g) at the considered DEET ratio compared to those of Al_2O_3 (119 mg/g) and ZrO_2 (63 mg/g) oxides, the amount of repellent diffusing from the surface of the polymeric film within 30 min was greater for the material containing the Al_2O_3 sorbent. Obviously, such materials would be able to repel faster, but SiO_2-phenyl-based materials could last longer.

In addition, the data show that less heat than that required to reach the boiling point of DEET itself (288 °C) can be used to release the repellent from the material. In this way, the targeting efficiency of DEET-based materials can be enhanced by thermal diffusion processes that allow the repellent concentration to be controlled and fully predictable.

To determine the parameters of the total flux of matter, the values of the heat flux J_f ($W \cdot m^{-1} \cdot K^{-1}$) were calculated for each polymeric material on the basis of the provisions of the Onsager theory and the Fourier law [22]. From the data obtained, the DEET repellent concentration determined by thermal diffusion was plotted against the heat flux acting on the materials after 30 min (Figure 6).

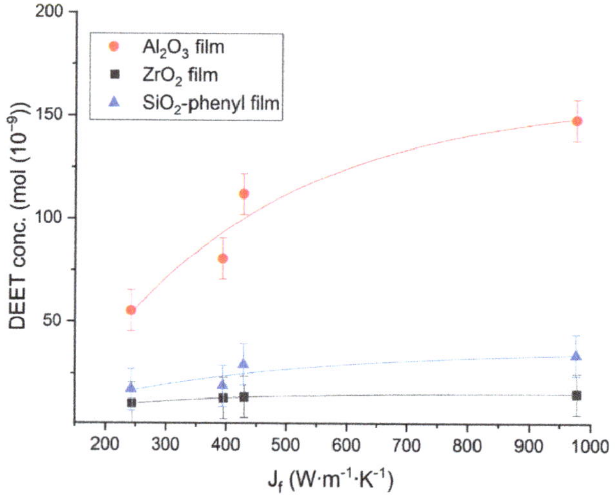

Figure 6. The dependence of the concentration of DEET (mol) as a result of thermal diffusion over the surface of polymer films containing the inorganic sorbents Al_2O_3, ZrO_2 and SiO_2-phenyl on the heat flux J_f ($W \cdot m^{-1} \cdot K^{-1}$) at 30 min (n = 10).

The heat flow on the polymer films containing Al_2O_3 could lead to DEET diffusion in air faster and more efficiently than when applied to the films containing the ZrO_2 or SiO_2 phenyl sorbents. Moreover, the repellent concentrations obtained from the polymer materials of different composition in this study showed only basic differences and could be increased by using more sorbent in the polyacrylate composition.

3.3. Study of Polymer Films Containing the Sorbents with the DEET Repellent, Applied to Textile Materials

The obtained polymer films containing Al_2O_3, ZrO_2 and SiO_2-phenyl with the DEET repellent, were coated on textiles via thermal polymerization of the polyacrylate copolymer at 170 °C. This approach not only allowed the repellent to be successfully incorporated into the surface of the textile but also resulted in materials that were resistant to environmental factors.

The structure of the samples obtained was examined by scanning electron microscopy to confirm that the inorganic sorbents with the DEET repellent successfully coated the textiles fibers (Figure 7). The resulting images showed that Al_2O_3, ZrO_2 and SiO_2-phenyl were all attached to the surface of the polymerized polyacrylate layer on the textile fibers, providing a repellent effect.

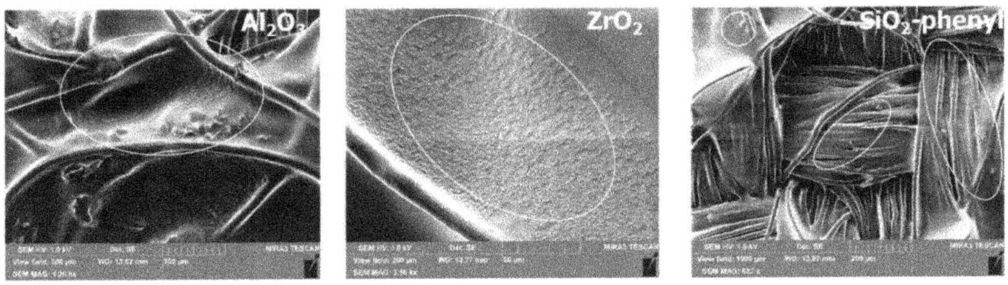

Figure 7. Scanning electron microscopy images of textile materials with incorporated Al_2O_3, ZrO_2 and SiO_2-phenyl inorganic sorbents containing DEET.

The concentration of DEET in the obtained textile samples was determined using HPLC- DAD after extraction of the repellent in ethyl acetate for 5 h. The experimental time was optimized by repeating the extraction under the same conditions. The results are summarized in Table 1.

Table 1. Results of the determination of the DEET repellent content in textile materials.

	Al_2O_3 Textile Material	ZrO_2 Textile Material	SiO_2-Phenyl Textile Material
DEET content, mg/m^2	66.7 ± 3.3	46.0 ± 2.3	85.6 ± 4.3
DEET content, mg/m^2 (re-extraction)	<0.12	<0.12	<0.12
Recovery rate, %		>99	

The results obtained fully correlated with the sorption capacity data of the sorbents themselves, with the highest DEET repellent content being a feature of SiO_2-phenyl oxide. A similar relationship was found for the sorption capacity of the textile materials containing Al_2O_3 and ZrO_2 oxides. The suitability of this method for the determination of repellent content in textile samples is shown by the DEET content on re-extraction under the given conditions (<0.12) and the extraction rate of this method (over 99%). It was shown that by creating functional polyacrylate-based coatings, it was possible to successfully deliver the difficult-to-dissolve repellent DEET to textiles, opening new possibilities for creating controlled-release repellent materials.

4. Conclusions

In this study, polymer films based on polyacrylate (2-(perfluorodecyl)ethyl acrylate and stearyl acrylate) containing the inorganic sorbents Al_2O_3, ZrO_2 and SiO_2-phenyl with the DEET repellent were prepared. The high sorption capacity of oxides allows them to be used as carrier materials for repellents in functional coatings for the targeted delivery of repellents to a variety of surfaces, such as textiles. The obtained scanning electron microscopy images and FTIR spectroscopy data confirmed that this approach can be used to obtain polymer films with repellent properties.

The thermal diffusion of DEET from the surface of polymeric materials was also the subject of detailed investigations. According to the results obtained, materials containing the Al_2O_3 sorbent are able to thermally diffuse the DEET repellent from their surface faster and at higher concentrations than materials containing the other ZrO_2 and SiO_2-phenyl oxides, despite the greater sorption capacity of phenyl group-modified silica. Release of the repellent from polymeric materials can also be achieved by applying less heat than is required to reach the boiling point of DEET. In this way, the resulting materials can be acted upon in a controlled manner by means of thermal diffusion processes, which helps to improve their target efficiency.

The possibility of using these polymer materials as functional coatings on textiles was demonstrated in the final step of the research. The SEM images and HPLC data obtained show that a polyacrylate formulation containing the inorganic sorbents Al_2O_3, ZrO_2 and SiO_2-phenyl with the DEET repellent was successfully applied to textiles, opening new possibilities for developing controlled-release materials.

Author Contributions: Conceptualization, S.Z. and S.A.; methodology, S.A.; software, S.Z. and M.R.; formal analysis, S.Z.; investigation, S.Z., EA. and V.M.; resources, S.Z., V.V. and M.R.; data curation, S.Z. and E.A.; writing—original draft preparation, S.Z.; writing—review and editing, S.A. and V.V.; visualization, S.Z.; supervision, S.A. and V.V.; project administration, S.A. All authors have read and agreed to the published version of the manuscript.

Funding: This research received no external funding.

Data Availability Statement: Data is contained within the article.

Acknowledgments: The authors are grateful to the staff of the MIREA—Russian Technological University (M.V. Lomonosov Institute of Fine Chemical Technologies) and Gubkin University for their assistance in conducting physicochemical studies.

Conflicts of Interest: The authors declare no conflicts of interest.

Abbreviations

The following abbreviations are used in this manuscript:

DEET	*N,N*-diethyl-3-methylbenzamide
SEM	Scanning electron microscopy
FTIR	Fourier transform infrared spectroscopy
ATR	Attenuated total reflectance
TGA	Thermogravimetric analysis
HPLC	High-performance liquid chromatography
DAD	Diode array detection

References

1. Tiwari, I.; Mahanwar, P.A. Polyacrylate/silica hybrid materials: A step towards multifunctional properties. *J. Dispers. Sci. Technol.* **2019**, *40*, 925–957. [CrossRef]
2. Shi, H.; Hong, L.; Pan, K.; Wei, W.; Liu, X.; Li, X. Biodegradable polyacrylate copolymer coating for bio-functional magnesium alloy. *Prog. Org. Coat.* **2021**, *159*, 106422. [CrossRef]
3. Wu, L.; Baghdachi, J. *Functional Polymer Coatings: Principles, Methods, and Applications*; John Wiley & Sons, Inc.: Hoboken, NJ, USA, 2015; pp. 1–369.
4. Coetzee, D.; Militky, J.; Venkataraman, M. Functional coatings by natural and synthetic agents for insect control and their applications. *Coatings* **2022**, *12*, 476. [CrossRef]
5. Sikder, A.; Pearce, A.K.; Parkinson, S.J.; Napier, R.; O'Reilly, R.K. Recent trends in advanced polymer materials in agriculture related applications. *ACS Appl. Polym. Mater.* **2021**, *3*, 1203–1217. [CrossRef]

6. Puoci, F.; Iemma, F.; Spizzirri, U.G.; Cirillo, G.; Curcio, M.; Picci, N. Polymer in agriculture: A review. *Am. J. Agric. Biol. Sci.* **2008**, *3*, 299–314. [CrossRef]
7. Mapossa, A.B.; Focke, W.W.; Tewo, R.K.; Androsch, R.; Kruger, T. Mosquito-repellent controlled-release formulations for fighting infectious diseases. *Malar. J.* **2021**, *20*, 165. [CrossRef]
8. Zverev, S.A.; Andreev, S.V.; Sakharov, K.A.; Akhmetshina, M.B.; Istomina, L.I.; Verzhutskaya, Y.A.; Shashina, N.I. Evaluation of the efficacy of permethrin-and cypermethrin-based textile against taiga tick, Ixodes persulcatus. *Exp. Appl. Acarol.* **2023**, *89*, 275–286. [CrossRef]
9. Duan, J.; Xu, Y.; Lei, Y.; Cheng, Y.; Tan, Y.; Zhu, S.; Xia, H.; Li, S.; Shi, P.; Tang, J. Preparation and properties of water-soluble coatings resistant to scratching on the surface of nuclear cladding. *Nucl. Eng. Des.* **2024**, *417*, 112831. [CrossRef]
10. Elsayed, G.A.; Hassabo, A.G. Insect repellent of cellulosic fabrics (a review). *Lett. Appl. NanoBioScience* **2022**, *11*, 3181–3190. [CrossRef]
11. Mapossa, A.B.; da Silva Júnior, A.H.; Mhike, W.; Sundararaj, U.; Silva de Oliveira, C.R. Electrospun Polymeric Nanofibers for Malaria Control: Advances in Slow-Release Mosquito Repellent Technology. *Macromol. Mater. Eng.* **2024**, *309*, 2400130. [CrossRef]
12. Fulton, A.C.; Thum, M.D.; Jimenez, J.; Camarella, G.; Cilek, J.E.; Lundin, J.G. Long-Term Insect Repellent Electrospun Microfibers from Recycled Poly (ethylene terephthalate). *ACS Appl. Mater. Interfaces* **2023**, *15*, 44722–44730. [CrossRef] [PubMed]
13. Chatha, S.A.S.; Asgher, M.; Asgher, R.; Hussain, A.I.; Iqbal, Y.; Hussain, S.M.; Bilal, M.; Saleem, F.; Iqbal, H.M. Environmentally responsive and anti-bugs textile finishes–Recent trends, challenges, and future perspectives. *Sci. Total Environ.* **2019**, *690*, 667–682. [CrossRef] [PubMed]
14. Peila, R.; Scordino, P.; Shanko, D.B.; Caldera, F.; Trotta, F.; Ferri, A. Synthesis and characterization of β-cyclodextrin nanosponges for N, N-diethyl-meta-toluamide complexation and their application on polyester fabrics. *React. Funct. Polym.* **2017**, *119*, 87–94. [CrossRef]
15. Bonadies, I.; Longo, A.; Androsch, R.; Jehnichen, D.; Göbel, M.; Di Lorenzo, M.L. Biodegradable electrospun PLLA fibers containing the mosquito-repellent DEET. *Eur. Polym. J.* **2019**, *113*, 377–384. [CrossRef]
16. Ap dos Santos, D.; Oliveira, A.M.; Cerize, N.N.; Manhani, K.C.; Santos, D.D.S.; de Araujo, H.C.; Costa, S.A.; Costa, S.M. Synthesis of polymeric particles with insect repellent for potential application on textile substrates. *Mater. Chem. Phys.* **2022**, *280*, 125662. [CrossRef]
17. Gilbert, I.H.; Gouck, H.K. Evaluation of repellents against mosquitoes in Panama. *Fla. Entomol.* **1955**, *38*, 153–163. [CrossRef]
18. Xiang, C.; Etrick, N.R.; Frey, M.W.; Norris, E.J.; Coats, J.R. Structure and properties of polyamide fabrics with insect-repellent functionality by electrospinning and oxygen plasma-treated surface coating. *Polymers* **2020**, *12*, 2196. [CrossRef]
19. Singh, A.; Sheikh, J. Development of multifunctional polyester using disperse dyes based through a combination of mosquito repellents. *J. Mol. Struct.* **2021**, *1232*, 129988. [CrossRef]
20. Akbarzadeh, A.; Mokhtari, J.; Kolkoohi, S.; Amin Sarli, M. Imparting insect repellency to nylon 6 fibers by means of a novel MCT reactive dye. *J. Appl. Polym. Sci.* **2012**, *126*, 1097–1104. [CrossRef]
21. Zverev, S.A.; Vinogradova, Y.V.; Selivanova, A.A.; Solovov, R.D.; Sakharov, K.A.; Ischenko, A.A.; Andreev, S.V. Study of sorption properties of zirconia, alumina, and silica in relation to repellents. *Colloid Polym Sci.* **2024**, *302*, 1259–1268. [CrossRef]
22. Sher, E.; Moshkovich-Makarenko, I.; Moshkovich, Y.; Cukurel, B. Implementation of the Onsager theorem to evaluate the speed of the deflagration wave. *Entropy* **2020**, *22*, 1011. [CrossRef]
23. Spearman, C. Correlation calculated from faulty data. *Br. J. Psychol.* **1910**, *3*, 271–295. [CrossRef]

surfaces

Article

Marangoni Flow-Driven Self-Assembly of Biomimetic Jellyfish-like Hydrogels for Spatially Controlled Enzyme Catalysis

Aoxiang Zhang [1,2], Huiying Zhou [2], Yanhui Guo [2] and Yu Fu [1,2,*]

1 School of Chemical and Environmental Engineering, Sichuan University of Science and Engineering, Zigong 643000, China; asongchueng@outlook.com
2 Department of Chemistry, College of Sciences, Northeastern University, Shenyang 110819, China; zhmoo@foxmail.com (H.Z.); gyh747474@163.com (Y.G.)
* Correspondence: yufu@suse.edu.cn

Abstract: Enzymatic catalysis has gained significant attention in green chemistry due to its high specificity and efficiency under mild conditions. However, challenges related to enzyme immobilization and spatial control often limit its practical applications. In this work, we report a Marangoni flow-driven strategy to fabricate a biomimetic jellyfish-like hydrogel with tunable tentacle-like structures. The formation process occurs entirely in an aqueous system without organic solvents or post-treatment, enabling the construction of ultra-thin, free-standing hydrogels through spontaneous interfacial self-assembly. The resulting structure exhibits high surface-area geometry and excellent biocompatibility, providing a versatile platform for localized enzyme loading. This method offers a simple and scalable route for engineering soft materials with complex morphologies, and expands the design space for bioinspired hydrogel systems.

Keywords: Marangoni flow; hydrogels; polyelectrolyte; biomimetic

1. Introduction

Enzymatic catalysis has become an important part of green chemistry [1], valued for its high specificity, efficiency, and sustainability under mild reaction conditions [2,3]. However, the practical application of free enzymes faces significant challenges. When dispersed directly in aqueous solutions, enzymes lack spatial control over their catalytic activity, leading to unregulated reaction zones and potential limitations in substrate diffusion. This spatial ambiguity not only reduces catalytic efficiency but also complicates process optimization, especially in industrial settings. Moreover, traditional enzyme immobilization strategies typically rely on rigid carriers (nanoparticles or metal–organic frameworks [4]) to stabilize enzymes and facilitate recycling. For example, Wei et al. reported a solid-state mechanochemical strategy to encapsulate enzymes within metal–organic frameworks (MOFs), effectively protecting the enzyme activity under biologically unfavorable conditions and enabling more stable catalytic performance [5]. While these materials improve enzyme stability, their inherent lack of biocompatibility and their non-degradable nature raise environmental concerns, particularly in sensitive ecosystems or biomedical applications. For instance, micro- or nano-sized substrates, despite enhancing mass transfer efficiency, are notoriously difficult to recover, resulting in secondary pollution and increased operational costs [6].

 https://doi.org/10.3390/surfaces8020028

To overcome these dual challenges of spatial control and environmental compatibility, bioinspired architectures have emerged as innovative enzyme immobilization platforms [7,8]. Li et al., inspired by the adhesive capability of mussel foot proteins, developed a bio-adhesive hydrogel with strong wet adhesion properties, offering an adaptable and robust interface for biological applications [9]. Drawing inspiration from natural systems, such as the predation mechanism of jellyfish [10–12], offers a promising approach to reconciling efficient catalysis with sustainable design. Jellyfish utilize flexible tentacles and dynamic fluid interactions to capture prey efficiently [13,14]—a process analogous to the need for enzymes to interact with substrates in a controlled spatial domain. Mimicking this strategy, we propose a biomimetic jellyfish-like hydrogel system formed by Marangoni flow at liquid–liquid interfaces [15–17]. The Marangoni effect refers to the mass transfer of fluid along an interface caused by a gradient in surface tension. This flow occurs when there is a difference in surface tension between two adjacent regions of a liquid [18,19]. This design leverages hydrodynamic forces to self-assemble a soft, biocompatible hydrogel matrix at the interface, which not only benefits the immobilization of enzymes (such as catalase) but also dynamically regulates their spatial distribution. The Marangoni effect, driven by interfacial tension gradients, enables the spontaneous formation of structured hydrogels without external energy input, making this design inherently aligned with the principles of green chemistry [20,21].

The resulting hydrogel mimics a jellyfish-like tentacle structure, providing an expanded surface area for enzyme attachment. The loose, porous structure of these tentacles enhances stable enzyme binding [22–24], while the use of green materials ensures biocompatibility [25–27]. Furthermore, the macroscopic size of the hydrogel facilitates easy recovery, overcoming the trade-off between the mass transfer efficiency and recyclability seen in micro-/nanosystems [28–30]. In this study, we demonstrate that the integration of Marangoni flow-driven self-assembly and biomimetic hydrogel design offers a novel pathway to engineering enzyme-loaded systems with tunable catalytic performance, environmental compatibility, and scalability. This approach not only overcomes the limitations of conventional enzyme immobilization but also establishes a sustainable framework with broad applications, from industrial biocatalysis to environmental remediation.

2. Materials and Methods

2.1. Materials

Sodium Alginate (Alg, AR, Macklin Biochemical Technology Co., Ltd., Shanghai, China), ε-Poly-L-lysine (ε-PL, MW < 5000, Macklin Biochemical Technology Co., Ltd., Shanghai, China), Fe_3O_4 (magnetite) nanoparticle dispersion (25% in H_2O, Macklin Biochemical Technology Co., Ltd., Shanghai, China), Anhydrous Calcium Chloride ($CaCl_2$, 99.99% (metals basis), Macklin Biochemical Technology Co., Ltd., Shanghai, China). The Catalase (CAT, 3500 μ/mg, Macklin Biochemical Technology Co., Ltd., Shanghai, China) and Hydrogen peroxide (H_2O_2, 3%, LIRCON Medical Technology Co., Ltd., Dezhou, China) were used without further purification. The deionized water was made in our laboratory.

2.2. Methods

A simple experimental setup is illustrated below. The sodium alginate solution was extruded at a rate of 3 μL/s, controlled by a microinjector pump (Figure 1). A needle with an inner diameter of 2 mm was chosen to control the droplet volume, approximately 0.05 mL. The needle was positioned 1 cm above the surface of the underlying solution to ensure that the initial kinetic energy of the droplet was minimal upon impact with the solution, thereby preventing droplet sinking or breakage.

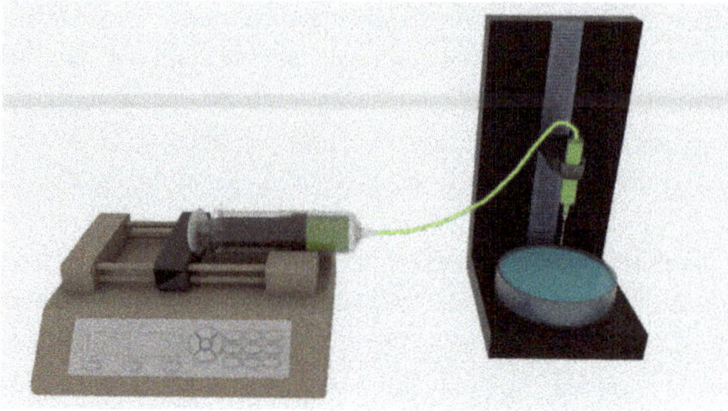

Figure 1. Schematic diagram of the experimental setup.

2.3. Characterization

Field Emission Scanning Electron Microscopy (FESEM): A field emission scanning electron microscope (SU8010, Hitachi High-Tech, Tokyo, Japan) was used to analyze the surface morphology of the samples. During standard scanning, an accelerating voltage of 5 kV was applied. For elemental composition analysis using energy-dispersive X-ray spectroscopy (EDS), the accelerating voltage was increased to 15 kV.

Fourier-Transform Infrared Spectroscopy (FT-IR): A VERTEX70 FT-IR spectrometer (Bruker, Ettlingen, Germany), with a resolution of 2 cm^{-1}, was used for the infrared spectral analysis.

Surface Tension Measurement: A KRÜSS contact angle instrument (Hamburg, Germany) was used to measure the surface tension of each liquid sample by the pendant drop method at 28 °C. Each measurement was repeated three times, and the average value was taken. The surface tension at the liquid–air interface creates an internal pressure difference within the droplet, causing an increase in internal pressure. This relationship between the pressure difference (Δp), the surface curvature radii (r_1 and r_2), and the surface tension (σ) can be described by the Young–Laplace equation:

$$\Delta p = \sigma \cdot (1/r_1 + 1/r_2)$$

By analyzing the droplet morphology to obtain r_1 and r_2, the surface tension of the sample can be calculated [31,32]. Further details on these procedures can be found in Supporting Information Figure S1.

Image Analysis: The areas and perimeters of the jellyfish-like droplets were determined from images using Python's OpenCV library. Briefly, each image was first converted to grayscale to eliminate background noise. Unwanted artifacts were removed, and the resulting image was filtered to reduce noise. Finally, the number of boundary pixels was recorded as the perimeter, and the total number of pixels identified as the droplet served as the area. For images with different scale bars, the conversion from pixels to actual dimensions was based on the number of pixels corresponding to the scale bar length. Further details on these procedures can be found in Supporting Information Figure S2.

MTT Assay: A cell suspension was prepared at a density of 1000–10,000 HIMECs (Human Intestinal Microvascular Endothelial Cells) per well in 100 μL, and the cells were allowed to adhere overnight. The next day, the cells were treated with varying concentrations of the test solutions, prepared by diluting a stock solution as follows:

Alg/SDS Solutions: Solutions of sodium alginate (Alg) containing 0.6 wt%, 0.5 wt%, 0.4 wt%, 0.3 wt%, 0.2 wt%, 0.1 wt%, 0.05 wt%, 0.01 wt%, 0.005 wt%, or 1 wt% SDS were configured.

Biomimetic Jellyfish-like Hydrogels: For samples containing 0.1–0.6 wt% SDS, the solid gel was ground into small particles and dispersed in PBS at 5 wt% for testing. For samples with SDS below 0.1 wt%, an equal volume of 1 wt% ε-polylysine (ε-pl) solution was mixed and stirred together with the corresponding Alg-SDS solution before testing.

After 48 h of incubation at 37 °C with 5% CO_2, the culture medium was removed. Then, 10 μL of 0.5% MTT solution was added to each well, and the plates were incubated for an additional 4 h at 37 °C. Subsequently, the MTT solution was discarded, and 100 μL of dimethyl sulfoxide (DMSO) was added to dissolve the formazan crystals formed by metabolically active cells. The optical density (OD) of each well was measured at 570 nm and 650 nm using a microplate reader. Corrected OD values were obtained by subtracting the OD at 650 nm from the OD at 570 nm.

The Bradford technique is used to calculate enzyme load: A 100 μL CAT/PBS solution with a concentration gradient of 0.05–0.12 mg/mL was placed in a 2 mL centrifuge tube, and 990 μL of Komas Brilliant Blue staining solution was added to stain the free CAT. The standard curve of concentration versus absorbance was plotted. Take 100 μL of the remaining enzyme solution of the biomimetic jellyfish-like hydrogel system in a 2 mL centrifuge tube and repeat the above operation. Standard curves were used to quantify enzyme concentration and the quantity of leftover enzymatic solution, and enzymatic loading was obtained by difference.

The relative enzyme activity calculation: A standard curve of H_2O_2 concentration versus absorbance at 240 nm was established using a UV–vis spectrophotometer with gradient concentrations of H_2O_2 ranging from 0.05% to 1.5%. A biomimetic jellyfish-like hydrogel was immersed in 15 mL of H_2O_2 solution at 25 °C and shaken at 200 rpm. The absorbance of the solution at 240 nm was measured, and the H_2O_2 concentration was calculated based on the standard curve. To rule out the effect of spontaneous H_2O_2 decomposition, a blank control experiment without enzyme addition was conducted. The actual amount of H_2O_2 decomposition catalyzed by the enzyme was determined by subtracting the decomposition observed in the blank control from that in the experimental group. The relative catalytic performance of catalase (CAT) under different conditions was evaluated using relative activity, defined as follows Equation (1):

$$\text{Relative activity (\%)} = \frac{H_2O_2 \text{ decomposed under each condition}}{H_2O_2 \text{ decomposed with free enzyme}} \times 100\% \quad (1)$$

The amount of free enzyme used in the calculation was determined based on the enzyme loading content of the immobilized system.

The catalytic activity of immobilized α-amylase and β-amylase was evaluated using a starch–iodine colorimetric method.

α-Amylase and β-amylase were immobilized onto the biomimetic jellyfish-like hydrogel using the same method, and the enzyme loading was determined accordingly. Either the immobilized enzymes or their free counterparts were immersed in a 0.5 wt% starch solution. After 30 min, 2 mL of the reaction mixture was withdrawn and mixed with 1 mL of 0.005 M iodine solution. After reacting for 20 min, the absorbance at 580 nm was measured using a UV–vis spectrophotometer. The relative enzymatic activity was calculated using the following equation:

$$\text{Relative activity(\%)} = \frac{\text{Starch decomposed under each condition}}{\text{Starch decomposed by free enzyme}} \times 100\% \quad (2)$$

Each catalytic cycle was repeated 12 times to evaluate the reusability of the immobilized enzymes.

3. Results and Discussion

As shown in Figure 2a, when sodium alginate droplets (0.2 wt% SDS, 2 wt% Alg) come into contact with the surface of a ε-PL solution (1 wt%), they rapidly form droplets with tentacle-like structures around the perimeter within approximately 10 s (the case without surfactants is in Supporting Information Figure S3). These tentacles grow exclusively in the two-dimensional plane at the liquid surface (Figure 2b) and do not appear at any other spatial locations. The carboxyl groups of Alg and the amino groups of ε-PL can form a complex through electrostatic interactions, creating a cohesive barrier that prevents further diffusion of solutes. As a result, no tentacles are formed at locations above or below the liquid surface (Figure 2c).

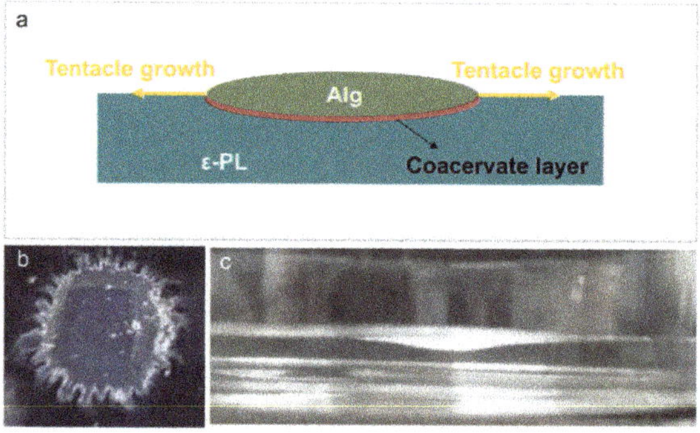

Figure 2. (**a**) A schematic diagram of the formation process of biomimetic jellyfish-like droplets; (**b**) a top-view photograph of the biomimetic jellyfish-like droplet; (**c**) a side-view photograph of the biomimetic jellyfish-like droplet.

By measuring the surface tension of the Alg droplets containing SDS and the ε-PL solution (Table 1), we hypothesize that the formation of tentacles is induced by Marangoni flow at the liquid–liquid interface. The surface tension gradient between the two solutions drives the Marangoni flow, causing fluid to move horizontally from the low surface tension interior of the droplet (35.72 mN/m) toward the high surface tension exterior solution (71.11 mN/m). This outward fluid movement creates favorable conditions for tentacle growth. In contrast, in the vertical direction, the presence of the coacervate layer prevents fluid from escaping, thereby inhibiting tentacle formation.

Table 1. Surface Tension Measurement.

Solution	Surface Tension (mN/m)
Alg	64.69 ± 1.74
Alg with SDS	35.72 ± 1.52
ε-PL	71.11 ± 1.48

By adjusting the surfactant concentration, we modulate the intensity of the Marangoni flow [33] and thus alter the tentacle growth rate. Meanwhile, changing the Alg concentration (which means adjusting the viscosity) regulates the speed of droplet deformation [34]. These two factors act antagonistically to jointly determine the growth process of the tentacles (Figure S4). Intuitively, with increasing surfactant concentration, the tentacles become longer and more densely distributed around the droplet (Figure 3a–c). In contrast, as the solution concentration decreases, the droplet's overall spreading area increases significantly, yet the tentacle morphology does not vary uniformly (Figure 3d). At certain concentrations, the tentacles appear much more elongated and denser than under other conditions (Figure 3e), whereas at lower concentrations, they look swollen (Figure 3f).

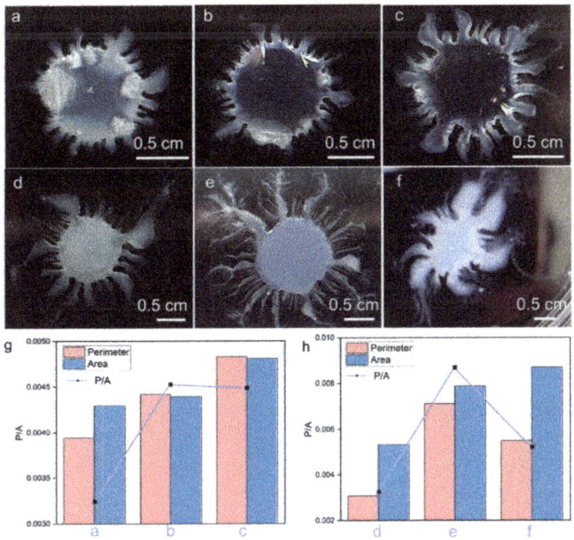

Figure 3. Photographs of biomimetic jellyfish-like droplets formed under different solution conditions: (**a**) 0.1 wt% SDS, 2 wt% Alg; (**b**) 0.2 wt% SDS, 2 wt% Alg; (**c**) 0.3 wt% SDS, 2 wt% Alg; (**d**) 0.2 wt% SDS, 1.5 wt% Alg; (**e**) 0.2 wt% SDS, 1.0 wt% Alg; (**f**) 0.2 wt% SDS, 0.5 wt% Alg. (**g**,**h**) Present the relative perimeter and area, as well as the perimeter-to-area ratio (P/A), calculated via image recognition for the droplets in (**a**–**c**) and (**d**–**f**), respectively.

In order to investigate these changes more precisely, we analyzed the area and perimeter of the jellyfish-like droplets through image processing and then calculated the perimeter-to-area ratio (P/A) to quantify tentacle growth. When the surfactant concentration reached 0.2 wt%, the morphological variations in the tentacles became relatively limited (Figure 3g). Although higher surfactant concentrations can accelerate tentacle growth, the increased solution viscosity limits fluid mobility, thereby constraining the growth rate. As the solution viscosity decreases, such constraints are alleviated, resulting in denser, longer tentacles. However, when the viscosity is too low, the droplet spreads so quickly that many tentacles are covered by the expanding liquid before they can fully form. This behavior manifests as an initial rise, followed by a subsequent decline, in the P/A value (Figure 3h).

Figure 4a shows the tentacle growth of the biomimetic jellyfish-like droplet as captured by a camera. Once the droplet contacts the solution surface, the carboxyl groups in Alg rapidly bind to the amino groups in ε-PL, forming a coacervate barrier through electrostatic interactions. This coacervate layer between the two aqueous phases prevents the further diffusion of solutes from the Alg droplet into the solution beneath [35]. Meanwhile, a strong Marangoni flow drives the droplet to spread outward across the horizontal liquid surface. Since the rate of droplet spread is slower than the rate of interfacial complexation, some peripheral regions form a dense coacervate layer earlier, leading to spatially nonuniform diffusion confinement. In contrast, regions with weaker complexation allow for continued solute transport, giving rise to protrusions and the eventual formation of "tentacle"—like structures. This process occurs within approximately six seconds and the tentacle morphology remains stable thereafter, as shown in the time-lapse images (Figure 4a). To further support the proposed formation mechanism, microscopic imaging of the edge regions (Figure 4b) revealed spatial variations in thickness across the tentacle structure. From left to right, the images represent the tentacle structure extending from its base toward the outer edge. A dense shadow adjacent to the root of the tentacle indicates the formation of a thick coacervate layer at the interface. As the tentacle extends outward, the microscopic images become progressively lighter, suggesting a gradual reduction in material content and concentration. Additionally, the accompanying video (Supporting Information Video S1) dynamically captures the formation and stabilization of the tentacle structures within seconds, further corroborating the proposed mechanism.

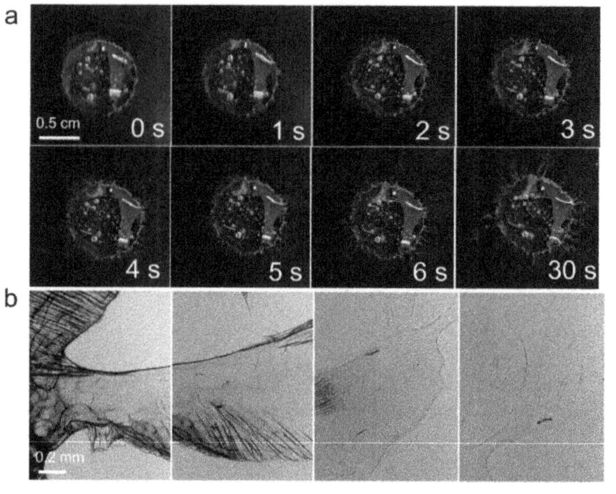

Figure 4. The tentacle formation process of the biomimetic jellyfish-like hydrogel. (**a**) Time-lapse images showing tentacle growth within 6 s after droplet deposition. (**b**) Microscopic images from tentacle root to tip (from **left** to **right**).

The SEM images of the central, transition, and peripheral regions of the jellyfish-like droplets (Figure 5a) reveal that the central region of the droplet is relatively dense (Figure 5b). As the structure extends towards the tentacle edges, it becomes increasingly porous and loose. This porous architecture facilitates the attachment of enzymes and other nanomaterials [36]. In the transition zone at the periphery of the central circular region, which corresponds to the base of the tentacles, wrinkle-like structures are observed (Figure 5c). These wrinkles are likely a result of the antagonistic forces between viscous forces and Marangoni flow. At positions closer to the tentacle edges, a sponge-like porous structure becomes evident (Figure 5d,e), and the tentacle itself becomes extremely thin, with a thickness of approximately 3.46 μm.

Further elemental composition analysis was performed on the three regions using EDS point scanning (Figure 5f–h, Table S1). The low sodium content indicates that the counterions (Na^+) from the Alg and ε-PL complex dissociate into the surrounding solution, providing the coacervate predominantly composed of polymer chains that have lost their coordinating counterions. A significant change in sulfur content was observed as the tentacles developed—the proportion of sulfur, originating from SDS, increased from 2.2% in the center to 5.3% at the edge. Additionally, the oxygen content from the sodium alginate decreased from 28.3% in the center to 11.1% at the periphery, indicating that alginate concentration is higher in the central region and gradually decreases towards the outer tentacles. Another notable change was in the nitrogen content from ε-PL, which increased from 11.8% in the center to 17.5% at the tentacle edge, supporting the idea that ε-PL interacts with SDS in the outer regions. Since the complexation between ε-PL and SDS is weaker than that with alginate, this likely contributes to the looser, more porous sponge-like structure observed at the tentacle periphery (Figure 5e). The coacervate forming the tentacles contains abundant hydroxyl, carboxyl, and amino groups, which facilitate enzyme attachment [37,38]. These findings confirm that as the position shifts from the central region to the tentacle edges, ε-PL/SDS complexes progressively replace ε-PL/Alg complexes. Due to the lower molecular weight and weaker complexation strength of ε-PL/SDS compared to ε-PL/Alg, the central region remains dense and compact, while the tentacle edges represent a looser, more porous structure.

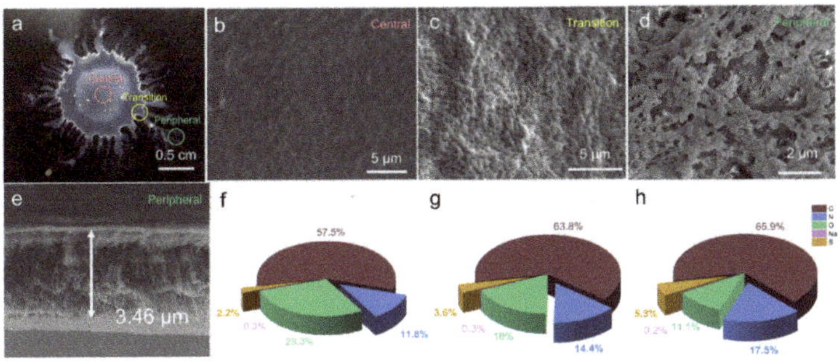

Figure 5. (**a**) An optical image of a biomimetic jellyfish-like droplet formed by depositing 0.2 wt% SDS and 1 wt% Alg into a 1 wt% ε-PL solution, with three distinct regions marked. (**b–d**) SEM images of the jellyfish-like droplet at different regions: (**b**) the central region, (**c**) the transition region, and (**d**) the peripheral region. (**e**) A cross-sectional SEM image of the tentacle edge. EDS point scanning results for elemental composition analysis of (**f**) the central region, (**g**) the transition region, and (**h**) the peripheral region, showing the distribution of C, N, O, Na, and S.

The FT-IR analysis shows peaks at 1407 cm^{-1} and 1630 cm^{-1} (carboxylate $-COO^-$ stretching from sodium alginate), 1036 cm^{-1} (S=O stretching from sulfate ester), and 1540 cm^{-1} (amide II band from N–H bending and C–N stretching). The 1540 cm^{-1} peak is notably stronger in the peripheral region than in the central region, consistent with EDS results, confirming the increased presence of ε-PL/SDS complexes at the periphery (Figure 6).

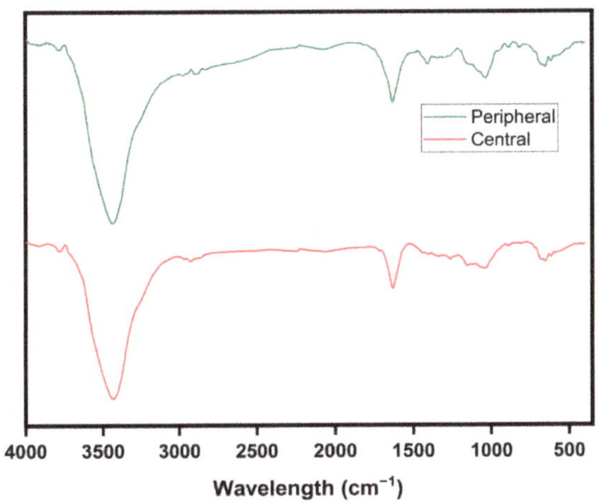

Figure 6. FT-IR spectra of the central and edge regions of a biomimetic jellyfish-like droplet formed by dropping 1 wt% Alg (with 0.2 wt% SDS) onto 1 wt% ε-PL solution.

As all of the materials used are biocompatible, and the porous tentacle structure facilitates enzyme binding, the biomimetic jellyfish-like hydrogel is well-suited for enzyme immobilization in catalytic applications. The porous morphology increases the surface area and provides physical confinement for enzyme molecules (Figure S5). Meanwhile, potential weak interactions—such as hydrogen bonding, electrostatic attraction, or hydrophobic interactions—may also contribute to the immobilization. Given its high perimeter-to-area ratio, we selected 1 wt% Alg with 0.2 wt% SDS for hydrogel fabrication. As shown in Figure 7a, a droplet was first deposited onto a 1 wt% ε-PL solution to form a biomimetic jellyfish-like droplet. After stabilization for 10 min, 0.1 mL of 25 wt% Fe_3O_4 nanoparticle dispersion was added to the center, allowing the nanoparticles to uniformly sediment at the bottom, and they were then left undisturbed for another 10 min. The droplet was then transferred to 0.1 M $CaCl_2$, inducing Alg crosslinking to encapsulate the Fe_3O_4 nanoparticles (10 min). After washing with deionized water, the hydrogel was immersed in 0.25 mg/mL catalase solution, gently stirred (5 h), and incubated (10 min) to allow enzyme attachment. The hydrogel was then washed and stored in PBS (Figure 7b). The immobilization likely involves a combination of physical entrapment facilitated by the hydrogel's porous tentacle structure, as well as possible non-covalent interactions (hydrogen bonding, electrostatic or hydrophobic interactions), similar to mechanisms reported in previous studies [39,40]. Zeta potential measurements further support this interpretation, showing that catalase carries a net negative charge while ε-polylysine (ε-PL) is positively charged under experimental conditions, indicating that electrostatic attraction plays a key role in the enzyme attachment process [41,42]. As the tentacle regions of the hydrogel are enriched in ε-PL, these regions exhibit higher local loading as a result of enhanced electrostatic interaction. MTT assays confirmed the biocompatibility of the system. PBS solutions containing 1 wt% Alg with varying SDS concentrations (Figure 7c), as well as the biomimetic jellyfish-like hydrogels formed from them (Figure 7d), showed no significant changes in corrected OD values compared to the blank group, indicating no cytotoxicity [43,44].

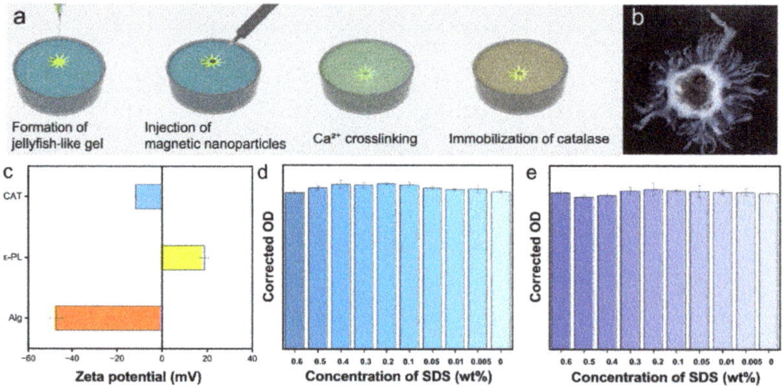

Figure 7. (**a**) A schematic illustration of the biomimetic jellyfish-like hydrogel fabrication process. (**b**) A photo of the fabricated biomimetic jellyfish-like hydrogel. (**c**) The zeta potential measurements of individual components used in the hydrogel system under the experimental conditions, CAT (-11.67 ± 0.42 mV), ε-PL (18.87 ± 1.84 mV), and Alg (-47.36 ± 2.72 mV). (**d**) The cytotoxicity assay results of Alg solutions with varying SDS concentrations. (**e**) The cytotoxicity assay results of biomimetic jellyfish-like hydrogels formed from Alg solutions with varying SDS concentrations.

In practical biocatalytic applications, enzymes are often subjected to environmental fluctuations, including extreme pH or temperatures conditions. Therefore, the chemical and catalytic robustness of the immobilized enzyme system is a key factor for its reusability and practical implementation. Compared with the hydrogel without tentacles, the jellyfish-like hydrogel exhibited higher catalytic efficiency (Figure 8a), which can be attributed to its enhanced enzyme loading capacity enabled by the tentacle structure (Figure S6). As shown in Figure 8b,c, the relative activity of catalase (CAT) immobilized within the biomimetic jellyfish-like hydrogel declined significantly more slowly than that of free CAT under both acidic (pH = 5) and alkaline (pH = 10) conditions, indicating better pH tolerance. This improved pH stability may be attributed to the porous tentacle network of the hydrogel, which physically confines the enzyme and provides a microenvironment that buffers against extreme pH-induced denaturation. In addition, the long-term thermal and storage stability of the immobilized enzyme system was investigated by incubating the jellyfish-like hydrogel-immobilized CAT at different temperatures (40 °C, 25 °C, and 5 °C) for 7 days (Figure 8d–f). Compared with the free CAT, the immobilized CAT retained a significantly higher proportion of its initial activity over time. At 5 °C, a common storage temperature, immobilized CAT maintained its activity better than the free enzyme, indicating that the hydrogel provides structural stabilization across diverse thermal conditions. Moreover, we tested the reusability of the hydrogel platform using other enzymes, including α-amylase and β-amylase. As illustrated in Figure 8g–i, all three enzymes retained over 50% of their initial activity after 12 catalytic cycles, highlighting the practical reusability of the jellyfish-like hydrogel framework for different enzyme systems. These results demonstrate that the jellyfish-like hydrogel not only enables effective enzyme immobilization but also enhances enzyme durability under multiple conditions, thereby confirming its potential for robust and reusable biocatalysis.

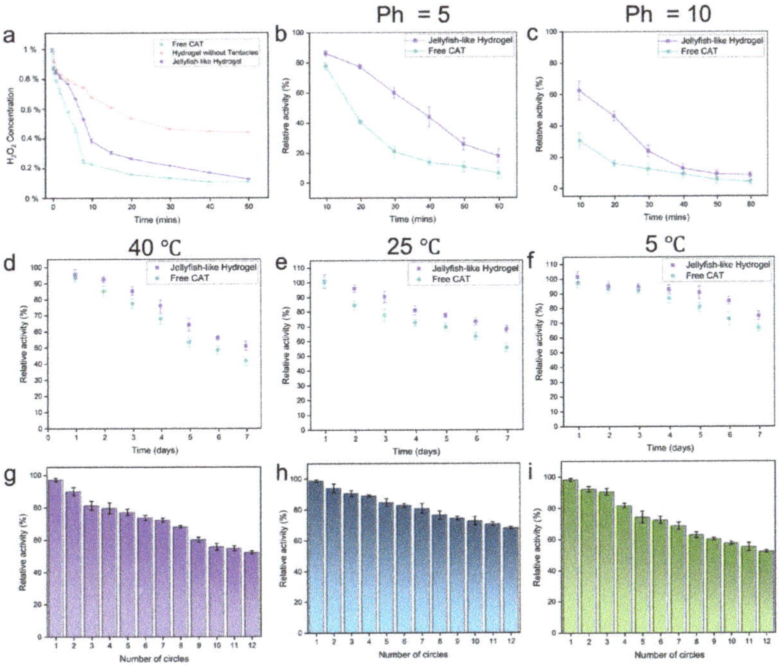

Figure 8. (**a**) Decomposition of H_2O_2 catalyzed by free CAT, CAT immobilized in hydrogel without tentacles, and CAT immobilized in jellyfish-like hydrogel over time. (**b**,**c**) Relative activity of free CAT and jellyfish-like hydrogel-immobilized CAT after incubation at pH = 5 (**b**) and pH = 10 (**c**) in PBS buffer for up to 60 min. (**d**–**f**) Relative activity of immobilized and free CAT incubated at 40 °C (**d**), 25 °C (**e**), and 5 °C (**f**) for 7 days, respectively. (**g**–**i**) Relative activity of catalase (**g**), α-amylase (**h**), and β-amylase (**i**) immobilized in jellyfish-like hydrogel over 12 catalytic cycles.

The biomimetic jellyfish-like hydrogel was used to simulate a spatially controlled enzymatic reaction in 0.2 wt% H_2O_2 solution (Supporting Information Video S2). Figure 9a–c demonstrate its controlled movement under non-contact magnetic actuation, forming a rectangular trajectory. In contrast, Figure 9d–f show a linear back-and-forth motion of a non-jellyfish-like hydrogel under identical conditions, suggesting the maneuverability of the biomimetic jellyfish-like hydrogel and the effective functionality of the magnetically loaded nanoparticles introduced via simple injection. The distinct motion trail left by the hydrogel is attributed to oxygen microbubbles adhering to the container bottom, whereas the non-jellyfish-like hydrogel leaves no visible trace. This suggests a significantly higher catalase loading capacity in the biomimetic jellyfish-like hydrogel, which is directly related to its high perimeter-to-area ratio and porous tentacle structure, providing an expanded enzyme immobilization space.

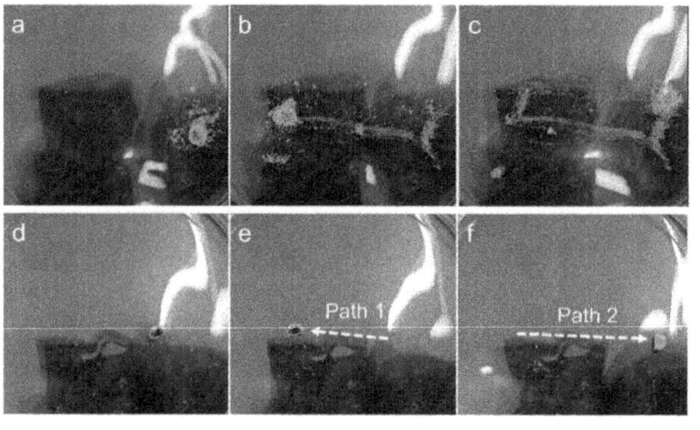

Figure 9. The motion states of the biomimetic jellyfish-like hydrogel at (**a**) 0 s, (**b**) 6 s, and (**c**) 12 s, respectively. (**d–f**) The motion states of the hydrogel without surfactant at (**d**) 0 s, (**e**) 3 s, and (**f**) 6 s, respectively.

4. Conclusions

In this study, we developed a biomimetic jellyfish-like hydrogel with a highly tunable structure and enhanced enzymatic catalytic performance by leveraging Marangoni flow-driven self-assembly at the liquid–liquid interface. The formation mechanism was systematically investigated, revealing that SDS-induced Marangoni flow drives tentacle formation, while Alg viscosity restricts outward extension, resulting in a high perimeter-to-area ratio and a tentacle-rich structure. SEM and EDS analyses confirmed the gradual transition from a dense central region to a porous tentacle periphery, which is attributed to the progressive replacement of ε-PL/Alg complexes by ε-PL/SDS complexes. FT-IR spectroscopy further validated the composition distribution, supporting the structural findings.

By incorporating Fe_3O_4 nanoparticles and catalase, the hydrogel demonstrated excellent spatial control and catalytic efficiency, as evidenced by its precise magnetically guided motion and distinct oxygen bubble trails in H_2O_2 solution. The high catalase loading capacity was directly correlated with the porous tentacle structure, which enhances enzyme immobilization. Furthermore, comparative studies of relative enzymatic activity under different conditions—including pH extremes, temperature variation, and repeated reuse—confirmed that the immobilized enzymes retained significantly higher activity than their free counterparts, demonstrating the hydrogel's practical potential in biocatalysis. Biocompatibility assessments via MTT assays confirmed that the biomimetic hydrogel exhibits no cytotoxicity, ensuring its suitability for biocatalytic and biomedical applications.

Overall, this study presents a simple and scalable strategy for fabricating biomimetic hydrogels with tunable morphology and high enzymatic catalytic efficiency. This approach not only advances the design of bioinspired catalytic platforms but also offers new opportunities for applications in biocatalysis, controlled drug delivery, and bioinspired material engineering.

Supplementary Materials: The following supporting information can be downloaded at: https://www.mdpi.com/article/10.3390/surfaces8020028/s1, Figure S1: Images of surface tension measurement results; Figure S2: Image processing of biomimetic jellyfish-like droplets; Figure S3: Photographs of Alg droplets without surfactants on the ε-PL solution surface; Figure S4: Biomimetic jellyfish-like droplets formed under different ε-PL concentrations; Table S1: Elemental weight percentages of C, N, O, Na, and S; Figure S5: Images of magnetic nanoparticle-loaded jellyfish-like and tentacle-free hydrogels; Figure S6: CAT loading capacity of the jellyfish-like hydrogel and tentacle-free hydrogel; Video S1: Biomimetic jellyfish-like hydrogel; Video S2: The time-lapse microscopic recording of tentacle formation in a jellyfish-like droplet.

Author Contributions: Investigation, A.Z. and Y.G.; data curation, A.Z., H.Z. and Y.G.; writing—original draft, A.Z.; validation, H.Z.; supervision, Y.F.; funding acquisition, Y.F.; project administration, Y.F.; writing—review and editing, Y.F. All authors have read and agreed to the published version of the manuscript.

Funding: This work was financially supported by the National Natural Science Foundation of China (22475032), and the Open Project of State Key Laboratory of Supramolecular Structure and Materials (sklssm 202406).

Data Availability Statement: Data is contained within the article or Supplementary Materials.

Acknowledgments: Thanks to the Equipment Support Team from the Analytical and Testing Center of Northeastern University.

Conflicts of Interest: The authors declare no conflicts of interest.

Abbreviations

The following abbreviations are used in this manuscript:

Alg	Sodium Alginate
ε-PL	ε-Poly-L-lysine

References

1. Bierbaumer, S.; Nattermann, M.; Schulz, L.; Zschoche, R.; Erb, T.J.; Winkler, C.K.; Tinzl, M.; Glueck, S.M. Enzymatic Conversion of CO_2: From Natural to Artificial Utilization. *Chem. Rev.* **2023**, *123*, 5702–5754. [CrossRef] [PubMed]
2. Woodley, J.M. Ensuring the Sustainability of Biocatalysis. *ChemSusChem* **2022**, *15*, e202102683. [CrossRef]
3. Savile, C.K.; Janey, J.M.; Mundorff, E.C.; Moore, J.C.; Tam, S.; Jarvis, W.R.; Colbeck, J.C.; Krebber, A.; Fleitz, F.J.; Brands, J.; et al. Biocatalytic Asymmetric Synthesis of Chiral Amines from Ketones Applied to Sitagliptin Manufacture. *Science* **2010**, *329*, 305–309. [CrossRef] [PubMed]
4. Liu, Y.; Wu, M.; Kang, A.; Zhang, X.; Xie, R.; Huang, Y.; Huang, J.; Chou, L.-Y.; Zhong, C. Living Seed Materials Made by Metal–Organic Framework-Encapsulated Bacillus Subtilis Spore. *Adv. Funct. Mater.* **2024**, *34*, 2309288. [CrossRef]
5. Wei, T.-H.; Wu, S.-H.; Huang, Y.-D.; Lo, W.-S.; Williams, B.P.; Chen, S.-Y.; Yang, H.-C.; Hsu, Y.-S.; Lin, Z.-Y.; Chen, X.-H.; et al. Rapid mechanochemical encapsulation of biocatalysts into robust metal–organic frameworks. *Nat. Commun.* **2019**, *10*, 5002. [CrossRef]
6. Ahmed, S.F.; Mofijur, M.; Rafa, N.; Chowdhury, A.T.; Chowdhury, S.; Nahrin, M.; Islam, A.B.M.S.; Ong, H.C. Green approaches in synthesising nanomaterials for environmental nanobioremediation: Technological advancements, applications, benefits and challenges. *Environ. Res.* **2022**, *204*, 111967. [CrossRef]
7. Zhu, X.; Du, C.; Gao, B.; He, B. Enhancing carbon capture efficiency with a large-sized bionic jellyfish-carbonic anhydrase complex. *Green Chem.* **2023**, *25*, 9926–9933. [CrossRef]
8. Wang, C.; Wang, Y.; Yan, M.; Zhang, W.; Wang, P.; Guan, W.; Zhang, S.; Yu, L.; Feng, J.; Gan, Z.; et al. Highly efficient self-floating jellyfish-like solar steam generators based on the partially carbonized Enteromorpha aerogel. *J. Colloid Interface Sci.* **2023**, *630*, 297–305. [CrossRef]

9. Li, K.; Xu, Z.; Liu, X.; He, Y.; Tian, X.; Xu, X.; Bo, G.; Yuan, S.; Xu, L.; Yang, M.; et al. Mussel foot inspired bionic adhesive material enhanced by a reconstructed in vitro system for interfacial adhesion. *Chem. Eng. J.* **2023**, *452*, 139580. [CrossRef]
10. Wang, Y.; Zhang, P.; Huang, H.; Zhu, J. Bio-Inspired Transparent Soft Jellyfish Robot. *Soft Robot.* **2023**, *10*, 590–600. [CrossRef]
11. Jordano, M.d.A.; Nagata, R.M.; Morandini, A.C. A review of the role played by cilia in medusozoan feeding mechanics. *Biol. Rev.* **2024**, *99*, 950–964. [CrossRef]
12. Gold, D.A.; Nakanishi, N.; Hensley, N.M.; Cozzolino, K.; Tabatabaee, M.; Martin, M.; Hartenstein, V.; Jacobs, D.K. Structural and Developmental Disparity in the Tentacles of the Moon Jellyfish Aurelia sp.1. *PLoS ONE* **2015**, *10*, e0134741. [CrossRef] [PubMed]
13. Tardent, P. The cnidarian cnidocyte, a high-tech cellular weaponry. *BioEssays* **1995**, *14*, 351–362. [CrossRef]
14. Karabulut, A.; McClain, M.; Rubinstein, B.; Sabin, K.Z.; McKinney, S.A.; Gibson, M.C. The architecture and operating mechanism of a cnidarian stinging organelle. *Nat. Commun.* **2022**, *13*, 3494. [CrossRef] [PubMed]
15. Suciu, D.G.; Smigelschi, O.; Ruckenstein, E. The spreading of liquids on liquids. *J. Colloid Interface Sci.* **1970**, *33*, 520–528. [CrossRef]
16. Slavtchev, S.; Kalitzova-Kurteva, P.; Mendes, M.A. Marangoni instability of liquid–liquid systems with a surface-active solute. *Colloids Surf. A Physicochem. Eng. Asp.* **2006**, *282*, 37–49. [CrossRef]
17. Iasella, S.V.; Sun, N.; Zhang, X.; Corcoran, T.E.; Garoff, S.; Przybycien, T.M.; Tilton, R.D. Flow regime transitions and effects on solute transport in surfactant-driven Marangoni flows. *J. Colloid Interface Sci.* **2019**, *553*, 136–147. [CrossRef]
18. Walls, D.J.; Haward, S.J.; Shen, A.Q.; Fuller, G.G. Spreading of miscible liquids. *Phys. Rev. Fluids* **2016**, *1*, 013904. [CrossRef]
19. Nikolov, A.D.; Wasan, D.T.; Wu, P. Marangoni flow alters wetting: Coffee ring and superspreading. *Curr. Opin. Colloid Interface Sci.* **2021**, *51*, 101387. [CrossRef]
20. Sheldon, R.A.; Brady, D. Green Chemistry, Biocatalysis, and the Chemical Industry of the Future. *ChemSusChem* **2022**, *15*, e202102628. [CrossRef]
21. Ardila-Fierro, K.J.; Hernández, J.G. Sustainability Assessment of Mechanochemistry by Using the Twelve Principles of Green Chemistry. *ChemSusChem* **2021**, *14*, 2145–2162. [CrossRef] [PubMed]
22. Jin, X.; Zhang, W.; Shan, J.; He, J.; Qian, H.; Chen, X.; Wang, X. Thermosensitive Hydrogel Loaded with Nickel–Copper Bimetallic Hollow Nanospheres with SOD and CAT Enzymatic-Like Activity Promotes Acute Wound Healing. *ACS Appl. Mater. Interfaces* **2022**, *14*, 50677–50691. [CrossRef] [PubMed]
23. Meyer, J.; Meyer, L.-E.; Kara, S. Enzyme immobilization in hydrogels: A perfect liaison for efficient and sustainable biocatalysis. *Eng. Life Sci.* **2022**, *22*, 165–177. [CrossRef]
24. Sobczak, M. Enzyme-Responsive Hydrogels as Potential Drug Delivery Systems—State of Knowledge and Future Prospects. *Int. J. Mol. Sci.* **2022**, *23*, 4421. [CrossRef]
25. Criado-Gonzalez, M.; Mijangos, C.; Hernández, R. Polyelectrolyte Multilayer Films Based on Natural Polymers: From Fundamentals to Bio-Applications. *Polymers* **2021**, *13*, 2254. [CrossRef] [PubMed]
26. Huang, M.-H.; Yang, M.-C. Swelling and biocompatibility of sodium alginate/poly(γ-glutamic acid) hydrogels. *Polym. Adv. Technol.* **2010**, *21*, 561–567. [CrossRef]
27. Orive, G.; Tam, S.K.; Pedraz, J.L.; Hallé, J.-P. Biocompatibility of alginate–poly-l-lysine microcapsules for cell therapy. *Biomaterials* **2006**, *27*, 3691–3700. [CrossRef]
28. Pang, Y.; Zeng, G.; Tang, L.; Zhang, Y.; Liu, Y.; Lei, X.; Li, Z.; Zhang, J.; Xie, G. PEI-grafted magnetic porous powder for highly effective adsorption of heavy metal ions. *Desalination* **2011**, *281*, 278–284. [CrossRef]
29. Khan, M.; Lo, I.M.C. A holistic review of hydrogel applications in the adsorptive removal of aqueous pollutants: Recent progress, challenges, and perspectives. *Water Res.* **2016**, *106*, 259–271. [CrossRef]
30. Deng, S.; Ting, Y.-P. Characterization of PEI-modified biomass and biosorption of Cu(II), Pb(II) and Ni(II). *Water Res.* **2005**, *39*, 2167–2177. [CrossRef]
31. Song, B.; Springer, J. Determination of Interfacial Tension from the Profile of a Pendant Drop Using Computer-Aided Image Processing: 1. Theoretical. *J. Colloid Interface Sci.* **1996**, *184*, 64–76. [CrossRef] [PubMed]
32. Song, B.; Springer, J. Determination of Interfacial Tension from the Profile of a Pendant Drop Using Computer-Aided Image Processing: 2. Experimental. *J. Colloid Interface Sci.* **1996**, *184*, 77–91. [CrossRef]
33. Sauleda, M.L.; Chu, H.C.W.; Tilton, R.D.; Garoff, S. Surfactant Driven Marangoni Spreading in the Presence of Predeposited Insoluble Surfactant Monolayers. *Langmuir* **2021**, *37*, 3309–3320. [CrossRef]
34. Cox, R.G. The deformation of a drop in a general time-dependent fluid flow. *J. Fluid Mech.* **1969**, *37*, 601–623. [CrossRef]
35. Xie, G.; Forth, J.; Zhu, S.; Helms, B.A.; Ashby, P.D.; Shum, H.C.; Russell, T.P. Hanging droplets from liquid surfaces. *Proc. Natl. Acad. Sci. USA* **2020**, *117*, 8360–8365. [CrossRef] [PubMed]
36. Imam, H.T.; Marr, P.C.; Marr, A.C. Enzyme entrapment, biocatalyst immobilization without covalent attachment. *Green Chem.* **2021**, *23*, 4980–5005. [CrossRef]
37. Liu, D.-M.; Chen, J.; Shi, Y.-P. Advances on methods and easy separated support materials for enzymes immobilization. *TrAC Trends Anal. Chem.* **2018**, *102*, 332–342. [CrossRef]

38. Hartmann, M.; Kostrov, X. Immobilization of enzymes on porous silicas—benefits and challenges. *Chem. Soc. Rev.* **2013**, *42*, 6277–6289. [CrossRef]
39. Zhang, Z.; Zhang, R.; Chen, L.; McClements, D.J. Encapsulation of lactase (β-galactosidase) into κ-carrageenan-based hydrogel beads: Impact of environmental conditions on enzyme activity. *Food Chem.* **2016**, *200*, 69–75. [CrossRef]
40. Fuenzalida, J.P.; Nareddy, P.K.; Moreno-Villoslada, I.; Moerschbacher, B.M.; Swamy, M.J.; Pan, S.; Ostermeier, M.; Goycoolea, F.M. On the role of alginate structure in complexing with lysozyme and application for enzyme delivery. *Food Hydrocoll.* **2016**, *53*, 239–248. [CrossRef]
41. Mukai, M.; Maruo, K.; Sasaki, Y.; Kikuchi, J.-i. Intermolecular Communication on a Liposomal Membrane: Enzymatic Amplification of a Photonic Signal with a Gemini Peptide Lipid as a Membrane-Bound Artificial Receptor. *Chem.—A Eur. J.* **2012**, *18*, 3258–3263. [CrossRef] [PubMed]
42. Adrar, N.; Ceylan, F.D.; Capanoglu, E. Hazelnut Protein and Sodium Alginate Complex Coacervates: An Effective Tool for the Encapsulation of the Hydrophobic Polyphenol Quercetin. *ACS Omega* **2024**, *9*, 37243–37252. [CrossRef] [PubMed]
43. Tam, S.K.; Dusseault, J.; Bilodeau, S.; Langlois, G.; Hallé, J.-P.; Yahia, L.H. Factors influencing alginate gel biocompatibility. *J. Biomed. Mater. Res. Part A* **2011**, *98*, 40–52. [CrossRef] [PubMed]
44. Broderick, E.; Lyons, H.; Pembroke, T.; Byrne, H.; Murray, B.; Hall, M. The characterisation of a novel, covalently modified, amphiphilic alginate derivative, which retains gelling and non-toxic properties. *J. Colloid Interface Sci.* **2006**, *298*, 154–161. [CrossRef]

surfaces

MDPI

Review

On the Formation and Characterization of Nanoplastics During Surface Wear Processes

Oguzhan Der [1], Hesam Khaksar [2] and Enrico Gnecco [2,*]

[1] Marine Engineering Department, Bandirma Onyedi Eylul University, 10200 Balikesir, Türkiye; oder@bandirma.edu.tr

[2] Marian Smoluchowski Institute of Physics, Jagiellonian University, 30-348 Krakow, Poland; hesam.khaksar@uj.edu.pl

* Correspondence: enrico.gnecco@uj.edu.pl

Abstract: The invasive presence of nanoplastics in various ecosystems makes them a significant environmental problem nowadays. One of the main production mechanisms of nanoplastics is mechanical wear. The combination of friction, abrasion, and shear forces can indeed lead to the progressive fragmentation of polymeric materials. The high surface area–volume ratio of the resulting nanoparticles not only alters the physicochemical properties of the polymers but also leads to increased interaction with biological systems, which raises questions about the persistence of nanoplastics in the environment and their potential toxicity. Despite the growing body of research on microplastics, studies specifically addressing the formation, characterization, and impact of wear-induced nanoplastics remain limited. This article describes current research on the formation mechanisms of nanoplastics generated by mechanical wear, highlighting the tribological processes underlying their release. Advanced characterization techniques used to identify the morphology and composition of these particles are also mentioned. The techniques include atomic force microscopy (AFM), scanning electron microscopy (SEM), and, to some extent, Raman spectroscopy. In the case of AFM, an example of application to the extrusion of nanoplastics from polystyrene surfaces subjected to repeated nanoscratching is also provided.

Keywords: nanoplastics; mechanical wear; environmental impact

1. Introduction

Microplastics, recognized as an important environmental pollutant of the Anthropocene, are widely studied due to their presence in marine and terrestrial ecosystems [1,2]. Defined as plastic particles less than 5 mm in size, microplastics can fragment due to mechanical wear and environmental degradation, leading to the formation of even smaller particles known as nanoplastics [3]. This definition is usually limited to plastic fragments usually below 100 nm in size [4], and definitely not above 1000 nm [5], as we assume in the following. Compared with microplastics, nanoplastics have different physicochemical properties, and they interact more strongly with biological systems [6]. Because of their size, nanoplastics can indeed penetrate deeply into organisms, raising concerns about their potential toxicological and environmental impact. Their detection and characterization require advanced techniques such as atomic force microscopy (AFM) [7], scanning electron microscopy (SEM) [8], or Raman spectroscopy [9]. In this work, we review the mechanisms of nanoplastic formation due to mechanical wear, briefly describe the aforementioned techniques, and exemplify the extrusion of nanoplastics in the case of a standard polymer surface scratched and imaged by AFM.

The wear of polymers is one of the most important mechanisms for the release of secondary micro- and nanoplastics into the environment [10,11]. Mechanical processes, such as friction, abrasion, and various shear forces, consume polymeric materials and release plastic particles from them into air, water, and soil [12,13]. These processes, which lead to progressive plastic fragmentation [14], are found in different environments: industrial machinery, vehicle tires, marine equipment, and consumer products. In this way, micro- and nanoplastics end up in oceans, fresh waters, and atmospheric samples [15]. Because of their tiny size, nanoplastics move easily in ecological systems and biological organisms [16]. For instance, they can enter the human body through inhalation or ingestion, potentially triggering inflammation, oxidative stress, and cellular damage [17]. New mitigation strategies for processes accompanied by polymer surface wear are therefore required to stem the flow of nanoplastics into the ecosystem and the resulting pollution [18]. To effectively reduce nanoplastic release associated with polymer surface wear, targeted mitigation strategies based on material science advancements are essential [19]. One approach involves surface modifications, such as low-friction coatings (e.g., diamond-like carbon coatings), which can significantly reduce mechanical degradation by minimizing direct polymer-to-surface contact [20,21]. Additionally, reinforced polymer composites, incorporating carbon fibers, glass fibers, or nanofillers, enhance structural integrity and wear resistance, thereby reducing fragmentation under stress [22,23]. Another promising strategy is the use of biodegradable or high-durability thermoplastics, such as polyetheretherketone (PEEK) or bio-based polylactic acid (PLA), which offer improved mechanical stability and reduced wear under repetitive loading [24,25]. Furthermore, processing techniques like plasma treatment or UV stabilization can enhance polymer resistance to environmental stressors, preventing premature degradation and subsequent nanoplastic generation [26,27]. By integrating these material science approaches, the release of nanoplastics due to wear can be significantly mitigated, contributing to a more sustainable polymer usage framework.

Studies on the environmental impacts of nanoplastics are rapidly increasing, with a strong focus on toxicological research. Extensive literature exists on the effects of nanoplastics on biological organisms, their accumulation in aquatic ecosystems, and their interactions with chemical pollutants [28–30]. However, there are relatively limited data on the formation mechanisms of nanoplastics via mechanical wear and the physical-mechanical dynamics of this process. Therefore, the fragmentation of plastics due to wear, the morphology of the resulting particles, and their environmental dispersion remain insufficiently understood.

The results outlined below emphasize that wear resistance and surface deformation characteristics significantly influence the amount and morphology of nanoplastic debris generated from polymer surfaces. Nevertheless, long-term dynamics under mechanical loads and the environmental transport of wear-generated nanoplastics remain largely unexplored. Future investigations should integrate tribological, spectroscopic, and computational modeling approaches to predict nanoplastic generation across various real-world applications. Moreover, expanding experimental studies on different polymer types and their degradation behavior under mechanical stress will be essential for a comprehensive understanding of nanoplastic pollution.

2. Recent Advances in Nanoplastics Research

2.1. Formation Mechanisms of Nanoplastics

The main mechanisms of nanoplastic generation by mechanical wear involve repeated mechanical stresses, including friction, shear forces, and abrasion, which lead to the gradual fragmentation of larger polymer structures [31]. In brittle polymers, such as polystyrene (PS), mechanical stresses exceeding the material strength threshold can result in the formation of microcracks, which propagate due to cyclic loading, fatigue, or external environmental factors such as temperature fluctuations and UV exposure [32]. These cracks may eventually lead to the detachment of nanometer-sized plastic particles [33,34]. In contrast, ductile polymers, such as polyethylene (PE), primarily undergo plastic deformation under mechanical stress, where material removal occurs through mechanisms such as tearing and plowing rather than crack formation [35]. This distinction is important in understanding how different polymer types contribute to nanoplastic generation under wear conditions.

Recent tribological studies have demonstrated that nanoplastics are released through different wear mechanisms, including abrasive wear, adhesive wear, and fatigue wear [11,36]. Pin-on-disk tribometry and reciprocating wear tests have been employed to simulate real-world plastic wear conditions, such as tire–road interactions and synthetic textile friction [37,38]. Findings from these studies indicate that thermoplastics with lower molecular weight and weaker intermolecular forces tend to release more nanoplastics under mechanical stress. Further, Alkhadra et al. [35] highlighted that tire wear particles (TWPs), a significant source of microplastic emissions, enter the environment through multiple pathways, including atmospheric deposition, wastewater effluents, and surface runoff. Their work demonstrated that smaller TWP fragments exhibit longer transport potential, contributing to the widespread dispersion of nanoplastics in marine ecosystems. The study also emphasized the role of pyrolysis-GC-MS in detecting benzothiazole as a molecular marker for TWPs, underscoring the need for more refined analytical methods to quantify nanoplastic release. Yu et al. [16] illustrated in Figure 1 the process of secondary micro- and nanoplastic (MNP) release through surface interactions, shear forces, compression, and repeated cycles of wear acting externally. Similar to natural beach settings, where wave action and sediment interactions accelerate plastic fragmentation, daily mechanical wear in industrial and consumer spaces intensifies nanoplastic production. While Figure 1 categorizes tire wear under adhesive wear, it is important to note that tire wear is not limited to adhesive wear but also involves abrasive wear. Adhesive wear occurs due to repeated friction between the tire and the road surface, leading to material detachment, whereas abrasive wear results from rough road textures progressively eroding the tire material. Furthermore, personal protective equipment (PPE) and personal care products, also depicted in Figure 1, contribute to micro- and possibly nano-plastic release through different wear mechanisms. PPE items, such as gloves and face masks, are subject to adhesive wear due to frequent frictional forces during use, leading to gradual degradation. On the other hand, personal care products, such as exfoliating scrubs, sponges, and brushes, experience abrasive wear, where continuous mechanical action causes material fragmentation [39].

The morphology of these fine particles differs depending on the direction of motion. Forces applied perpendicularly to the polymer surface (normal forces) create localized stress concentrations, initiating cracks that propagate and lead to tile-like secondary fragments. In contrast, shear forces acting parallel to the surface induce material fatigue and peeling, leading to flake-like detachments. This differentiation is consistent with reported fragmentation scenarios, where vertically applied forces tend to create cracks and sharp-edged particles, while shear forces promote surface peeling and flake-like structures, as noted in studies investigating particle formation during plastic wear and environmental stress [16]. Such processes are critical in effective mitigation strategies, especially in high-friction environments such as those associated with vehicle tires, synthetic textiles, and plastic packaging.

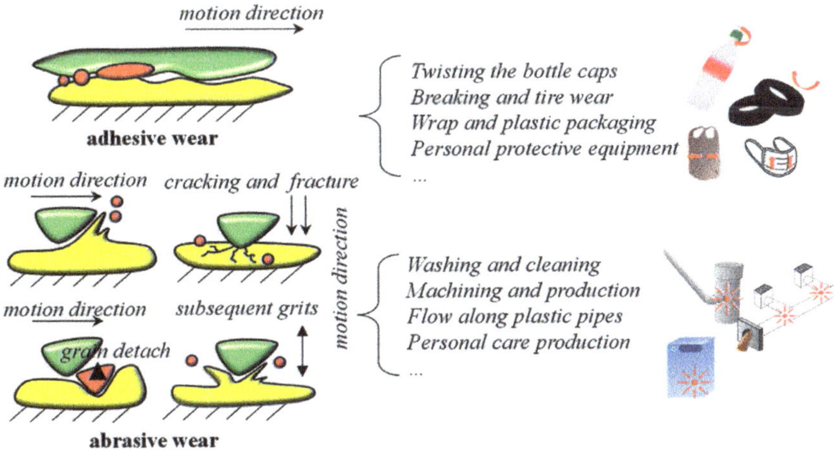

Figure 1. The formation of secondary MNPs through adhesive and abrasive wear processes and their corresponding activities. Reprinted with permission from Ref. [16]. Copyright 2024, MDPI.

Particulate matter arising from tire road wear (TRWP) is considered one of the significant factors in micro- and nanoplastic pollution (in Figure 2). According to recent estimations, about 10% of the plastic pollution stemming from tire wear may be found in the world's oceans. Although direct evidence is still lacking for the formation of nanoplastics from TRWPs, these particles are comprised of heavy metals and organic additives that may increase their risk to the environment and toxicologically. Nonetheless, importantly, there is still little investigation into the nanoscale dynamics of tire wear, hence the need for new hypotheses and modeling approaches for a much better understanding of the fragmentation processes taking place [40,41].

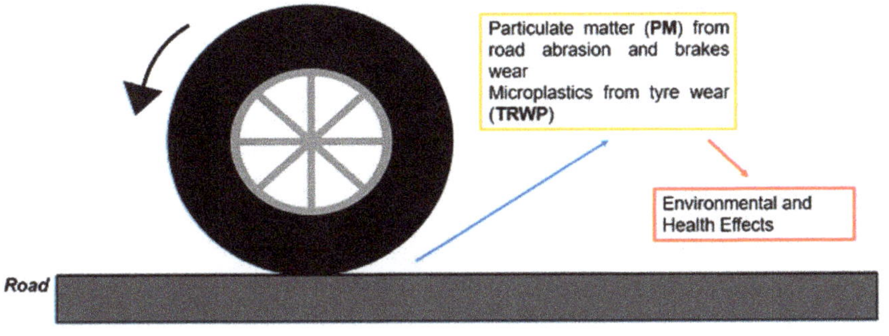

Figure 2. Diagram illustrating the formation and effects of TRWPs. Reprinted with permission from Ref. [40]. Copyright 2024, MDPI.

It is assumed that under actual road conditions, abraded tires actually produce nanoplastics through localized stress concentration and interfacial debonding at rubber–filler interfaces [41]. Repeated mechanical contact, high-pressure loading, and lateral shear between the tire tread and road surface may create microscale fatigue, leading to nanoscale fragments being detached from the viscoelastic matrix itself [42]. This process might be further affected by fillers such as carbon black or silica, which are points of stress concentration facilitating the development of cracks and release of nanoparticles. Finally, we also recognize the fact that nanoplastic particles can be caused by ultraviolet-induced destruction of pre-existing microplastic tire fragments. This photodegradation pathway is not abrasion-dependent but could significantly contribute to secondary nanoplastic pollution in sun-exposed environments. Therefore, both mechanical and photo-induced fragmentation pathways could be treated as complementary contributors to tire-derived nanoplastics.

The events associated with plastic weathering in the natural environment are dynamic and complex, since the degradation of polymers from UV irradiation and mechanical weathering leads progressively to the formation of microplastics and nanoplastics and the release of chemical additives [43]. Yang et al. showed that extended exposure to environmental stressors oxidizes the surface of the polymer, rendering it brittle and initiating crack or flake formation, which allows the detachment of nanoscale plastic particles [44]. The presence of reactive oxygen species (ROS) during photo-oxidation further exacerbates nanoplastic release, emphasizing the importance of environmental conditions in wear-related pollution. Nanoplastic release is closely linked to the aging of plastics, with smaller sized fragments becoming more predominant over time. While infrared and Raman analyses indicate that aging also promotes the leaching of polymer additives, this process is distinct from mechanical fragmentation. Additive leaching occurs due to polymer oxidation and degradation, whereas nanoplastic formation primarily results from surface cracking and flake detachment under mechanical stress. Both phenomena contribute to environmental pollution, but nanoplastic formation is directly tied to the physical breakdown of polymer structures. These findings are also relevant for tire road wear particles, as their ongoing mechanical wear against road surfaces similarly leads to progressive micro- and nanoplastic generation. Understanding these degradation mechanisms is essential for predicting nanoplastic pollution and designing effective mitigation strategies.

Figure 3 illustrates the surface morphological changes in aged polystyrene (PS) plastics, as observed through SEM imaging. The images show the progressive development of surface roughness, granular oxidation, and ultimately the formation of cracks and lamellar projections. As aging advances, cracks widen, increasing the likelihood of nanoplastic detachment. These observations support the idea that surface degradation mechanisms—such as crack formation and flake detachment—are primary contributors to the release of nanoplastics from weathered polymer surfaces. While this process is particularly evident in PS, similar degradation patterns have been reported for other thermoplastics under prolonged environmental exposure [44].

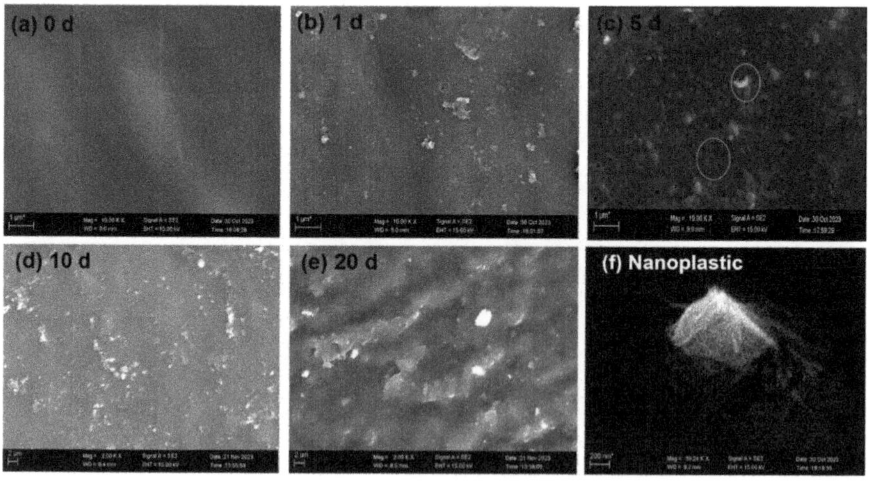

Figure 3. SEM images of polystyrene (PS) at various stages of weathering (**a**–**e**) along with an SEM image depicting the resulting nanoparticles (**f**). Note the formation of holes after 1 day and of granular protrusions after 5 days (highlighted by red circles and ovals). Adapted with permission from Ref. [44]. Copyright 2024, Elsevier.

In this context, it is also worth mentioning the multi-scale abrasion mechanics model developed by Li et al. [45] to describe the formation of particulate matter during wear processes. Their findings indicate that nanoplastic generation is directly linked to fracture mechanics, where fatigue-induced microcrack propagation plays a critical role. Their model quantitatively relates abrasion-induced emissions to material properties, revealing that materials with lower fracture toughness tend to release higher concentrations of smaller plastic fragments. This approach provides a robust theoretical foundation for predicting and mitigating nanoplastic emissions from polymeric surfaces under repeated stress. In addition to mechanical wear, environmental factors also contribute to the degradation of TWPs over time.

As highlighted by Gnecco et al. [46], nanoripples can emerge due to periodic instabilities occurring under shear stress of polymeric surfaces. Their formation and evolution are governed by the interplay between lateral forces and the plastic response of the material. This process is particularly significant for PS, a commonly found plastic in the environment, and repeated mechanical stresses can trigger nanoparticle detachment. As shown by Hennig et al. [47], nanoplastics are nucleated from the crests of the ripples and easily displaced by the scratching tool (i.e., an AFM tip, see below). Similar results were recently obtained by Khaksar et al. on polymeric blends [12]. In both cases, the release of nanoplastics is possibly the result of crazing mechanisms in the polymer surfaces under tension. Although these studies were conducted using AFM under controlled nanoscratching conditions (constant loading forces in the sub μN range and constant scan velocities of few tens of μm/s at most), it is not excluded that similar mechanisms of nanoparticle detachment and surface instabilities occur in everyday life. For instance, tire abrasion on roads introduces continuous and fluctuating mechanical stresses, leading to progressive polymer degradation and nanoplastic detachment [48]. Processes similar to those observed under rather idealized laboratory conditions may also occur, although, at the time of writing, we are not aware of any systematic investigations extending the aforementioned studies in this direction. The same can be said for synthetic polymer coatings on ship hulls and industrial equipment, which experience mechanical erosion due to abrasive contact with sediment contributing to nanoplastic release, and for the wear of polymer fibers in textiles during washing cycles.

Altogether, these studies illustrate that nanoplastic formation during mechanical wear is influenced by intrinsic material properties, wear mechanics, and external environmental factors. A comprehensive understanding of these parameters is essential for developing sustainable materials and wear-resistant polymers to mitigate nanoplastic pollution. Despite these advancements, significant research gaps remain in fully elucidating the relationship between mechanical wear parameters and nanoplastic release rates. In addition to mechanical wear, environmental aging mechanisms—particularly UV radiation and elevated temperatures—play a crucial role in the progressive degradation of polymeric materials [27]. Studies have shown that prolonged exposure to sunlight and heat promotes surface oxidation and polymer chain scission, resulting in embrittlement and enhanced fragmentation potential [49,50]. This photo-oxidative degradation pathway facilitates the detachment of nanoscale fragments, even in the absence of direct mechanical abrasion. Notably, Yang et al. [44] demonstrated that the generation of reactive oxygen species (ROS) during UV-induced aging exacerbates crack formation and flake detachment, further contributing to nanoplastic release. These aging effects can also interact with wear mechanisms: for example, aged and oxidized surfaces are typically more susceptible to fragmentation under mechanical stress. Therefore, a comprehensive understanding of nanoplastic generation must account for the synergistic effects between environmental aging and tribological wear.

2.2. Mechanical Properties and Wear Behavior

Nanoplastics exhibit distinct mechanical properties that influence their formation, durability, and environmental persistence [51]. The elastic modulus, hardness, and wear resistance of plastic films determine how easily they are fragmented under mechanical stress. Different polymers respond to mechanical wear differently, affecting the amount and morphology of nanoplastics they generate [52].

Polymers with high brittleness and low tensile strength release more nanoplastics under wear tests [11,53]. Thermoplastics such as PE, PP, and PS exhibit varying degrees of mechanical properties and wear mechanisms. PE and PP are widely regarded as ductile polymers with moderate tensile strength, exhibiting plowing wear behavior, characteristic of ductile materials [13,54,55]. While they are mechanically softer than high-performance engineering polymers, they do not possess high brittleness. In contrast, PS and other amorphous thermoplastics demonstrate greater brittleness and are more prone to cutting wear behavior [56]. The distinction in wear mechanisms is crucial for selecting appropriate materials in tribological applications. Additionally, studies have demonstrated that mechanical degradation and fragmentation of thermoplastics like PE, PP, and PS can lead to the formation of microplastics and nanoplastics under environmental stressors [57–59]. A key aspect in this context is the mechanical breakdown of the materials, which is influenced by the molecular weight and structural stiffness of the polymers [60]. Environmental stressors also include ultraviolet (UV) light, which weakens polymer chains and increases fragmentation susceptibility, and consequently worsens the degradation of PE and other thermoplastics [61].

Frictional forces and roughness also play a crucial role in wear-induced nanoplastic generation [62]. Pin-on-disk tribological tests have demonstrated that increased sliding speed and applied load results in higher nanoplastic release rates. Surface modification techniques, such as adding fillers or reinforcing fibers, have been explored to reduce polymer wear and mitigate nanoplastic generation [63]. Despite these insights, challenges remain in accurately quantifying nanoplastic generation under real-world conditions. Many studies focus on controlled laboratory settings [64], whereas real-world factors such as temperature variations, humidity, and UV exposure can significantly alter polymer wear behavior [65,66].

2.3. Characterization Techniques

The accurate detection and characterization of nanoplastics are essential for understanding their formation, environmental fate, and toxicity [67], and a few analytical techniques can be used to identify and quantify nanoplastics.

Atomic force microscopy (AFM) is one of the most important tools for studying the morphology of polymer surfaces on the nanoscale [68]. It also plays a crucial role in understanding the nanoscale fragmentation process [69]. As an example, Figure 4 shows the formation of surface ripples as observed while scratching initially flat polystyrene surfaces with an AFM tip. Scratching was repeated one, three, or five times along a series of parallel lines, left to right only in the so-called "hover mode" [70]. The resulting particles had diameters in the order of 100 nm.

As seen in the previous example, AFM provides very accurate images of surface changes caused by wear and allows the process of generating individual nanoplastics to be observed. In addition, AFM plays a key role in quantifying nanoplastic removal mechanisms [71]. AFM-based nanoscratching is a powerful technique for investigating wear mechanisms at the nanoscale, offering valuable insights into how polymeric materials respond to mechanical interactions under controlled conditions [72]. The method employs diamond-coated tips under precisely controlled normal loads, effectively replicating abrasive wear commonly observed in real-world applications [73]. For instance, in marine environments, ship coatings endure continuous exposure to sand and debris [74]; in biomedical applications, polymeric implants experience frictional wear [75]; and in industrial settings, polymer seals and gaskets undergo tribological degradation due to repetitive mechanical contact [76]. One of the significant advantages of AFM-based nanoscratching is its ability to provide precise control over normal force, sliding speed, and environmental conditions such as humidity and temperature, which are crucial factors affecting polymer wear behavior [77,78]. These controlled experiments allow for a systematic evaluation of how material properties—such as cross-linking density, molecular structure, and the presence of reinforcing fillers—impact wear resistance in conditions that resemble those encountered in real-world applications.

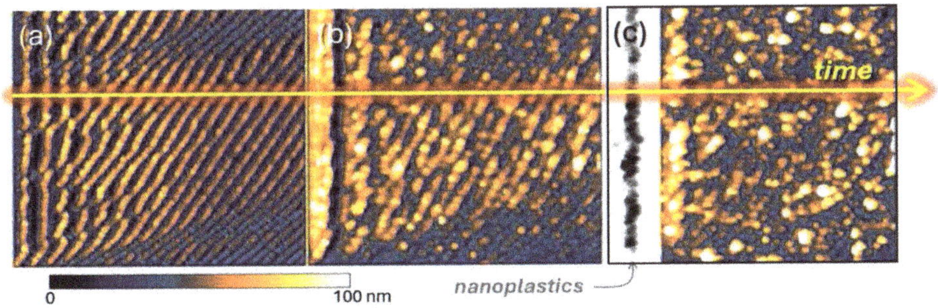

Figure 4. **(a–c)** Topography images of three polystyrene surfaces previously scratched in ambient condition (one, three, and five times) by a silicon probe (NSG-01 NT-MDT, spring constant k = 5.1 N/m) with a normal force F_N = 100 nN and a scan velocity v = 15 μm/s. Frame sizes: 5 × 4 μm^2. Reprinted with permission from Ref. [47]. Copyright 2021, Elsevier.

Furthermore, studies have shown that wear behavior observed through AFM nanoscratching correlates well with macroscopic tribological tests, reinforcing its effectiveness as a predictive tool [79,80]. By enabling the analysis of plastic deformation, material transfer, and debris formation at the nanoscale, this method provides valuable data on the early-stage wear mechanisms that contribute to the long-term degradation of polymeric materials. Thus, AFM-based nanoscratching serves as a bridge between fundamental nanoscale wear analysis and the broader understanding of wear mechanisms in practical applications. AFM plays a crucial role in understanding nanoscale fragmentation processes and mechanical wear behavior [81]. However, obtaining high-resolution images and conducting well-controlled nanoscratching experiments require extensive fine-tuning [82]. Key factors include tip characterization, contact mechanics, and force modulation, all of which influence the accuracy of wear track analysis [83]. Tip radius, material properties, and cantilever stiffness must be carefully selected to ensure reproducibility in nanoplastic detection. Additionally, advanced AFM modes, such as PeakForce Tapping mode, provide further advantages in force control and nanoscale mechanical property mapping [84]. Unlike conventional contact-mode AFM, PeakForce mode dynamically modulates the tip–sample interaction forces, minimizing sample damage while simultaneously capturing quantitative mechanical properties such as adhesion, modulus, and deformation at a nanometer resolution [85]. In this context, PeakForce mode is particularly relevant for nanoplastic characterization, as it allows for real-time tracking of dynamic wear processes. Furthermore, AFM-IR (Atomic Force Microscopy-Infrared Spectroscopy) has emerged as a powerful tool for simultaneously obtaining morphological and chemical characterization of nanoplastics [86]. This hybrid technique overcomes AFM's limitation in chemical identification by integrating IR absorption mapping, which can detect polymer-specific spectral fingerprints with nanometer-scale spatial resolution [87]. Given its ability to combine topographical imaging with molecular identification, AFM-IR provides a more comprehensive approach to nanoplastic analysis, particularly in distinguishing different polymer types in mixed samples. Altogether, these AFM-based characterizations provide fundamental insights into the mechanical processes governing the formation of nanoplastics and their potential dispersion in the environment. A limitation of AFM remains, however, in the inability to chemically identify the surfaces under investigation, a limitation that can be mitigated by complementary spectroscopic methods [88]. It is worth noting, for instance, that the integration of AFM with infrared (IR) and Raman spectroscopy allows for simultaneous morphological and chemical characterization of surfaces [86].

Nanoplastics' size and shape can also be investigated using SEM [89] and TEM [90]. While SEM provides high-resolution surface imaging [91], TEM offers superior resolution for sub-micron polymer debris, making it particularly suitable for detecting nanoplastics smaller than 100 nm [92]. Additionally, Energy-Dispersive X-ray Spectroscopy (EDS) can be combined with both SEM and TEM to determine elemental composition, offering valuable insights into the chemical characteristics of nanoplastics [93]. High-resolution transmission-mode SEM (T-SEM) has also been employed (in Figure 5) to enhance the detection of nanoplastic particles at the nanoscale [94,95]. However, the quality of SEM images can be affected by various factors, such as accelerating voltage, beam current, and sample charge [96]. The physical state of some polymeric materials can be modified by the vacuum conditions used for SEM, which is especially critical for nanoplastics, as a high vacuum might distort polymeric structures or cause volatile components to evaporate [97]. To address these issues, low-voltage SEM methods and cryo-SEM are often used to preserve the natural structure of sensitive materials [98]. Furthermore, adjusting the working distance, detection angle, and beam energy can further improve SEM imaging quality while minimizing sample damage. Advances in SEM-EDS technology, particularly the use of high-sensitivity silicon drift detectors (SDDs), have significantly improved elemental mapping at the nanoscale. Similarly, TEM-EDS allows for precise elemental characterization of nanoplastics at much smaller dimensions, making it a complementary technique for detecting wear-induced plastic debris [99,100].

Raman spectroscopy and Fourier Transform Infrared (FTIR) spectroscopy are also useful for determining the chemical composition of polymeric fragments, although their application is, admittedly, limited to microplastics [101]. Raman spectroscopy allows for the analysis of those particles non-destructively and for distinguishing between various polymer components. However, it can occasionally be difficult to interpret Raman spectra due to fluorescence interference. Functional groups and polymer-specific spectral fingerprints can be easily identified using FTIR spectroscopy, especially for weakly Raman-active polymers [102]. Because of its sensitivity to surface characteristics, the attenuated total reflectance (ATR-FTIR) mode is widely utilized and a crucial tool for microplastic investigation. Furthermore, Raman spectroscopy can be improved by employing metal nanoparticles for surface-enhanced Raman scattering (SERS), which provides ultra-sensitive detection and broadens its uses in biomedical domains [103]. To fully characterize microplastics in environmental samples, both approaches are frequently applied in tandem, and for best results, they should be combined.

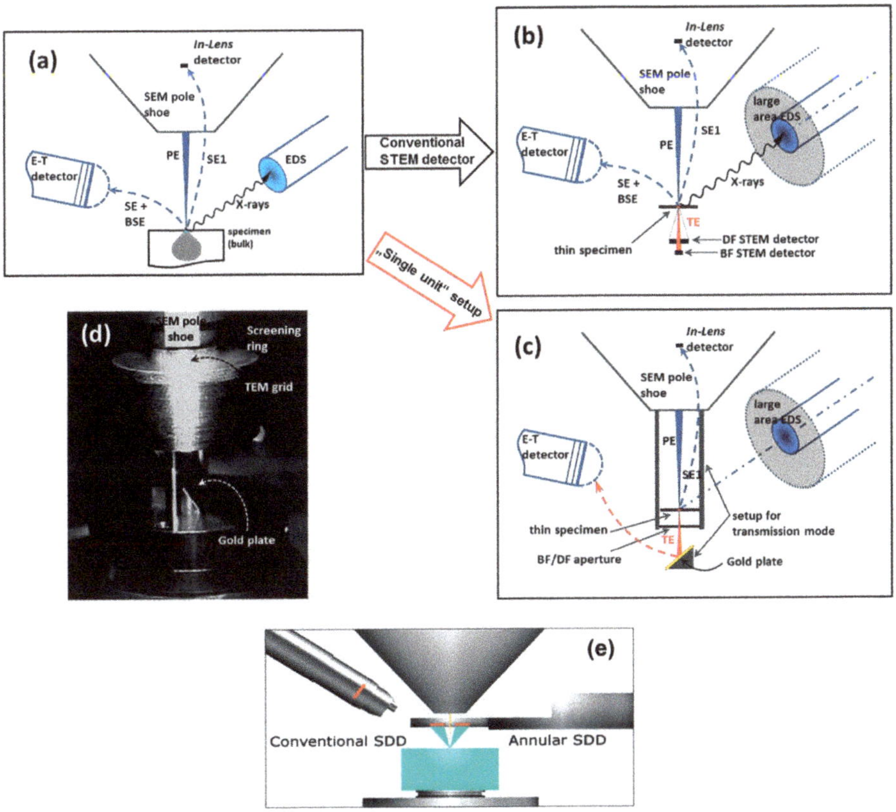

Figure 5. Schemes of SEM/EDS technique: (**a**) Bulk sample analysis in an SEM. (**b**) Thin sample analysis. (**c**) Thin sample analysis in transmission mode (T-SEM). (**d**) Photograph of the T-SEM setup in working position (close to the pole shoe). (**e**) Geometry of the annular SDD EDS. Reprinted with permission from Ref. [94]. Copyright 2016, IOP.

Besides mapping techniques based on AFM and SEM, Raman imaging has recently emerged as a powerful technique to identify and visualize nanoplastics down to 100 nm. As shown by Sobhani et al. [104], Raman mapping enables nanoplastics' chemical characterization by detecting polymer-specific vibrational fingerprints at the nanoscale. This study involved the identification of polystyrenic nanoplastics with a size of down to 100 nm using a confocal Raman microscope with high-resolution pixel stepping (100 nm × 100 nm). Figure 6 clearly demonstrates that even a single nanoplastic particle smaller than the laser spot (~300 nm) can be visualized through pixel-specific spectral analysis. The immense advantages of this imaging method compared to conventional ones, especially in finding nanoplastics from complex environmental samples like polishing dust from vehicle paints, cannot be overstated. The ability to discriminate between microplastics and nanoplastics through their Raman spectra is opening new gateways in environmental risk assessment. Thus, Raman characterization supports AFM and SEM in carrying out simultaneous chemical identification and spatial resolution of plastic debris on the nanoscale.

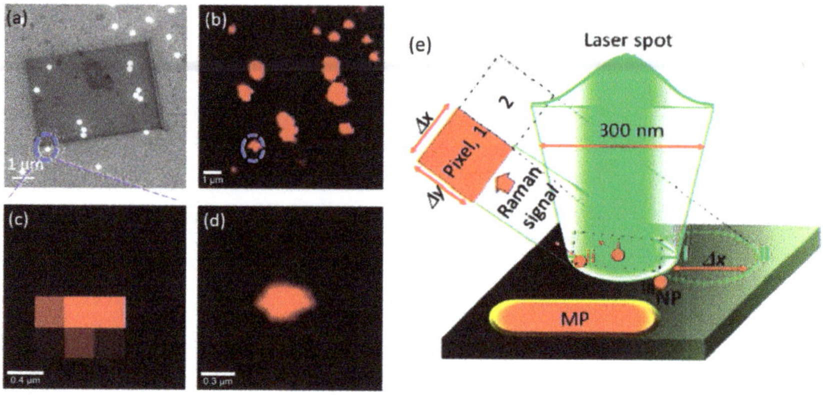

Figure 6. Raman technique for characterization of nanoplastics (NPs) or microplastics (MPs) on a gold surface: (**a**) SEM image, (**b**) Raman mapping image, (**c**,**d**) single nanoplastic without/with pixel color interpolation, (**e**) laser spot scans and excitation of the NP. Reprinted with permission from Ref. [104]. Copyright 2020, Elsevier.

Although nanoscale experiments, such as AFM-based nanoscratching, enhance our understanding of surface instabilities and the development of nanoplastic under very well-defined conditions, it must be noted that neither the type nor the extent of results would be in any way comparable to those formed with engineered or natural environments. Parameters such as contact pressure, asperity size, and force magnitudes vary greatly in real-world scenarios—ranging from nN at the tip–sample interface in AFM to N or kN ranges in applications such as tire–road interactions or marine equipment wear. Thus, direct quantitative extrapolation remains challenging. Nonetheless, the ripple formation and nanoparticle detachment observed at the nanoscale represent fundamental surface phenomena that may also manifest during the early stages of macroscale wear, especially under cyclic or repetitive loading. To effectively close the scale gap, such future endeavors will combine nanoscale understanding with multiscale computational modeling approaches and upscaled experimental work to predict wear behavior for various conditions. Also, we acknowledge that definitions of abrasive wear in natural environments have not yet gained a quantitative foundation; thus, much more research on reliable metrics—such as energy dissipation, particle emission rates, and force distributions under uncontrolled environmental stress—is needed to serve as an input for further developments.

3. Conclusions

Because of their distinct physicochemical characteristics and tiny size, nanoplastics cause serious environmental and health risks. These particles are commonly found in different ecosystems because of the fragmentation of larger plastic materials caused by friction, abrasion, and shear pressures. Their high surface area–volume ratio makes them more prone to adsorb contaminants and interact with biological systems. Determining the long-term environmental fate and potential hazards of nanoplastics requires an understanding of the mechanisms behind their generation, but investigations focusing on wear-induced nanoplastic formation remain limited so far. Thus, comprehensive studies are needed to establish direct correlations between material properties and fragmentation behavior. Integrating tribological analyses with spectroscopic and microscopic techniques can enhance the accuracy of nanoplastic detection and provide a deeper understanding of their release dynamics under real-world conditions. In this context, AFM is particularly useful for investigating the wear-related mechanisms leading to nanoplastic release, while

SEM, Raman spectroscopy, and IR spectroscopy provide crucial insights into the size, morphology, and chemical composition of the generated nanoplastics.

To sum up, nanoplastics produced by mechanical wear processes are a new environmental problem that needs to be addressed urgently. Developing successful mitigation techniques will require filling in the current information gaps regarding the toxicity, characterization, and generation of nanoplastics. To fully evaluate nanoplastic contamination, future studies should concentrate on interdisciplinary techniques that integrate environmental chemistry, material science, and toxicology. To minimize the release of nanoplastics into the environment, it will also be essential to promote sustainable material alternatives and enhance waste management techniques.

A final remark is necessary. Although our focus remained on nanoplastics, it is important to note that the number of direct studies on their wear-induced formation is still limited compared to the literature on microplastics. Consequently, findings from microplastic research cannot be downscaled a priori. However, recent evidence by Jurkschat et al. [105] demonstrated that vehicle tires—historically known to release microplastics—can also contribute to nanoplastic pollution. In their study, TWPs accounted for the largest proportion of nanoplastic mass in remote Alpine snow samples. These results justify the inclusion of tire wear in discussions of nanoplastic generation, even if mechanistic studies are still developing in this area.

Author Contributions: Conceptualization, O.D., H.K. and E.G.; writing—original draft preparation, O.D.; writing—review and editing, E.G.; visualization, H.K.; funding acquisition, E.G. All authors have read and agreed to the published version of the manuscript.

Funding: This research was partially funded by the Strategic Program Excellence Initiative at the Jagiellonian University 'SciMat', grant number U1U/P05/NO/01.05.

Data Availability Statement: No new data were created or analyzed in this study. Data sharing is not applicable to this article.

Conflicts of Interest: The authors declare no conflicts of interest. The funders had no role in the design of the study; in the collection, analyses, or interpretation of data; in the writing of the manuscript; or in the decision to publish the results.

References

1. Hurley, R.; Woodward, J.; Rothwell, J.J. Microplastic Contamination of River Beds Significantly Reduced by Catchment-Wide Flooding. *Nat. Geosci.* **2018**, *11*, 251–257. [CrossRef]
2. Yildiz, R.O.; Koc, E.; Der, O.; Aymelek, M. Unveiling the Contemporary Research Direction and Current Business Management Strategies for Port Decarbonization Through a Systematic Review. *Sustainability* **2024**, *16*, 10959. [CrossRef]
3. Thompson, R.C.; Olsen, Y.; Mitchell, R.P.; Davis, A.; Rowland, S.J.; John, A.W.G.; McGonigle, D.; Russell, A.E. Lost at Sea: Where Is All the Plastic? *Science* **2004**, *304*, 838. [CrossRef] [PubMed]
4. Koelmans, A.A.; Besseling, E.; Shim, W.J. Nanoplastics in the Aquatic Environment. Critical Review. In *Marine Anthropogenic Litter*; Springer International Publishing: Cham, Switzerland, 2015; pp. 325–340.
5. Frias, J.P.G.L.; Nash, R. Microplastics: Finding a Consensus on the Definition. *Mar. Pollut. Bull.* **2019**, *138*, 145–147. [CrossRef]
6. Mitrano, D.M.; Wick, P.; Nowack, B. Placing Nanoplastics in the Context of Global Plastic Pollution. *Nat. Nanotechnol.* **2021**, *16*, 491–500. [CrossRef]
7. Li, W.; Luo, Y.; Pan, X. Identification and Characterization Methods for Microplastics Basing on Spatial Imaging in Micro-/Nanoscales. In *Microplastics in Terrestrial Environments*; Springer: Berlin/Heidelberg, Germany, 2020; pp. 25–37.
8. Wagner, J.; Wang, Z.-M.; Ghosal, S.; Rochman, C.; Gassel, M.; Wall, S. Novel Method for the Extraction and Identification of Microplastics in Ocean Trawl and Fish Gut Matrices. *Anal. Methods* **2017**, *9*, 1479–1490. [CrossRef]
9. Schiavi, S.; Parmigiani, M.; Galinetto, P.; Albini, B.; Taglietti, A.; Dacarro, G. Plasmonic Nanomaterials for Micro- and Nanoplastics Detection. *Appl. Sci.* **2023**, *13*, 9291. [CrossRef]

10. Adetuyi, B.O.; Mathew, J.T.; Inobeme, A.; Falana, Y.O.; Adetunji, C.O.; Shahnawaz, M.; Oyewole, O.A.; Eniola, K.I.T.; Yerima, M.B.; Popoola, O.A. Tyres, Bitumen Wear and Plastic Bottles, Other Single Used Plastic as Major Sources of Microplastic. In *Microplastic Pollution*; Springer Nature: Singapore, 2024; pp. 141–160.
11. Rahman, E.; BinAhmed, S.; Keyes, P.; Alberg, C.; Godfreey-Igwe, S.; Haugstad, G.; Xiong, B. Nanoscale Abrasive Wear of Polyethylene: A Novel Approach To Probe Nanoplastic Release at the Single Asperity Level. *Environ. Sci. Technol.* **2024**, *58*, 13845–13855. [CrossRef]
12. Khaksar, H.; Ma, C.; Janiszewska, N.; Awsiuk, K.; Budkowski, A.; Gnecco, E. Nanoscale Wear Evolution on a Polystyrene/Poly (n-Butyl Methacrylate) Blend. *Wear* **2024**, *536–537*, 205160. [CrossRef]
13. Ordu, M.; Der, O. Polymeric Materials Selection for Flexible Pulsating Heat Pipe Manufacturing Using a Comparative Hybrid MCDM Approach. *Polymers* **2023**, *15*, 2933. [CrossRef]
14. Sipe, J.M.; Bossa, N.; Berger, W.; von Windheim, N.; Gall, K.; Wiesner, M.R. From Bottle to Microplastics: Can We Estimate How Our Plastic Products Are Breaking Down? *Sci. Total Environ.* **2022**, *814*, 152460. [CrossRef] [PubMed]
15. Musa, I.O.; Auta, H.S.; Ilyasu, U.S.; Aransiola, S.A.; Makun, H.A.; Adabara, N.U.; Abioye, O.P.; Aziz, A.; Jayanthi, B.; Maddela, N.R.; et al. Micro- and Nanoplastics in Environment: Degradation, Detection, and Ecological Impact. *Int. J. Environ. Res.* **2024**, *18*, 1. [CrossRef]
16. Yu, Y.; Craig, N.; Su, L. A Hidden Pathway for Human Exposure to Micro- and Nanoplastics—The Mechanical Fragmentation of Plastic Products during Daily Use. *Toxics* **2023**, *11*, 774. [CrossRef] [PubMed]
17. da Silva Antunes, J.C.; Sobral, P.; Branco, V.; Martins, M. Uncovering Layer by Layer the Risk of Nanoplastics to the Environment and Human Health. *J. Toxicol. Environ. Health Part B* **2025**, *28*, 63–121. [CrossRef]
18. Suzuki, G.; Uchida, N.; Tuyen, L.H.; Tanaka, K.; Matsukami, H.; Kunisue, T.; Takahashi, S.; Viet, P.H.; Kuramochi, H.; Osako, M. Mechanical Recycling of Plastic Waste as a Point Source of Microplastic Pollution. *Environ. Pollut.* **2022**, *303*, 119114. [CrossRef]
19. Bulut, M.S.; Ordu, M.; Der, O.; Basar, G. Sustainable Thermoplastic Material Selection for Hybrid Vehicle Battery Packs in the Automotive Industry: A Comparative Multi-Criteria Decision-Making Approach. *Polymers* **2024**, *16*, 2768. [CrossRef]
20. Nilsson, D.; Stavlid, N.; Lindquist, M.; Hogmark, S.; Wiklund, U. The Role of Aluminum Additions in the Oxidation and Wear of a TaC:C Low-Friction Coating. *Surf. Coat. Technol.* **2009**, *203*, 2989–2994. [CrossRef]
21. Yoshida, K.; Horiuchi, T.; Kano, M.; Kumagai, M. Or0404—Effect of a Tribochemical Reacted Film on Friction and Wear Properties of DLC Coatings. *Plasma Process. Polym.* **2009**, *6*, S96–S101. [CrossRef]
22. Fan, X.; Li, G.; Guo, Y.; Zhang, L.; Xu, Y.; Zhao, F.; Zhang, G. Role of Reinforcement Types and Silica Nanoparticles on Tribofilm Growth at PTFE-Steel Interface. *Tribol. Int.* **2020**, *143*, 106035. [CrossRef]
23. Venkatesan, M.; Palanikumar, K.; Boopathy, S.R. Experimental Investigation and Analysis on the Wear Properties of Glass Fiber and CNT Reinforced Hybrid Polymer Composites. *Sci. Eng. Compos. Mater.* **2018**, *25*, 963–974. [CrossRef]
24. Reyes Acosta, Y.K.; Cruz Martinez, W.E.; Reyes Acosta, A.V.; Cepeda Tovar, V.A.; Contreras Esquivel, J.C.; Aguilar Gonzales, C.N.; Narro Cespedes, R.I.; Martinez, R.R. Thermal Properties by Adding Natural Oils, Foods, Organic Materials, Fibers, and Nanocomposites in the PLA, and Applications in 3D Printing. In *Engineering Principles for Food Processing Technology and Product Realization*; Apple Academic Press: New York, NY, USA, 2024; pp. 87–110.
25. Reyes Acosta, Y.K.; Martinez, E.C.; Reyes Acosta, A.V.; Torre, L.S.; Aguilar Gonzales, C.N.; Narro Cespedes, R.I. Release Studies to Improve the Mechanical Properties of the Biopolymer, Polylactic Acid (PLA) for Food Packaging Applications. In *Engineering Principles for Food Processing Technology and Product Realization*; Apple Academic Press: New York, NY, USA, 2024; pp. 67–85.
26. Levchenko, I.; Xu, S.; Baranov, O.; Bazaka, O.; Ivanova, E.; Bazaka, K. Plasma and Polymers: Recent Progress and Trends. *Molecules* **2021**, *26*, 4091. [CrossRef]
27. Maraveas, C.; Kyrtopoulos, I.V.; Arvanitis, K.G.; Bartzanas, T. The Aging of Polymers under Electromagnetic Radiation. *Polymers* **2024**, *16*, 689. [CrossRef]
28. Yong, C.; Valiyaveettil, S.; Tang, B. Toxicity of Microplastics and Nanoplastics in Mammalian Systems. *Int. J. Environ. Res. Public Health* **2020**, *17*, 1509. [CrossRef] [PubMed]
29. Shi, C.; Liu, Z.; Yu, B.; Zhang, Y.; Yang, H.; Han, Y.; Wang, B.; Liu, Z.; Zhang, H. Emergence of Nanoplastics in the Aquatic Environment and Possible Impacts on Aquatic Organisms. *Sci. Total Environ.* **2024**, *906*, 167404. [CrossRef]
30. Zhang, L.; García-Pérez, P.; Muñoz-Palazon, B.; Gonzalez-Martinez, A.; Lucini, L.; Rodriguez-Sanchez, A. A Metabolomics Perspective on the Effect of Environmental Micro and Nanoplastics on Living Organisms: A Review. *Sci. Total Environ.* **2024**, *932*, 172915. [CrossRef] [PubMed]
31. Gulati, S.; Amar, A.; Olihan, S. Environmental Fate, Behavior, and Risk Management Approaches of Nanoplastics in the Environment. In *Solid Waste Treatment Technologies*; CRC Press: Boca Raton, FL, USA, 2024; pp. 148–172.
32. Vettegren', V.I.; Lyashkov, A.I.; Savitskii, A.V.; Shcherbakov, I.P.; Vasil'ev, K.D. Microcrack Dynamics in a Polymer Composite Material during Friction. *Tech. Phys.* **2012**, *57*, 1445–1448. [CrossRef]
33. Ito, M.M.; Gibbons, A.H.; Qin, D.; Yamamoto, D.; Jiang, H.; Yamaguchi, D.; Tanaka, K.; Sivaniah, E. Structural Colour Using Organized Microfibrillation in Glassy Polymer Films. *Nature* **2019**, *570*, 363–367. [CrossRef]

34. Zhou, H.; Zhang, Z. Evolution of Silicon Particle Damage on Fatigue Crack Initiation and Early Propagation in an Aluminum Alloy. *Rare Met.* **2023**, *42*, 2470–2476. [CrossRef]
35. Alkhadra, M.A.; Root, S.E.; Hilby, K.M.; Rodriquez, D.; Sugiyama, F.; Lipomi, D.J. Quantifying the Fracture Behavior of Brittle and Ductile Thin Films of Semiconducting Polymers. *Chem. Mater.* **2017**, *29*, 10139–10149. [CrossRef]
36. Su, Y.; Yang, C.; Wang, S.; Li, H.; Wu, Y.; Xing, B.; Ji, R. Mechanochemical Formation of Poly(Melamine-Formaldehyde) Microplastic Fibers During Abrasion of Cleaning Sponges. *Environ. Sci. Technol.* **2024**, *58*, 10764–10775. [CrossRef]
37. Solasa, K.C.; Venkataraman, N.V.; Choudhury, P.R.; Schueller, J.K.; Bhattacharyya, A. On the Use of Alternative Measurement Methods in the Estimation of Wear Rates in Rotary-Pin-on-Disk Tribometry. *Tribol. Lett.* **2024**, *72*, 45. [CrossRef]
38. Orgeldinger, C.; Rosnitschek, T.; Tremmel, S. Unveiling an Additively Manufactured Open Hardware Pin-on-Disc Tribometer Considering Its High Reproducibility. *Wear* **2024**, *552–553*, 205437. [CrossRef]
39. Budinski, K.G. Adhesive Transfer to Abrasive Particles in Abrasion Testing. *Wear* **2011**, *271*, 1258–1263. [CrossRef]
40. Morreale, M.; La Mantia, F.P. Current Concerns about Microplastics and Nanoplastics: A Brief Overview. *Polymers* **2024**, *16*, 1525. [CrossRef] [PubMed]
41. Giechaskiel, B.; Grigoratos, T.; Mathissen, M.; Quik, J.; Tromp, P.; Gustafsson, M.; Franco, V.; Dilara, P. Contribution of Road Vehicle Tyre Wear to Microplastics and Ambient Air Pollution. *Sustainability* **2024**, *16*, 522. [CrossRef]
42. Yu, L.; Hu, J.; Li, R.; Yang, Q.; Guo, F.; Pei, J. Tire-Pavement Contact Pressure Distribution Analysis Based on ABAQUS Simulation. *Arab. J. Sci. Eng.* **2022**, *47*, 4119–4132. [CrossRef]
43. Cao, Y.; Lin, H.; Zhang, K.; Xu, S.; Yan, M.; Leung, K.M.Y.; Lam, P.K.S. Microplastics: A Major Source of Phthalate Esters in Aquatic Environments. *J. Hazard. Mater.* **2022**, *432*, 128731. [CrossRef]
44. Yang, N.; Men, C.; Zhang, Y.; Xie, Z.; Zuo, J. Exploring Polystyrene Weathering Behavior: From Surface Traits to Micro(Nano)Plastics and Additives Release. *J. Environ. Manag.* **2024**, *367*, 121880. [CrossRef]
45. Li, K.; Zhang, Y.; Yu, K.; Du, H.; Sioutas, C.; Wang, Q. Mechanics of Abrasion-Induced Particulate Matter Emission. *J. Mech. Phys. Solids* **2024**, *188*, 105661. [CrossRef]
46. Gnecco, E.; Pedraz, P.; Nita, P.; Dinelli, F.; Napolitano, S.; Pingue, P. Surface Rippling Induced by Periodic Instabilities on a Polymer Surface. *New J. Phys.* **2015**, *17*, 032001. [CrossRef]
47. Hennig, J.; Litschko, A.; Mazo, J.J.; Gnecco, E. Nucleation and Detachment of Polystyrene Nanoparticles from Plowing-Induced Surface Wrinkling. *Appl. Surf. Sci. Adv.* **2021**, *6*, 100148. [CrossRef]
48. Heinrich, G.; Klüppel, M. Basic Mechanisms and Predictive Testing of Tire-Road Abrasion. In *Degradation of Elastomers in Practice, Experiments and Modeling*; Springer: Berlin/Heidelberg, Germany, 2022; pp. 1–14.
49. Lisci, C.; Sitzia, F.; Pires, V.; Mirão, J. Building Stones Durability by UVA Radiation, Moisture and Spray Accelerated Weathering. *J. Build. Pathol. Rehabil.* **2022**, *7*, 60. [CrossRef]
50. Gu, X.; Michaels, C.A.; Drzal, P.L.; Jasmin, J.; Martin, D.; Nguyen, T.; Martin, J.W. Probing Photodegradation beneath the Surface: A Depth Profiling Study of UV-Degraded Polymeric Coatings with Microchemical Imaging and Nanoindentation. *J. Coat. Technol. Res.* **2007**, *4*, 389–399. [CrossRef]
51. El Hadri, H.; Gigault, J.; Maxit, B.; Grassl, B.; Reynaud, S. Nanoplastic from Mechanically Degraded Primary and Secondary Microplastics for Environmental Assessments. *NanoImpact* **2020**, *17*, 100206. [CrossRef]
52. Arinstein, A.; Zussman, E. Electrospun Polymer Nanofibers: Mechanical and Thermodynamic Perspectives. *J. Polym. Sci. B Polym. Phys.* **2011**, *49*, 691–707. [CrossRef]
53. Wohlleben, W.; Bossa, N.; Mitrano, D.M.; Scott, K. Everything Falls Apart: How Solids Degrade and Release Nanomaterials, Composite Fragments, and Microplastics. *NanoImpact* **2024**, *34*, 100510. [CrossRef]
54. Mourad, A.-H.I. Impact of Blending Ratio and Thermal Treatment on the Mechanical Behaviour of Polyethylene/Polypropylene Blends. In *Volume 6: Materials and Fabrication, Parts A and B, Proceedings of the ASME 2008 Pressure Vessels and Piping Conference, Chicago, IL, USA, 27–31 July 2008*; ASMEDC: New York, NY, USA, 2008; pp. 145–155.
55. Mourad, A.-H.I. Thermo-Mechanical Characteristics of Thermally Aged Polyethylene/Polypropylene Blends. *Mater. Des.* **2010**, *31*, 918–929. [CrossRef]
56. Bermúdez, M.D.; Brostow, W.; Carrión-Vilches, F.J.; Cervantes, J.J.; Pietkiewicz, D. Wear of Thermoplastics Determined by Multiple Scratching. *e-Polymers* **2005**, *5*, 1–9. [CrossRef]
57. Kadac-Czapska, K.; Knez, E.; Gierszewska, M.; Olewnik-Kruszkowska, E.; Grembecka, M. Microplastics Derived from Food Packaging Waste—Their Origin and Health Risks. *Materials* **2023**, *16*, 674. [CrossRef]
58. Mattsson, K.; Björkroth, F.; Karlsson, T.; Hassellöv, M. Nanofragmentation of Expanded Polystyrene Under Simulated Environmental Weathering (Thermooxidative Degradation and Hydrodynamic Turbulence). *Front. Mar. Sci.* **2021**, *7*, 578178. [CrossRef]
59. Song, Y.K.; Hong, S.H.; Jang, M.; Han, G.M.; Jung, S.W.; Shim, W.J. Corrections to "Combined Effects of UV Exposure Duration and Mechanical Abrasion on Microplastic Fragmentation by Polymer Type". *Environ. Sci. Technol.* **2018**, *52*, 3831–3832. [CrossRef] [PubMed]

60. Kierkels, J.T.A.; Dona, C.-L.; Tervoort, T.A.; Govaert, L.E. Kinetics of Re-embrittlement of (Anti)Plasticized Glassy Polymers after Mechanical Rejuvenation. *J. Polym. Sci. B Polym. Phys.* **2008**, *46*, 134–147. [CrossRef]
61. Sun, J.; Zheng, H.; Xiang, H.; Fan, J.; Jiang, H. The Surface Degradation and Release of Microplastics from Plastic Films Studied by UV Radiation and Mechanical Abrasion. *Sci. Total Environ.* **2022**, *838*, 156369. [CrossRef] [PubMed]
62. Wang, P.; Liang, H.; Jiang, L.; Qian, L. Effect of Nanoscale Surface Roughness on Sliding Friction and Wear in Mixed Lubrication. *Wear* **2023**, *530–531*, 204995. [CrossRef]
63. Chelliah, N.; Kailas, S.V. Synergy between Tribo-Oxidation and Strain Rate Response on Governing the Dry Sliding Wear Behavior of Titanium. *Wear* **2009**, *266*, 704–712. [CrossRef]
64. Randhawa, K.S.; Patel, A. The Effect of Environmental Humidity/Water Absorption on Tribo-Mechanical Performance of Polymers and Polymer Composites—A Review. *Ind. Lubr. Tribol.* **2021**, *73*, 1146–1158. [CrossRef]
65. Kalin, M.; Kupec, A. The Dominant Effect of Temperature on the Fatigue Behaviour of Polymer Gears. *Wear* **2017**, *376–377*, 1339–1346. [CrossRef]
66. Korku, M.; İlhan, R.; Feyzullahoğlu, E. Investigation of Effects of Environmental Conditions on Wear Behaviors of Glass Fiber Reinforced Polyester Composite Materials. *Polym. Compos.* **2025**, *46*, 355–371. [CrossRef]
67. Lai, Y.; Zhou, X.; Liu, J. Separation and Enrichment of Nanoplastics from Samples. In *Analysis of Microplastics and Nanoplastics*; Elsevier: Amsterdam, The Netherlands, 2025; pp. 281–293.
68. Shere, L.; Zhang, Z.J.; Preece, J.A. Application of Atomic Force Microscopy in Formulation Engineering. *Johns. Matthey Technol. Rev.* **2018**, *62*, 438–452. [CrossRef]
69. Ishida, N. Atomic Force Microscopy. In *Non-Destructive Material Characterization Methods*; Elsevier: Amsterdam, The Netherlands, 2024; pp. 89–125.
70. Kammoun, M.; Dupres, V.; Landoulsi, J.; Subramaniam, M.; Hawse, J.; Bensamoun, S.F. Transversal Elasticity of TIEG1 KO Muscle and Tendon Fibers Probed by Atomic Force Microscopy. *Comput. Methods Biomech. Biomed. Eng.* **2019**, *22*, S308–S310. [CrossRef]
71. Lv, S.; Wang, Q.; Li, Y.; Gu, L.; Hu, R.; Chen, Z.; Shao, Z. Biodegradation of Polystyrene (PS) and Polypropylene (PP) by Deep-Sea Psychrophilic Bacteria of Pseudoalteromonas in Accompany with Simultaneous Release of Microplastics and Nanoplastics. *Sci. Total Environ.* **2024**, *948*, 174857. [CrossRef] [PubMed]
72. Shen, N.; Feigenbaum, E.; Suratwala, T.; Steele, W.; Wong, L.; Feit, M.D.; Miller, P.E. Nanoplastic Removal Function and the Mechanical Nature of Colloidal Silica Slurry Polishing. *J. Am. Ceram. Soc.* **2019**, *102*, 3141–3151. [CrossRef]
73. Pöhl, F.; Hardes, C.; Theisen, W. Deformation Behavior and Dominant Abrasion Micro Mechanisms of Tempering Steel with Varying Carbon Content under Controlled Scratch Testing. *Wear* **2019**, *422–423*, 212–222. [CrossRef]
74. Soon, Z.Y.; Tamburri, M.N.; Kim, T.; Kim, M. Estimating Total Microplastic Loads to the Marine Environment as a Result of Ship Biofouling In-Water Cleaning. *Front. Mar. Sci.* **2024**, *11*, 1502000. [CrossRef]
75. Modica, F.; Basile, V.; Surace, R.; Fassi, I. Replication Study of Molded Micro-Textured Samples Made of Ultra-High Molecular Weight Polyethylene for Medical Applications. *Micromachines* **2023**, *14*, 523. [CrossRef]
76. Zhang, Y.; Yang, X.; Wang, S.; Lv, G.; Gao, Y.; Chen, K.; Yang, H. Research Status and Prospect of Wear and Aging on Hydraulic Rubber Sealing Materials. *Plast. Rubber Compos.* **2023**, *52*, 249–266. [CrossRef]
77. He, Y.; Yan, Y.; Geng, Y.; Hu, Z. Fabrication of None-Ridge Nanogrooves with Large-Radius Probe on PMMA Thin-Film Using AFM Tip-Based Dynamic Plowing Lithography Approach. *J. Manuf. Process.* **2017**, *29*, 204–210. [CrossRef]
78. Colaço, R. An AFM Study of Single-Contact Abrasive Wear: The Rabinowicz Wear Equation Revisited. *Wear* **2009**, *267*, 1772–1776. [CrossRef]
79. Reichelt, M.; Cappella, B. Wear Volume of Self-Mated Steel at the Submicron-Scale: An Atomic Force Microscopy Study. *J. Tribol.* **2022**, *144*, 061702. [CrossRef]
80. Singh, V.P.; Kumar, A.; Kumar, D.; Kuriachen, B. Effect of Welding Speed on Microstructure, Nano-Mechanical and Nano-Tribological Characteristics of Dissimilar Friction Stir Welded AA6061-T6 and AZ31 Alloy. *J. Adhes. Sci. Technol.* **2024**, *38*, 2295–2317. [CrossRef]
81. Stan, G.; King, S.W. Atomic Force Microscopy for Nanoscale Mechanical Property Characterization. *J. Vac. Sci. Technol. B* **2020**, *38*, 060801. [CrossRef]
82. Lin, C.-T.; Liao, W.-C.; Chen, J.-Y.; Su, H.-C.; Chiang, K.-N. Design and Analysis of a Nano-Probe of the AFM Based on the Small/Large Deflection Theory. In *Electronic and Photonic Packaging, Electrical Systems Design and Photonics, and Nanotechnology, Proceedings of the ASME 2004 International Mechanical Engineering Congress and Exposition, Anaheim, CA, USA, 13–19 November 2004*; ASMEDC: New York, NY, USA, 2004; pp. 337–344.
83. Krasnoborodko, S.Y.; Vysokikh, Y.E.; Bulatov, M.F.; Churikov, D.V.; Smagulova, S.A.; Shevyakov, V.I. Defocused Ion Beam Etching of the Silicon Probes for High Resolution Atomic-Force Microscopy. In Proceedings of the 2019 PhotonIcs & Electromagnetics Research Symposium—Spring (PIERS-Spring), Rome, Italy, 17–20 June 2019; IEEE: New York, NY, USA, 2019; pp. 1063–1066.
84. Xu, K.; Sun, W.; Shao, Y.; Wei, F.; Zhang, X.; Wang, W.; Li, P. Recent Development of PeakForce Tapping Mode Atomic Force Microscopy and Its Applications on Nanoscience. *Nanotechnol. Rev.* **2018**, *7*, 605–621. [CrossRef]

85. Pittenger, B.; Erina, N.; Su, C. Mechanical Property Mapping at the Nanoscale Using PeakForce QNM Scanning Probe Technique. In *Nanomechanical Analysis of High Performance Materials*; Springer: Berlin/Heidelberg, Germany, 2014; pp. 31–51.
86. Kochan, K.; Peleg, A.Y.; Heraud, P.; Wood, B.R. Atomic Force Microscopy Combined with Infrared Spectroscopy as a Tool to Probe Single Bacterium Chemistry. *J. Vis. Exp.* **2020**, *163*, e61728. [CrossRef]
87. Cho, H.; Felts, J.R.; Yu, M.-F.; Bergman, L.A.; Vakakis, A.F.; King, W.P. Improved Atomic Force Microscope Infrared Spectroscopy for Rapid Nanometer-Scale Chemical Identification. *Nanotechnology* **2013**, *24*, 444007. [CrossRef] [PubMed]
88. Zhou, L.; Cai, M.; Tong, T.; Wang, H. Progress in the Correlative Atomic Force Microscopy and Optical Microscopy. *Sensors* **2017**, *17*, 938. [CrossRef]
89. Vladár, A.E.; Hodoroaba, V.-D. Characterization of Nanoparticles by Scanning Electron Microscopy. In *Characterization of Nanoparticles*; Elsevier: Amsterdam, The Netherlands, 2020; pp. 7–27.
90. Konomi, B.A.; Dhavala, S.S.; Huang, J.Z.; Kundu, S.; Huitink, D.; Liang, H.; Ding, Y.; Mallick, B.K. Bayesian Object Classification of Gold Nanoparticles. *Ann. Appl. Stat.* **2013**, *7*, 640–668. [CrossRef]
91. Tsiper, S.; Dicker, O.; Kaizerman, I.; Zohar, Z.; Segev, M.; Eldar, Y.C. Sparsity-Based Super Resolution for SEM Images. *Nano Lett.* **2017**, *17*, 5437–5445. [CrossRef] [PubMed]
92. John Mardinly, A. Electron Tomography and Three-Dimensional Aspects of Transmission Electron Microscopy. *EDFA Tech. Artic.* **2005**, *7*, 6–12. [CrossRef]
93. Girão, A.V.; Caputo, G.; Ferro, M.C. Application of Scanning Electron Microscopy–Energy Dispersive X-Ray Spectroscopy (SEM-EDS). *Compr. Anal. Chem.* **2017**, *75*, 153–168.
94. Hodoroaba, V.-D.; Rades, S.; Salge, T.; Mielke, J.; Ortel, E.; Schmidt, R. Characterisation of Nanoparticles by Means of High-Resolution SEM/EDS in Transmission Mode. *IOP Conf. Ser. Mater. Sci. Eng.* **2016**, *109*, 012006. [CrossRef]
95. Buhr, E.; Senftleben, N.; Klein, T.; Bergmann, D.; Gnieser, D.; Frase, C.G.; Bosse, H. Characterization of Nanoparticles by Scanning Electron Microscopy in Transmission Mode. *Meas. Sci. Technol.* **2009**, *20*, 084025. [CrossRef]
96. Pretorius, E. Influence of Acceleration Voltage on Scanning Electron Microscopy of Human Blood Platelets. *Microsc. Res. Tech.* **2010**, *73*, 225–228. [CrossRef] [PubMed]
97. Albalak, R.J. (Ed.) *Polymer Devolatilization*; Routledge: Oxfordshire, UK, 2017; ISBN 9780203742914.
98. Bell, D.C.; Mankin, M.; Day, R.W.; Erdman, N. Successful Application of Low Voltage Electron Microscopy to Practical Materials Problems. *Ultramicroscopy* **2014**, *145*, 56–65. [CrossRef]
99. Zheng, K.; Gao, Y.; Bai, X.; Che, R.; Zhang, Z.; Han, X.; Bando, Y.; Yang, S.; Wang, E.; Cao, Q. In Situ TEM: Theory and Applications. In *Progress in Nanoscale Characterization and Manipulation*; Springer: Berlin/Heidelberg, Germany, 2018; pp. 381–477.
100. Slater, T.J.A.; Lewis, E.A.; Haigh, S.J. Energy Dispersive X-Ray Tomography for 3D Elemental Mapping of Individual Nanoparticles. *J. Vis. Exp.* **2016**, *113*, e52815. [CrossRef]
101. Kotula, A.P.; Orski, S.V.; Brignac, K.C.; Lynch, J.M.; Heilala, B.M.J. Time-Gated Raman Spectroscopy of Recovered Plastics. *Mar. Pollut. Bull.* **2022**, *181*, 113894. [CrossRef] [PubMed]
102. Käppler, A.; Fischer, D.; Oberbeckmann, S.; Schernewski, G.; Labrenz, M.; Eichhorn, K.-J.; Voit, B. Analysis of Environmental Microplastics by Vibrational Microspectroscopy: FTIR, Raman or Both? *Anal. Bioanal. Chem.* **2016**, *408*, 8377–8391. [CrossRef] [PubMed]
103. Gaur, R.; Manikandan, P.; Manikandan, D.; Umapathy, S.; Padhy, H.M.; Maaza, M.; Elayaperumal, M. Noble Metal Ion Embedded Nanocomposite Glass Materials for Optical Functionality of UV–Visible Surface Plasmon Resonance (SPR) Surface-Enhanced Raman Scattering (SERS) X-Ray and Electron Microscopic Studies: An Overview. *Plasmonics* **2021**, *16*, 1461–1493. [CrossRef]
104. Sobhani, Z.; Zhang, X.; Gibson, C.; Naidu, R.; Megharaj, M.; Fang, C. Identification and Visualisation of Microplastics/Nanoplastics by Raman Imaging (i): Down to 100 Nm. *Water Res.* **2020**, *174*, 115658. [CrossRef]
105. Jurkschat, L.; Gill, A.J.; Milner, R.; Holzinger, R.; Evangeliou, N.; Eckhardt, S.; Materić, D. Using a Citizen Science Approach to Assess Nanoplastics Pollution in Remote High-Altitude Glaciers. *Sci. Rep.* **2025**, *15*, 1864. [CrossRef] [PubMed]

surfaces

MDPI

Article

Tagua Powder as a Bio-Based Filler for Waterborne Acrylic Coatings: Enhancing Performances and Sustainability

Andrea Cristoforetti, Luca Donati and Stefano Rossi *

Department of Industrial Engineering, University of Trento, via Sommarive n. 9, 38123 Trento, Italy; andrea.cristoforetti@unitn.it (A.C.); luca.donati@studenti.unitn.it (L.D.)
* Correspondence: stefano.rossi@unitn.it; Tel.: +39-0461-282403

Abstract: The increasing demand for sustainable materials in the coatings industry is driving the replacement of synthetic components with bio-based alternatives. In this study, Tagua powder, derived from the seeds of *Phytelephas macrocarpa*, was incorporated as a filler in a waterborne acrylic-based coating to evaluate its effects on abrasion and protective properties. Two different particle size ranges (40–63 μm and ≤40 μm) and concentrations (1 wt% and 3 wt%) were tested. Morphological analyses confirmed a homogeneous dispersion of the filler within the coating matrix, with larger particles inducing surface roughness. The results demonstrated that the addition of Tagua powder significantly improved abrasion resistance, with the coating containing 3 wt% of larger particles (40–63 μm), reducing mass loss by 24.5% after 1000 Taber abrasion cycles compared to the reference coating. However, due to its lignocellulosic nature, the filler increased water uptake, leading to a decrease in barrier properties. Coatings with 3 wt% filler exhibited a reduction in electrochemical impedance modulus by approximately one order of magnitude after 670 h of immersion in a 3.5 wt% NaCl solution, indicating lower corrosion protection. Despite this, the performance in filiform corrosion resistance remained comparable to the reference, suggesting that Tagua powder does not critically affect adhesion properties. These findings highlight the potential of Tagua powder as a functional bio-based filler, offering enhanced mechanical durability while requiring a strategic coating design, such as a multilayer system, to mitigate moisture sensitivity. This study provides valuable insights into the development of environmentally friendly coatings with improved wear resistance.

Keywords: organic coatings; waterborne paint; bio-based filler; tagua powder; natural pigment

1. Introduction

During the past century, the prices of raw materials have been consistently climbing as a result of the unsustainable linear economic model. This system has made resources less available and more expensive while also inflicting irreversible damage on the environment and leading to a growing accumulation of waste [1–3]. One possible solution to this problem is a production aimed at the minimization of the impact on the environment. In this scenario, sustainable sources, such as bio-based materials, are applied to replace the products derived from fossil fuels in order to optimize waste management systems [4]. For example, in the paints field, researchers are increasingly focused on ecological and multifunctional alternatives to standard synthetic fillers because their production typically does not consider characteristics of environmental sustainability [5]. A sustainable approach is to use pigments that can be produced without chemical reactions, with reduced energy consumption and zero waste.

Recent research indicates that fillers derived from food and natural waste materials can significantly enhance the value of composite materials by reducing production costs and repurposing recycled resources effectively [6]. For example, the use of bio-based feedstock in material production grew from 5% in 2004 to 12% in 2010, reaching around 18% in 2020. Forecasts predict this could rise to 25% by 2030 [7]. These aspects have driven the research and the industrial sector toward the adoption of polymeric matrices added with bio-based fillers, in particular, those sourced from natural materials or derived from agricultural and food waste. One promising example of a natural bio-based filler is $CaCO_3$ powder from chicken eggshells tested in several organic coatings [8–10]. All these studies highlighted that thanks to their high content of $CaCO_3$, about 95%, chicken eggshells can replace commercial $CaCO_3$ without compromising the protective performances of the coating. Similarly, seashell waste from various mollusks provides an abundant supply of $CaCO_3$ that can be used to enhance the mechanical and thermal properties of bio-composites. In coatings, this natural filler also improves flame retardancy and reduces smoke emissions [11]. Additionally, agro-industrial waste offers a low-cost and valuable option as a filler. Materials like pistachio nutshells [12], cherry seeds, olive pits [6], peanut shells [13], avocado seeds [14], and groundnut and coconut shells [15] have been studied and utilized to improve mechanical, thermal, and wear-resistant properties of polymer composites.

An intriguing alternative filler material emerges in the form of Tagua, often referred to as "natural ivory". This remarkable material is derived from the seeds found within the fruits of the *Phytelephas Macrocarpa* tree, a palm species native to tropical regions of South America [16,17]. Known for its exceptional hardness and dense, ivory-like texture, Tagua has been prized for centuries in both artisanal and industrial applications [18]. Historically, it has been extensively used in the production of jewelry, chess pieces, billiard balls, buttons, and various decorative items [19]. During its peak industrial usage, this material constituted up to 20% of all buttons manufactured in the United States, underscoring its importance as a sustainable and versatile resource. It is naturally available and only needs to be pulverized. Its production does not lead to deforestation; on the contrary, it supports the sustainable development of local communities. Tagua is harvested in Ecuador when it naturally falls from wild trees, contributing to forest conservation and preventing deforestation risks. The collectors are independent farmers who supply Tagua to local wholesalers. Drying and initial processing stages are carried out locally, further supporting the regional economy [20].

Despite its long history and recognized potential, the use of Tagua in advanced materials remains underexplored. In particular, no studies have investigated the incorporation of Tagua powder as a functional filler in organic coatings. Leveraging its inherent hardness and unique physical properties, this work seeks to fill that gap by evaluating the impact of Tagua powder on the aesthetic qualities, durability, and mechanical performance of waterborne paints. In the current context of environmental concerns and stricter volatile organic compounds (VOCs) regulations, the coatings industry has been driven toward more sustainable alternatives, with waterborne paints emerging as a key solution. Widely adopted in the DIY and residential markets, these coatings offer advantages such as low VOC content [21], reduced odor, non-flammability [22], and easy cleanup with water. Their formulation has evolved to replicate the application and processing properties of traditional solvent-based systems through the use of water-dispersible or -emulsifiable resins.

The challenges of water sensitivity in waterborne paints, primarily linked to the use of hydrophilic additives, highlight the need for innovative solutions to enhance their protective properties [21]. Natural fillers such as Tagua could be interesting as sustainable alternatives to synthetic components, offering potential benefits in improved wear performance of the outer layer of the coating system.

This study explores the potential of Tagua as a filler in waterborne acrylic paints, evaluating its impact on water absorption, film integrity, and overall coating performance. The findings aim to advance the development of eco-friendly, high-performance waterborne coatings on steel substrates while leveraging the sustainable attributes of natural materials.

The paint was prepared by adding two different amounts of Tagua powder for two different particle size ranges to assess how the filler concentration and the granulometry influence the coatings' performance. The morphology and the aesthetic characteristics of the coatings were characterized by scanning electron microscope (SEM), optical microscope, gloss and roughness analysis, and contact angle to study the influence of the Tagua powder. The influence of bio-filler on the corrosion resistance and performance of composite coatings in aggressive environments was investigated through neutral salt spray chamber exposure (NSST), humidostatic chamber, and electrochemical impedance spectroscopy (EIS). Moreover, it was possible to study the promising mechanical wear improvement of the different coatings by exploiting the Taber test [23] and the Scrub test [24].

2. Materials and Methods

2.1. Samples Preparation

Natural ivory in the form of Tagua buttons was supplied by Mabo (Telgate, BG, Italy). The buttons were ground using a batch mill (IKA® WERKE M20, Staufen, Germany) to produce a fine powder. The resulting powder was sieved, yielding two distinct particle size distributions: one containing particles with sizes between 63 μm and 40 μm and the other containing particles smaller than 40 μm. The white waterborne 2K acrylic-based paint, AQUACRIL SATIN DTM, was supplied by FRANCHI & KIM (Castenedolo, BS, Italy). The Tagua powder was added to the paint at concentrations of 1 wt% and 3 wt% for each particle size range. The resulting formulations were mixed for 5 min using a high-speed homogenizer (IKA® ULTRA TURRAX T18, Staufen, Germany) to ensure homogeneous dispersion of the bio-filler particles. The paints were then applied onto the carbon steel substrates using a spray deposition technique. Five different coating types were prepared, including four bio-filler-enriched paints (two particle size ranges at two concentrations each) and the reference paint, free of bio-filler. Table 1 summarizes the coating types. Carbon steel substrates (Q-panel type R; composition: 0.15% C, Fe bal.) were used in three different dimensions: 76 mm × 152 mm × 0.81 mm, 102 mm × 102 mm × 0.81 mm, and 102 mm × 203 mm × 0.81 mm. These substrates were provided by Q-Lab (Westlake, OH, USA). After deposition, the coated samples were dried at room temperature for 15 min, followed by oven curing at 60 °C for 40 min.

Table 1. Coatings formulation considered. Different particle content and particle size were evaluated.

Sample	Particle Content [wt%]	Particle Size [μm]
C0	0	-
C1S63	1	$40 < d \leq 63$
C3S63	3	$40 < d \leq 63$
C1S40	1	$d \leq 40$
C3S40	3	$d \leq 40$

2.2. Characterization Methods

The morphology and composition of the Tagua powder and acrylic-based white paint, as well as the cross-sections of the coatings, were analyzed using a low-vacuum scanning electron microscope (SEM, JEOL IT 300, Tokyo, Japan). Powder size distribution was determined from SEM images using ImageJ software 1.53t. The gloss at 60° incident light rays of the coatings, influenced by different amounts of Tagua powder, was measured with an Erichsen 503 glossmeter (Milan, Italy) according to ASTM D523-14 [25]. Measurements were performed on three specimens per coating type, with five measurements per specimen for a total of 15 readings per coating. Coating roughness was assessed using a MAHR MarSurf PS1 profilometer (Mahr Gruppe, Göttingen, Germany) to evaluate R_a and R_z parameters. The coating thickness was measured with a Phynix Surfix FN thickness gauge (Phynix Sensortechnik, Neuss, Germany). To evaluate the effect of the filler on the surface wettability of the coatings, contact angle measurements were performed following the ASTM D7334/08 standard [26]. A Nikon 60 mm lens f/2.8 (Nikon Instruments, Tokyo, Japan) for macro pictures was used, and the wetting angle was measured by using the NIS-Elements Microscope Imaging software. The drops were formed with a syringe and dropped from a height of about 2 cm. Once the drop was focused, the picture was captured, and the wetting angle was measured by imaging software. In total, 10 measurements were performed for each sample to obtain statistical validity. Abrasion resistance was evaluated using the Taber and Scrub tests. The Taber test, performed following ASTM D4060–19 [23], utilized a Taber Abraser 5135 (Taber Industries, Tonawanda, NY, USA) with CS10 abrasive wheels loaded with 250 g each. Samples underwent 1.000 cycles at 60 rpm, with mass loss measured every 100 cycles. The wheels were cleaned and resurfaced every 250 cycles as per the standard. The Scrub test was conducted with an Elcometer 1720 Abrasion and Washability Tester (Elcometer, Manchester, U.K.) in accordance with ASTM D4213-24 [24]. Samples were subjected to 2000 cycles using an abrasive pad (90.0 ± 0.5 mm $\times$ 39.0 ± 0.5 mm) weighing 400 g, moving at 37 cycles per minute. Weight loss was evaluated every 500 cycles. The test was performed in dry mode to prevent solution absorption into the coatings. The NSST was performed to assess the corrosion protection provided by the coatings in accordance with ASTM B117-19 [27]. Images of the tested samples before aging is reported in Appendix A. Samples were exposed to a 5 wt.% sodium chloride solution in a salt spray chamber (Ascott Analytical Equipment Limited, Tamworth, U.K.) for 1.000 h. Since the NSST does not provide a complete characterization of the common failure mechanism [28], filiform corrosion (FFC) resistance was performed at a constant humidity of 80% at 40 °C after a 5 h chlorides contamination stage in the salt spray chamber [29]. Observations were made periodically to identify defects such as pores, blisters, and rust [30]. EIS was employed to evaluate coating defect levels and resistance to water and aggressive ion absorption. Measurements were conducted using an Autolab PGSTAT302N potentiostat (Metrohm, Herisau, Switzerland) controlled by Nova 1.11 software. A sinusoidal signal with a 20 mV peak-to-peak amplitude was applied over a frequency range of 100 kHz to 10 mHz. The setup included an electrochemical cell with a 15 cm^2 area in contact with a 3.5 wt.% sodium chloride solution. The cell housed an Ag/AgCl/3M KCl reference electrode and a platinum counter electrode. Impedance measurements were taken at different immersion periods.

3. Results

3.1. Filler and Coating Characterization

The initial analysis conducted was Energy-Dispersive X-ray Spectroscopy to determine the composition of the Tagua powder. The results, presented in Table 2, reveal that the powder is primarily composed of carbon (51 wt.%) and oxygen (46 wt.%), with trace amounts of other elements, including Al, K, Cl, Cu, Ca, P, S, Mg, and Si.

Table 2. Composition of Tagua powder.

Sample	Normalize Mass [%]
C	51.04
O	42.28
F	0.04
Al	0.98
P	0.15
Cl	0.32
K	0.57
Ca	0.21
Cu	0.25
S	0.07
Mg	0.07
Si	0.03

The SEM images of the two powders differently sieved are reported in Figure 1a and 1c, respectively, and for both images, the size distribution of the powders, determined with the aid of ImageJ software, was extracted (Figure 1b,d). The particle size was measured considering the maximum dimension of the particles. As a result, elongated particles with a major dimension larger than the sieve aperture may still pass through it due to their shape. However, the dimension of the maximum dimension for 90% of the particles (D90) for the coarse powder, 60.3 μm, indicates that nearly all granules are smaller than the sieve aperture size (63 μm). For the other type of powder, D90 is calculated at around 40.2 μm.

In addition to evaluating the morphology and particle size distribution, their adhesion to the matrix was also observed. To evaluate the compatibility between the paint matrix and the bio-based filler, the samples underwent a brittle fracture process by means of liquid nitrogen, allowing investigation of the cross-sectional structure through SEM analysis. In Figure 2, the cross-section of samples C3S63 and C3S40 is shown, and the images reveal the presence of Tagua granules well embedded within the paint layer. Despite the brittle fracture process, the bio-based filler remains firmly integrated into the matrix. This observation confirms strong compatibility between the filler and the polymeric matrix for both the bigger and the smaller particles. Interestingly, larger particles, as shown in Figure 2a, induce the formation of surface bulges, which increase surface roughness. In contrast, smaller particles, depicted in Figure 2b, result in a less pronounced effect on the surface texture.

Regarding the coating surface properties, the gloss values of the coating change not only as a function of the filler concentration but also as a function of its granulometry. The presence of the Tagua determines a decrease in the gloss value, providing a matt effect. This feature is in agreement with the higher roughness detected on the loaded paint with respect to the reference one (C0) in Table 3. Furthermore, a greater amount of Tagua added determines lower gloss together with higher roughness regardless of the particle size. The most relevant change is reached by adding 3.0 wt% of powder in coating type C3S63, which shows a decrease in gloss of around 30% with respect to the reference sample C0. Therefore, the introduction of the bio-based filler causes a gradual opacification of the coating. This trend is more evident when increasing the dimension of the filler. This bio-based filler influences both the surface morphology and texture of the coating, demonstrating its potential to provide aesthetic multifunctionality. Regarding the modification of the wettability of the coating, the contact angle is reported in Table 3 as being hydrophilic, which leads to an increase in the contact angle. So, increasing the content and the dimension of the particles increases the contact angle of the droplets deposited on the coating.

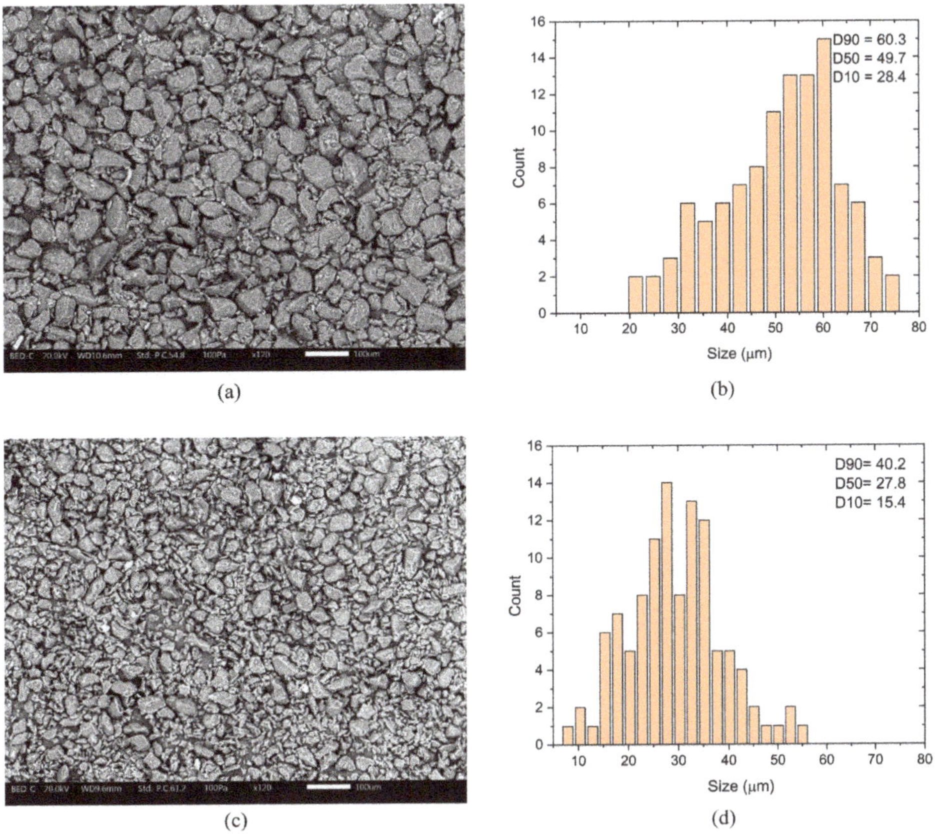

Figure 1. SEM-BED images of Tagua particles sieved through 60 μm mesh (**a**) and its particle size distribution (**b**). SEM-BED image of Tagua particles sieved through 40 μm mesh (**c**) at 120 and its particle size distribution (**d**).

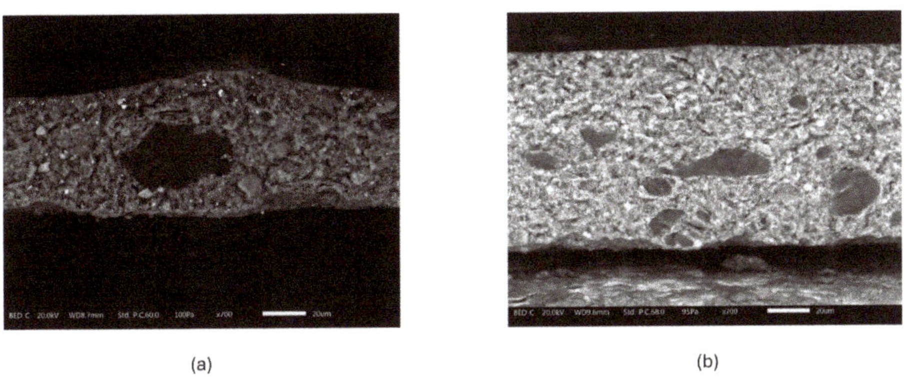

Figure 2. SEM-BED-C images of cross-sections for (**a**) C3S63 and (**b**) C3S40.

Table 3. Thickness and surface properties of the analyzed coatings.

Title 1	Thickness [μm]	Gloss 60° [GU]	Roughness Ra [μm]	Roughness Rz [μm]	Contact Angle [°]
C0	70 ± 7	41 ± 2	0.26 ± 0.02	1.62 ± 0.16	50 ± 5
C3S63	64 ± 11	28 ± 1	1.00 ± 0.12	5.61 ± 0.72	62 ± 2
C1S63	55 ± 2	35 ± 1	0.52 ± 0.10	3.40 ± 0.76	55 ± 4
C1S40	56 ± 5	34 ± 1	0.37 ± 0.03	2.56 ± 0.26	55 ± 5
C3S40	67 ± 8	33 ± 1	0.64 ± 0.10	3.95 ± 0.72	60 ± 1

3.2. Wear Resistance

The graph in Figure 3a illustrates the mass loss recorded for each coating type as a function of the number of Taber test abrasion cycles. As the test progresses, the mass loss consistently increases due to the continuous removal of the surface wear by the rotating wheels. The hardness of the Tagua particles in all the concentrations and granulometries leads to a reduction in mass loss. However, the paint added with the 3 wt% of particles having bigger particles appears to be the best choice for optimizing the abrasion resistance of the paint. Indeed, C3S63 shows a saved mass loss of 24.5% after 1000 cycles compared to the reference C0. The behavior during abrasion of the coating charged with smaller particles in higher amounts gives a similar result to the one of the coating charged with bigger particles but in lower amounts. A slight improvement in wear resistance is observed when tripling the amount of smaller particles; however, their tendency to remain embedded within the matrix reduces their effectiveness in reinforcing the coating. In contrast, larger particles protrude more from the surface, providing greater mechanical resistance against abrasion, while smaller particles are more easily detached during wear. On the other hand, the difference in mass loss between C3S63 and C3S40 is much more pronounced, indicating that the influence of filler size is less significant when particle content is small but becomes considerably more impactful as particle content increases.

It is noticeable that the damage after 1000 cycles on the C3S63 surface in the right region of Figure 3b. In contrast, the left portion of the image, which represents the region of the sample not subjected to abrasion, displays bulges caused by particles coated with paint. A detail of the same specimen is displayed in Figure 3c.

It can be observed that the motion of the wheels during the 1000 Taber cycles has removed the outer layer of the coating composed of paint, and it seems that the Tagua particles are well anchored to the polymer matrix, and no signs of pull-off were observed. During the cycles, the surrounding polymer was removed, while the filler remained attached to the paint, leading to a reduction in material loss thanks to the good wear resistance of the granules. Referring to Figure 3c, it is possible to see a sort of cut in the particle, so it is clear that the Tagua also gets abraded, and it is unable to provide complete shielding to the paint. However, by observing the edges of the particle, it is possible to see that the particle is slightly elevated compared to the matrix. These kinds of "steps" created between these granules may serve as resistance points against the grinding wheel's movement. As a result, the wheels tend to glide over these steps, reducing the removal of the polymeric matrix in the coating. Indeed, it can be seen how the mark left by the wheel stops above the particle and then starts again in a region slightly distant from it. Therefore, for this type of superficial abrasive process, Tagua powder effectively reduces the removal of the polymer matrix, thereby enhancing the durability of the coating in repeated abrasive contact.

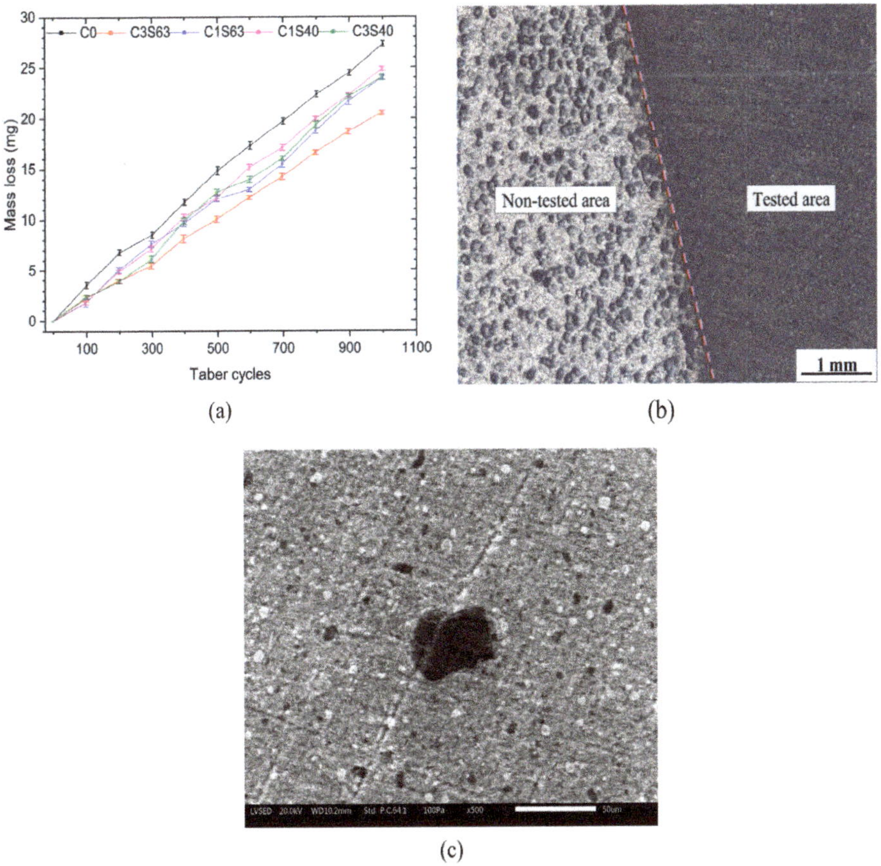

Figure 3. Mass loss evolution during 1000 cycles of Taber test (**a**). SEM-BED images (**b**,**c**) highlight the different morphologies on the tested coating surface for the C3S63 sample at different magnifications.

The Scrub test is one reliable method exploited for the characterization of the mechanical properties of organic-based composite coatings. It provides valuable insights into the reinforcing effects of pigments and both inorganic and organic fillers on the mechanical strength of coatings. The abrasion resistance of the coatings was assessed by subjecting the samples to wear cycles caused by an abrasive pad. The weight loss of the samples was monitored every 500 abrasion cycles, and then it is necessary to calculate L, which is the loss in coating mass per unit area, following Equation (1) in which it is possible to find the sample's initial weight (m_0) and the weight after the nth cycle (m_n). Additionally, A is the area of the sample occupied by the pad during the test [31].

$$\mathrm{L} = \frac{m_0 - m_n}{A} \quad (1)$$

By observing the trend of mass loss per unit area (L) as a function of abrasion cycles in the graph shown in Figure 4a, it is possible to evaluate the contribution of the concentration and the granulometry of the filler on the abrasion of the composite coating. It is noted that under such testing conditions, the addition of Tagua particles in all granulometries and concentrations increases the mass loss during the entire sequence of abrasion cycles compared to the sample coated with paint free of filler. In particular, for coatings containing the same concentrations of particles, the bigger size of the filler causes an increase in mass loss. In addition, also the measures of roughness reported in Table 3 corroborate the outcome of the mass losses. C3S63 contains big particles, resulting in higher roughness than the other samples. This means offering many protruding points where the pad can exert its abrasive action in a severe manner by removing the paint that covers the particles, and sometimes, it is also able to remove some of them, which results in a significant mass loss. The explanation of these data can be found by comparing the images of the area subjected to the abrasion for C3S63 and C0 taken after 2000 cycles and reported in Figure 4b,c. The C3S63 surface shows that the pad performs strong abrasive action on the raised areas corresponding to the places where are present the particles covered by paint, while the surface of C0 is smoother than C3S63, and it seems that the pad exerted less abrasive action. However, for this type of test, as was found for the Taber test, no cases of pull-off were detected.

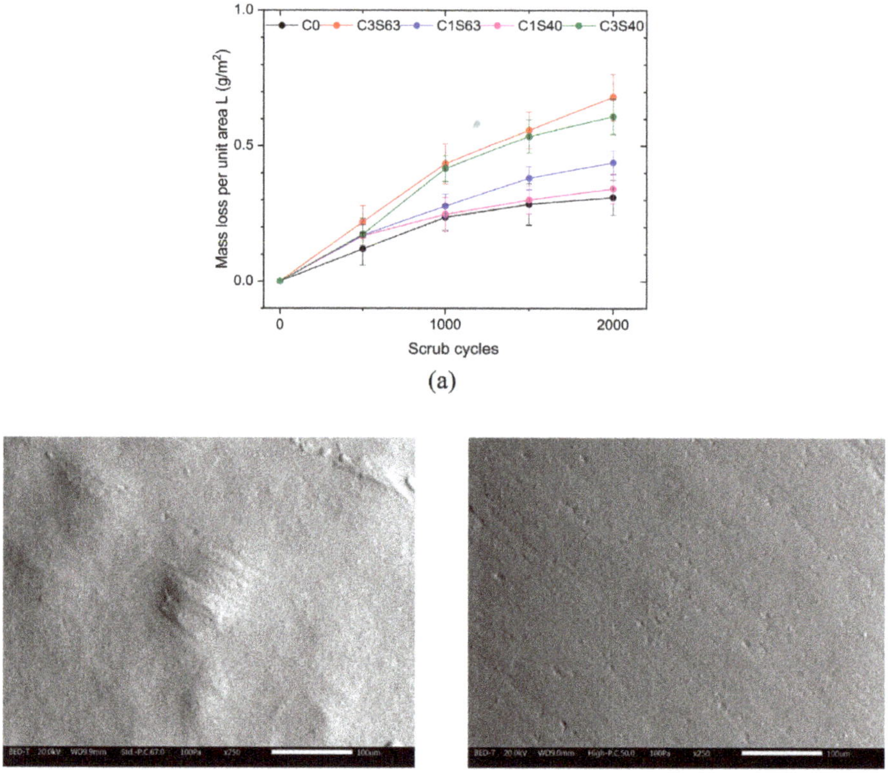

(a)

(b)

(c)

Figure 4. Mass loss evolution during 2000 cycles of the Scrub test (**a**). SEM-BED images (**b**) and (**c**) highlight the different roughness morphologies on the tested coating surface for C3S63 and C0 samples, respectively.

Lowering the particle concentration to 1 wt% leads to smaller mass losses after 2000 Scrub cycles, likely because the elevated points are less numerous. Additionally, the lower amount of filler reduces the contribution of cut particle pieces to the mass loss. This test involved a different type of abrasive action compared to the Taber test, and as a result, the findings differ significantly from those of the Taber test, which identified C3S63 as the best option.

3.3. Aging Tests

The coated samples were placed in the NSST for a total duration of 1000 h, and the samples were observed after 24, 72, 150, 300, 500, and 1000 h. The blister formation caused by water uptake was visually inspected during each monitoring session. After 300 h, some blisters became visible in the area for samples C3S40 and C3S63. The appearance of the coated steel panels after 1000 h is displayed in Figure 5. The inclusion of the bio-based filler does not enhance the protective properties of the coating; rather, it appears to influence them by promoting water uptake by its hygroscopic tendency. Tagua is composed of lignocellulose and is characterized by a hygroscopic behavior, which is already highlighted in several works [32–35].

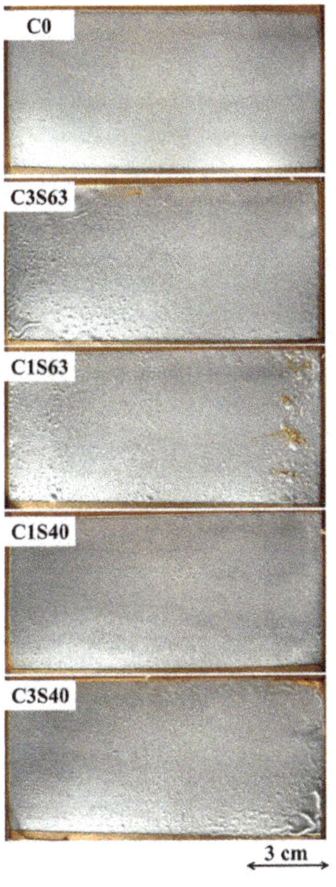

Figure 5. Coated samples after 1000 of NSST exposure.

After 1000 h in the salt spray chamber, the samples loaded with 3% Tagua powder (C3S63 and C3S40) exhibit blisters evenly distributed across the tested surface, accompanied by visible delamination. In the case of sample C1S40, the blisters are fewer and smaller in size, with better durability. Conversely, sample C1S63 might initially appear to be the most damaged. According to [36] ISO 4628-2, a blistering grade can be assigned based on the density and size of the blisters for each sample, focusing on the area reasonably far from the edges. These evaluations are summarized in Table 4. A higher amount of filler translates to a wider area damaged by blisters, while bigger particles cause a reduced number of blisters but have bigger dimensions. In the NSST, the higher absorption of water caused by the presence of Tagua particles is aggravated by the humidity set at 100 r.h.%, which may not be fully representative of the real durability performances.

Table 4. Percentage of area covered by blisters, number of blisters, and blistering grade.

Title 1	Area Covered by Blisters [%]	Number of Blisters	ISO 4628-2 Blister Density	ISO 4628-2 Blister Size
C0	/	/	/	/
C3S63	6.3	58	3	4
C1S63	1.4	28	3	3
C1S40	1.1	36	2	3
C3S40	3.1	103	4	3

Furthermore, the comparison between C3S63 and C3S40 underscores the impact of particle size. Larger particles appear to exacerbate water absorption, potentially creating pathways that facilitate water transport through capillary action. Meanwhile, for the samples containing 1 wt% of particles, it is observed that the blistering is less noticeable, and the effect of particle size is less pronounced. The barrier properties of the composite layers are weakened by the presence of these fillers, as their interface with the polymer matrix creates pathways that allow moisture and aggressive ions to penetrate. Larger particles, due to their hydrophilic nature and increased likelihood of surface exposure with incomplete matrix coverage (see Figure 2), create preferential pathways for electrolyte infiltration. This effect, combined with the expanded interface between the polymer matrix and the filler, promotes higher water uptake and reduces the overall barrier performance of the coating. Similar behavior has been reported in previous studies [6], where lignocellulose-based fillers increased water absorption by forming percolating pathways that facilitate water transport through capillary action within the filler.

A similar aging campaign was performed on similar samples where a longitudinal scratch in the coating was introduced. The results are not reported here since no differences have been detected between the coating loaded with Tagua and the reference. In the case of a damaged coating, the effect of the filler results is hidden by the macroscopic source of electrolyte from scratch, and the paint wet adhesion seems not altered.

To evaluate durability not only against cathodic delamination but also in relation to a similarly common yet less frequently considered failure mechanism, FFC resistance was tested in a humidostatic chamber at 80 r.h.% (FFC test). This setup favors the electrochemical processes driving FFC over those responsible for cathodic delamination. In this scenario, closer to natural exposure in a continental environment, low chloride contamination and relatively lower humidity compared to NSST conditions were simulated. Under these conditions, delamination driven by the nucleation and propagation of filaments through an advancing anodic front showed less critical results regarding the use of Tagua powder. In fact, no significant worsening was observed with the introduction of the filler with respect to the reference (Figure 6a). As already reported in the literature [28,37,38], different aging conditions could lead to opposite performance in coatings durability. In the case at hand, the sample with the higher amount of particles loaded and the biggest size (C3S63) displays after 500 h of cabinet FFC test the minor degradation in terms of fewer filaments nucleated (Figure 7b), focusing on the average and maximum threads length this sample presents feature fully comparable with the blank one (Figure 7a). The increased water uptake due to the presence of Tagua particles does not exacerbate FFC as observed with cathodic disbonding. Instead, it appears to alter the typical propagation pattern, leading to fewer but wider threads rather than the characteristic fine filaments. This suggests that the presence of the bio-based filler and the resulting modification of the barrier layer may disrupt the formation of the differential aeration cells necessary to sustain the conventional anodic undermining mechanism. As a general consideration, the FFC degradation was comparable to the reference and aligned with propagation rates reported in the literature [39]. Thus, in the tested humidity range, it is estimated to be a more frequent exposure condition rather than the NSST [40,41].

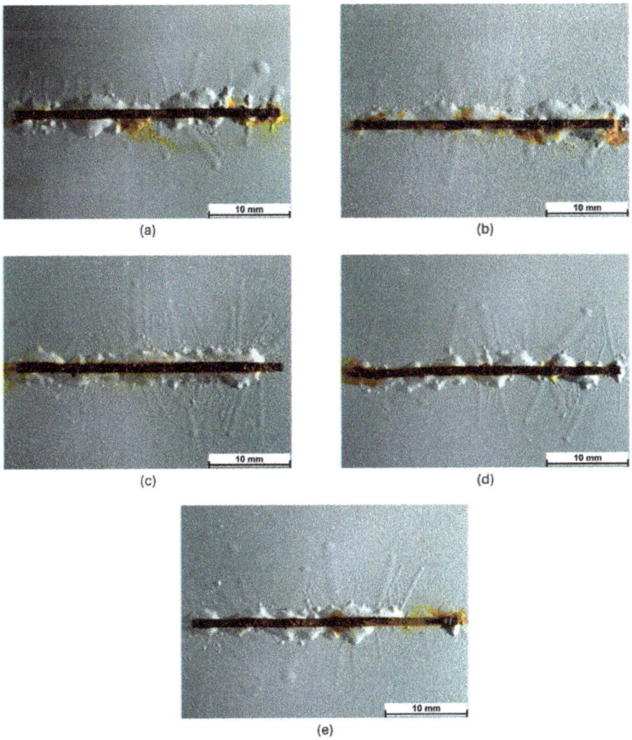

Figure 6. Coated samples after 500 h of FFC test. Samples C0 (**a**), C3S63 (**b**), C1S63 (**c**), C1S40 (**d**), and C3S40 (**e**) are displayed.

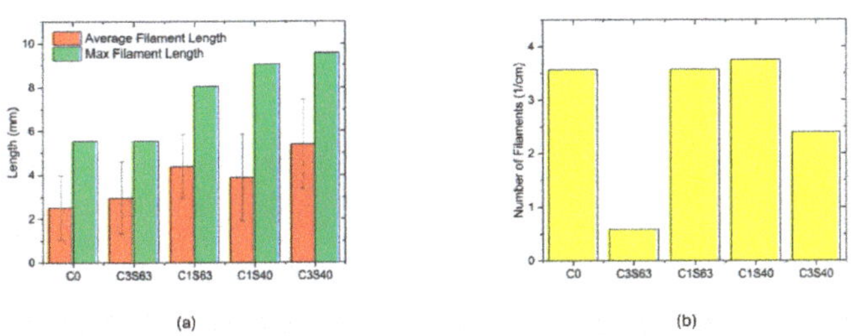

Figure 7. FFC degradation parameters after 500 h of humidostatic test at 80 r.h.% and 40 °C. (**a**) reports the mean and maximum filament length, while (**b**) displays the specific number of nucleated threads per unit length.

Moreover, considering the results obtained, Tagua powder emerges as an excellent filler for enhancing abrasion resistance. Its strategic application in the topcoat layer, where wear resistance is crucial, could improve the durability of the coating. When combined with a protective system design featuring a thinner primer layer, this approach may offer a well-balanced solution, ensuring both mechanical durability and effective corrosion protection.

3.4. Electrochemical Impedance Spectroscopy

EIS monitoring during electrolyte immersion is commonly employed to evaluate the protective performance of paints, including their adhesion properties, barrier properties, and potential defects. The impedance modulus in the Bode plot at low frequencies (10^{-2} Hz), denoted as $|Z|_{(0.01)}$, in some cases, could serve as a key parameter for approximating the overall coating's level of protection. This is commonly recognized as a potential source of misinterpretation in the case of organic coatings with a high level of barrier properties [42]. However, in the present study, the waterborne nature of the paints and the increased permeability introduced by the addition of Tagua powder provide suitable boundary conditions for a reliable evaluation of these parameters as a representative index of the coating's protective performance. According to various authors [43,44], a $|Z|_{(0.01)}$ value of 10^6 $\Omega\cdot cm^2$ is generally considered the minimum threshold for classifying a coating as "protective". The performance of the coatings was assessed by monitoring the evolution of their impedance modulus, $|Z|_{(0.01)}$, over time. The results, depicted in Figure 8, show the variation in $|Z|_{(0.01)}$ throughout the 670 h exposure to the test solution. Initially, all samples exhibited similar values for $|Z|_{(0.01)}$ around 10^{10} $\Omega\cdot cm^2$. In the following 8 h after, the low-frequency impedance continues to decrease significantly reaching a value around 10^8 $\Omega\cdot cm^2$ for all the samples, and this is due to the water uptake typical of the waterborne paint [45]. This type of coating containing additives like wetting and dispersing agents, stabilizers, and thickeners having hydrophilic characteristics makes the films more susceptible to water absorption. However, after 24 h of immersion, all samples appear to stabilize, with for $|Z|_{(0.01)}$ remaining relatively constant for the remainder of the test. As indicated in the NSST, the presence of Tagua particles in the paint, due to their hydrophilic tendency, leads to a general decrease in the impedance modulus at the end of the test, compared to the paint without particles, which retains the highest impedance modulus. C3S63 shows the lowest for $|Z|_{(0.01)}$ after 670 h of exposure to the electrolyte, while C3S40 containing the same amount of filler shows a higher value, also highlighting this time the negative effect of bigger particles, which favor the formation of detrimental pathways that allow the electrolyte to reach the substrate.

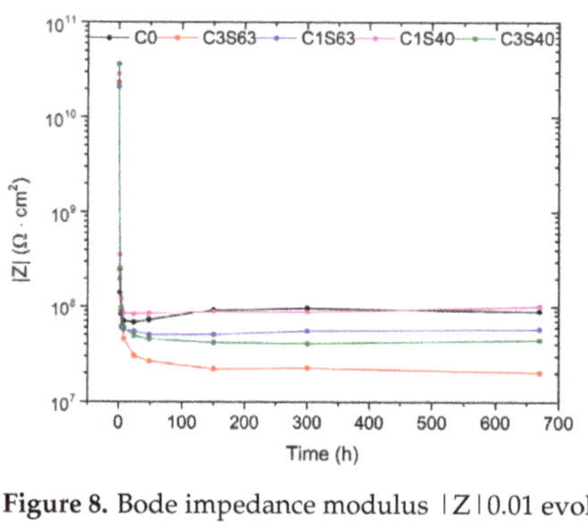

Figure 8. Bode impedance modulus $|Z|0.01$ evolution with time.

Figure 9 displays the Bode spectra of the monitored samples in time. In Figure 9d, it is possible to see that after 24 h for the extreme case of C3S63, a shoulder at a low frequency (0.1 Hz) develops, and the impedance drops at higher frequencies at increasing times. The outcome at low frequencies is representative of a second low-frequency time constant appearance that is typical of the electrochemical reactions at the coating–substrate interface. Instead, for sample C3S40 (Figure 9j), there is the evolution of the Bode phase spectrum, and the second low-frequency time constant is less visible in the sampled range and likely starts at lower frequencies. Regarding the other samples, reference C0, low-loaded C1S40, and

C1S63, their spectra exhibit a single time constant curve at high frequencies that gradually shifts toward higher frequencies over time.

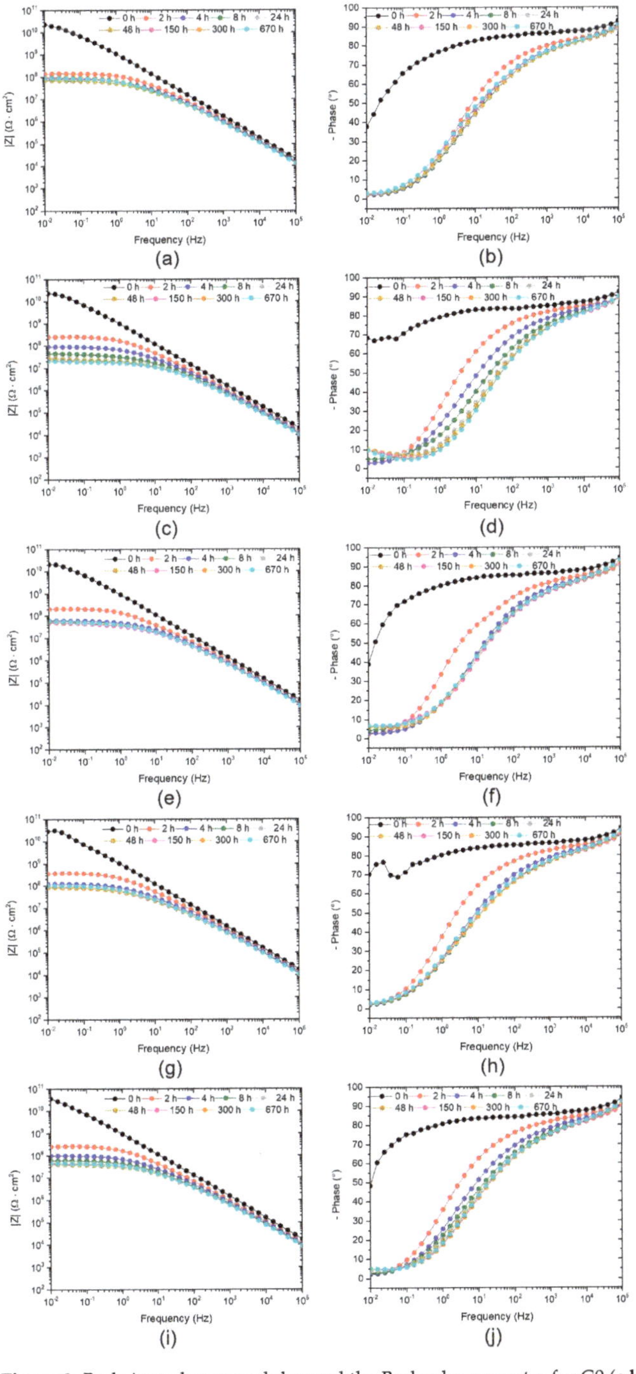

Figure 9. Bode impedance modulus and the Bode phase spectra for C0 (**a**,**b**), C3S63 (**c**,**d**), C1S63 (**e**,**f**), C1S40 (**g**,**h**), and C3S40 (**i**,**j**).

4. Discussion

The incorporation of Tagua powder as a bio-based filler significantly influences the aesthetic, mechanical, and protective properties of the waterborne acrylic paint. The addition of this filler effectively reduces gloss and increases surface roughness compared to the unfilled paint. Notably, filler concentration plays a more significant role in altering these properties than particle size. While variations in particle size do impact gloss and roughness, their effect is less pronounced compared to differences in filler content. When comparing coatings with the same filler concentration but different particle sizes, the changes in surface characteristics are minimal, whereas coatings with different filler concentrations exhibit more substantial variations in gloss and roughness.

Mechanical performance results show contrasting trends depending on the type of abrasion stress. In the Taber test, which evaluates resistance to abrasive forces, Tagua powder enhances durability due to its compact structure. A higher filler concentration and larger particle size contribute to lower mass loss, with the best performance observed in the coating containing 3 wt% of the largest particles, showing a 24.5% reduction in mass loss compared to the unmodified acrylic matrix. This highlights the potential of Tagua powder as an effective reinforcement for applications requiring high abrasion resistance. However, in the Scrub test, which involves another type of abrasion, the results differ. Here, the best-performing coating is the one with the lower concentration of the smallest particles. The increase in surface roughness caused by the filler creates more protruding points where the abrasive pad can exert force, leading to greater mass loss at higher filler concentrations. While the Taber test provides valuable insights into the mechanical wear resistance of coatings, it does not always fully reflect real-world performance across all applications. For instance, in the automotive industry, discrepancies have been observed between Taber test results and actual service conditions, where factors such as delamination and chipping play a significant role [46]. This highlights the importance of considering multiple evaluation methods. Scrub tests, for example, can offer complementary data, particularly for coatings exposed to frequent cleaning cycles. Therefore, selecting the most appropriate test method requires a thorough understanding of the specific environmental and mechanical stresses the coating will face in its intended application.

Regarding protective properties, despite good compatibility between the filler and the polymer matrix, the lignocellulosic nature of Tagua powder introduces challenges related to water absorption. Both NSST and EIS measurements indicate that the addition of the filler negatively affects the barrier properties of the coating in saturated, humid environments. Higher filler concentrations and larger particle sizes exacerbate this issue, leading to reduced protective performance. This effect is less pronounced when the coating is exposed to lower humidity levels, where FFC may become the primary failure mechanism for this type of coating [47]. The hydrophilic nature of lignocellulosic materials represents a key industrial concern, and ongoing research aims to address this limitation [48]. Promising advancements have been reported by successfully enhancing the water resistance of bio-filled coatings by functionalizing olive pit powder with silane and wax [6]. The functionalization processes applied to such fillers can significantly alter their hydrophobic properties, potentially mitigating issues related to water absorption and enhancing barrier performance in high-humidity environments.

The introduction of Tagua powder into the waterborne acrylic binder significantly increases the interfacial area between the filler and the matrix. Despite this, no voids or delamination were observed, indicating good compatibility between the two components. However, this interfacial region can represent a potential weak point in terms of coating durability, particularly concerning barrier properties and cohesion. It is important to note that the size of the Tagua particles plays a crucial role in determining the behavior of the filler within the coating system. Smaller particles provide a greater interfacial area, leading to better dispersion and stronger interactions with the binder. In contrast, larger particles often result in incomplete coverage by the binder, leaving the filler partially exposed to the environment. This effect may negatively influence the coating's protective performance, especially in terms of moisture and corrosion resistance. Thus, particle size is a key factor in optimizing the interfacial properties and overall performance of bio-based filled coatings.

This technology could also be implemented in the case at hand. However, the most promising strategy appears to be the design of a multilayer organic coating consisting of an unfilled primer ensuring strong adhesion to the substrate, a thicker barrier layer, and a topcoat functionalized with Tagua particles to enhance wear resistance [49]. Furthermore, the increased water uptake resulting from the porous or hydrophilic nature of bio-fillers used for wear resistance can be mitigated by integrating hybrid fillers. A dual-filler strategy, combining bio-fillers to enhance mechanical properties with hydrophobic additives to improve water resistance, offers a balanced solution [50,51]. This synergistic approach not only enhances durability in humid environments but may also contribute to improved fire resistance, making it a promising strategy for high-performance protective coatings.

This paper, for the first time, explores the potential of Tagua as a coating additive. To assess its fundamental properties, a single-layer system was adopted. At this stage, the objective was not to develop a fully formulated multilayer coating system but rather to establish a fundamental understanding of the effects of Tagua powder as a bio-based filler in coatings. By adopting a single-layer configuration, we aimed to accurately assess its influence on key properties such as abrasion resistance, gloss, and water uptake without the interference of additional layers. The findings confirmed that while Tagua powder improves wear resistance, its hydrophilic nature increases water absorption, making it unsuitable for direct application in a layer in contact with a metallic substrate. Based on these results, a more practical and effective two-layer coating system is proposed, where a primer ensures adhesion and corrosion protection while a Tagua-containing topcoat enhances mechanical durability. This study lays the groundwork for further optimization, aligning with the growing interest in sustainable coatings for environmentally oriented engineering applications.

5. Conclusions

This study demonstrated the potential of Tagua powder as a bio-based additive for acrylic coatings, particularly in improving abrasion durability. One of the key findings is its significant contribution to wear resistance, a fundamental parameter for the long-term performance of organic coatings. The incorporation of Tagua particles effectively reduced mass loss in abrasion tests, confirming its role in enhancing the mechanical strength of the polymeric matrix. In particular, coatings containing 3 wt% of larger particles (40–63 μm) exhibited a 24.5% reduction in mass loss after 1000 Taber abrasion cycles compared to the reference coating, while coatings with smaller particles ($\leq$40 μm) at the same concentration showed a more modest improvement of approximately 10%.

However, the lignocellulosic nature of Tagua also introduces hydrophilic characteristics, leading to increased water uptake. EIS revealed that coatings with 3 wt% Tagua powder, especially with larger particles, experienced a reduction in impedance modulus by approximately one order of magnitude after 670 h of immersion in a 3.5 wt% NaCl solution, indicating a decrease in barrier properties. Despite this, filiform corrosion resistance tests showed comparable performance to the reference, suggesting that the adhesion properties of the coatings were not significantly affected by the presence of the bio-based filler.

These results highlight the need for an optimized coating design to balance the benefits of increased wear resistance with the challenges of water sensitivity. A multilayer approach, where a topcoat reinforced with Tagua powder is combined with a dedicated barrier layer, could provide a viable solution for applications requiring both durability and environmental sustainability. This study provides a foundation for further research aimed at optimizing formulation strategies and evaluating long-term performance in real-world conditions.

Author Contributions: Conceptualization, S.R. and A.C.; methodology, S.R.; validation, S.R. and A.C.; investigation, L.D.; resources, S.R.; data curation, L.D. and A.C.; writing—original draft preparation, A.C. and L.D.; writing—review and editing, S.R. and A.C.; supervision, S.R.; project administration, S.R. All authors have read and agreed to the published version of the manuscript.

Funding: This research received no external funding.

Institutional Review Board Statement: Not applicable.

Informed Consent Statement: Not applicable.

Data Availability Statement: Data available upon request.

Acknowledgments: The authors would like to thank Mabo, Telgate (BG), for providing the Tagua, and Stefano Tonini from Estalia, Castenedolo (BS), for supplying the waterborne coating.

Conflicts of Interest: The authors declare no conflicts of interest.

Appendix A

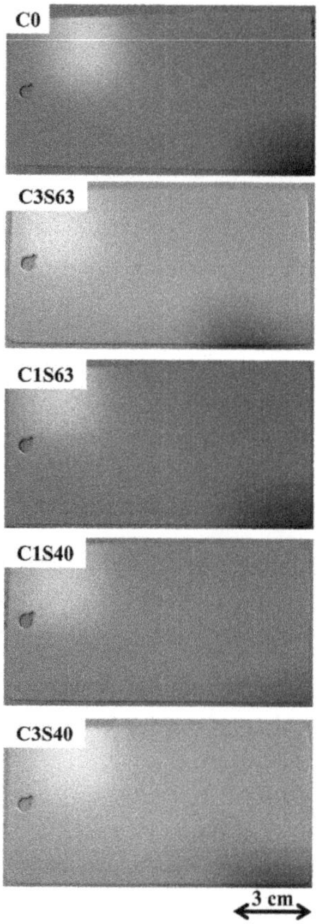

Figure A1. Coated samples before NSST exposure.

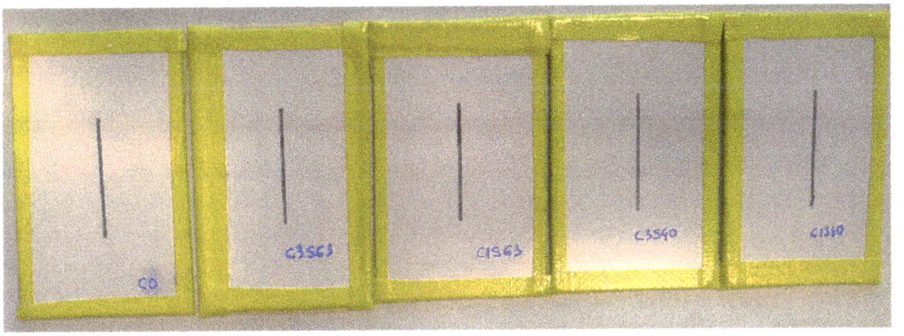

Figure A2. Coated samples before the FFC test. A 3 cm longitudinal scratch is introduced in the organic layer.

References

1. Huttmanová, E.; Chovancová, J.; Petruška, I.; Sallaku, K. Circular Economy in Action: Examining the Decoupling of Economic Growth and Material Use across EU Countries. *EJSD* **2024**, *13*, 13. [CrossRef]
2. Fellner, J.; Lederer, J.; Scharff, C.; Laner, D. Present Potentials and Limitations of a Circular Economy with Respect to Primary Raw Material Demand. *J. Ind. Ecol.* **2017**, *21*, 494–496. [CrossRef]
3. Schüpfer, D.; Wagner-Wenz, R.; Hendrich, K.; Weidenkaff, A. Materials Come around and Go around: Adapting to Nature's Circularity. *MRS Bull.* **2023**, *48*, 1184–1189. [CrossRef]
4. Proskurina, S. Carbon Neutrality in the Finnish Energy Sector: Prospects for a Fossil-Fuel Phase Out. *Biofuels Bioprod. Biorefining* **2024**, *18*, 1065–1076. [CrossRef]
5. Calovi, M.; Rossi, S. Synergistic Contribution of Bio-Based Additives in Wood Paint: The Combined Effect of Pigment Deriving from Spirulina and Multifunctional Filler Based on Carnauba Wax. *Prog. Org. Coat.* **2023**, *182*, 107713. [CrossRef]
6. Calovi, M.; Rossi, S. Functional Olive Pit Powders: The Role of the Bio-Based Filler in Reducing the Water Uptake Phenomena of the Waterborne Paint. *Coatings* **2023**, *13*, 442. [CrossRef]
7. Gurunathan, T.; Mohanty, S.; Nayak, S.K. A Review of the Recent Developments in Biocomposites Based on Natural Fibres and Their Application Perspectives. *Compos. Part A Appl. Sci. Manuf.* **2015**, *77*, 1–25. [CrossRef]
8. Yew, M.C.; Ramli Sulong, N.H.; Yew, M.K.; Amalina, M.A.; Johan, M.R. The Formulation and Study of the Thermal Stability and Mechanical Properties of an Acrylic Coating Using Chicken Eggshell as a Novel Bio-Filler. *Prog. Org. Coat.* **2013**, *76*, 1549–1555. [CrossRef]
9. Toro, P.; Quijada, R.; Yazdani-Pedram, M.; Arias, J.L. Eggshell, a New Bio-Filler for Polypropylene Composites. *Mater. Lett.* **2007**, *61*, 4347–4350. [CrossRef]
10. Yew, M.C.; Yew, M.K.; Saw, L.H.; Ng, T.C.; Durairaj, R.; Beh, J.H. Influences of Nano Bio-Filler on the Fire-Resistive and Mechanical Properties of Water-Based Intumescent Coatings. *Prog. Org. Coat.* **2018**, *124*, 33–40. [CrossRef]
11. Sienkiewicz, N.; Dominic, M.; Parameswaranpillai, J. Natural Fillers as Potential Modifying Agents for Epoxy Composition: A Review. *Polymers* **2022**, *14*, 265. [CrossRef] [PubMed]
12. Balasundar, P.; Narayanasamy, P.; Senthil, S.; Abdullah Al-Dhabi, N.; Prithivirajan, R.; Shyam Kumar, R.; Ramkumar, T.; Subrahmanya Bhat, K. Physico-Chemical Study of Pistachio (Pistacia Vera) Nutshell Particles as a Bio-Filler for Eco-Friendly Composites. *Mater. Res. Express* **2019**, *6*, 105339. [CrossRef]
13. Prabhakar, M.N.; Shah, A.U.R.; Rao, K.C.; Song, J.-I. Mechanical and Thermal Properties of Epoxy Composites Reinforced with Waste Peanut Shell Powder as a Bio-Filler. *Fibers Polym.* **2015**, *16*, 1119–1124. [CrossRef]
14. Zuluaga-Parra, J.D.; Ramos-deValle, L.F.; Sánchez-Valdes, S.; Torres-Lubián, J.R.; Rodriguez-Fernadez, O.S.; Hernández-Hernández, E.; da Silva, L.; Rodríguez-Gonzalez, J.A.; Borjas-Ramos, J.J.; Vázquez-Rodríguez, S.; et al. Phosphorylated Avocado Seed: A Renewable Biomaterial for Preparing a Flame Retardant Biofiller. *Fire Mater.* **2022**, *46*, 968–980. Available online: https://onlinelibrary-wiley-com.ezp.biblio.unitn.it/doi/10.1002/fam.3044 (accessed on 13 January 2025). [CrossRef]
15. Suthan, R.; Jayakumar, V.; Bharathiraja, G. Wear Analysis of Bio-Fillers Reinforced Epoxy Composites. *Mater. Today Proc.* **2020**, *22*, 793–798. [CrossRef]
16. Smith, N. Phytelephas Macrocarpa. In *Palms and People in the Amazon*; Smith, N., Ed.; Springer International Publishing: Cham, Switzerland, 2015; pp. 429–444. ISBN 978-3-319-05509-1.
17. Acosta-Solis, M. Tagua or Vegetable Ivory: A Forest Product of Ecuador. *Econ. Bot.* **1948**, *2*, 46–57. [CrossRef]
18. Cooper, D. On the Structure of the Nut Known as Vegetable Ivory. *Trans. Microsc. Soc. J.* **1842**, *1*, 97–99. [CrossRef]

19. Barfod, A.S.; Bergmann, B.; Pedersen, H.B. The Vegetable Ivory Industry: Surviving and Doing Well in Ecuador. *Econ. Bot.* **1990**, *44*, 293–300. [CrossRef]
20. MABO Website. Available online: https://mabo.it/it (accessed on 15 February 2025).
21. Chek, Y.W.; Ang, D.T.-C. Progress of Bio-Based Coatings in Waterborne System: Synthesis Routes and Monomers from Renewable Resources. *Prog. Org. Coat.* **2024**, *188*, 108190. [CrossRef]
22. Zhan, W.; Li, L.; Chen, L.; Kong, Q.; Chen, M.; Chen, C.; Zhang, Q.; Jiang, J. Biomaterials in Intumescent Fire-Retardant Coatings: A Review. *Prog. Org. Coat.* **2024**, *192*, 108483. [CrossRef]
23. *ASTM F1978-22*; Standard Test Method for Measuring Abrasion Resistance of Metallic Thermal Spray Coatings by Using the Taber Abraser. ASTM International: West Conshohocken, PA, USA, 2022.
24. D01 Committee ASTM International. *Test Method for Scrub Resistance of Paints by Abrasion Weight Loss*; ASTM International: West Conshohocken, PA, USA, 2024. [CrossRef]
25. E12 Committee ASTM International. *Test Method for Specular Gloss*; ASTM International: West Conshohocken, PA, USA, 2018. [CrossRef]
26. D01 Committee ASTM International. *Practice for Surface Wettability of Coatings, Substrates and Pigments by Advancing Contact Angle Measurement*; ASTM International: West Conshohocken, PA, USA, 2022. [CrossRef]
27. *ASTM B117-19*; Standard Practice for Operating Salt Spray (Fog) Apparatus. ASTM International: West Conshohocken, PA, USA, 2019. [CrossRef]
28. Cristoforetti, A.; Deflorian, F.; Rossi, S.; Fedel, M. On the Occurrence of Filiform Corrosion on Organic Coated Carbon Steel Exposed to Cyclic Aging Test. *Corrosion* **2023**, *79*, 1339–1344. [CrossRef] [PubMed]
29. *ASTM D2803-09*; Standard Guide for Testing Filiform Corrosion Resistance of Organic Coatings on Metal. ASTM International: West Conshohocken, PA, USA, 2020. [CrossRef]
30. D01 Committee ASTM International. *Test Method for Evaluating Degree of Blistering of Paints*; ASTM International: West Conshohocken, PA, USA, 2025. [CrossRef]
31. Calovi, M.; Rossi, S.; Deflorian, F.; Dirè, S.; Ceccato, R.; Guo, X.; Frankel, G.S. Effects of Graphene-Based Fillers on Cathodic Delamination and Abrasion Resistance of Cataphoretic Organic Coatings. *Coatings* **2020**, *10*, 602. [CrossRef]
32. Gašparovič, L.; Koreňová, Z.; Jelemenský, Ľ. Kinetic Study of Wood Chips Decomposition by TGA. *Chem. Pap.* **2010**, *64*, 174–181. [CrossRef]
33. Mansour, O.Y. Lignocellulose–Polymer Composite. III. *J. Appl. Polym. Sci.* **1993**, *47*, 839–846. [CrossRef]
34. Najafi, S.K.; Tajvidi, M.; Chaharmahli, M. Long-Term Water Uptake Behavior of Lignocellulosic-High Density Polyethylene Composites. *J. Appl. Polym. Sci.* **2006**, *102*, 3907–3911. [CrossRef]
35. Yang, X.F.; Vang, C.; Tallman, D.E.; Bierwagen, G.P.; Croll, S.G.; Rohlik, S. Weathering Degradation of a Polyurethane Coating. *Polym. Degrad. Stab.* **2001**, *74*, 341–351. [CrossRef]
36. *ISO 4628-2:2016*; Paints and Varnishes—Evaluation of Degradation of Coatings—Designation of Quantity and Size of Defects, Part 2: Designation of Degree of Blistering. International Organization for Standardization: Geneva, Switzerland, 2016.
37. Cambier, S.M.; Posner, R.; Frankel, G.S. Coating and Interface Degradation of Coated Steel, Part 1: Field Exposure. *Electrochim. Acta* **2014**, *133*, 30–39. [CrossRef]
38. Cambier, S.M.; Frankel, G.S. Coating and Interface Degradation of Coated Steel, Part 2: Accelerated Laboratory Tests. *Electrochim. Acta* **2014**, *136*, 442–449. [CrossRef]
39. Cristoforetti, A.; Rossi, S.; Deflorian, F.; Fedel, M. Recent Progress in Understanding Filiform Corrosion on Organic Coated Steel: A Comprehensive Review. *Prog. Org. Coat.* **2024**, *192*, 108469. [CrossRef]
40. Cristoforetti, A.; Rossi, S.; Deflorian, F.; Fedel, M. An Electrochemical Study on the Mechanism of Filiform Corrosion on Acrylic-Coated Carbon Steel. *Prog. Org. Coat.* **2023**, *179*, 107525. [CrossRef]
41. Cristoforetti, A.; Rossi, S.; Deflorian, F.; Fedel, M. Comparative Study between Natural and Artificial Weathering of Acrylic-coated Steel, Aluminum, and Galvanized Steel. *Mater. Corros.* **2023**, *74*, 1429–1438. [CrossRef]
42. Cristoforetti, A.; Deflorian, F.; Rossi, S.; Fedel, M. On the Limits of the EIS Low-Frequency Impedance Modulus as a Tool to Describe the Protection Properties of Organic Coatings Exposed to Accelerated Aging Tests. *Coatings* **2023**, *13*, 598. [CrossRef]
43. Amirudin, A.; Thieny, D. Application of Electrochemical Impedance Spectroscopy to Study the Degradation of Polymer-Coated Metals. *Prog. Org. Coat.* **1995**, *26*, 1–28. [CrossRef]
44. Rossi, S.; Deflorian, F.; Fontanari, L.; Cambruzzi, A.; Bonora, P.L. Electrochemical Measurements to Evaluate the Damage Due to Abrasion on Organic Protective System. *Prog. Org. Coat.* **2005**, *52*, 288–297. [CrossRef]
45. Lendvay-Győrik, G.; Pajkossy, T.; Lengyel, B. Water Uptake of Water-Borne Paint Resin Films as Studied by Impedance Spectroscopy and Gravimetry. *Prog. Org. Coat.* **2007**, *59*, 95–99. [CrossRef]
46. Sun, J.; Mukamal, H.; Liu, Z.; Shen, W. Analysis of the Taber Test in Characterization of Automotive Side Windows. *Tribol. Lett.* **2002**, *13*, 49–54. [CrossRef]
47. Bautista, A. Filiform Corrosion in Polymer-Coated Metals. *Prog. Org. Coat.* **1996**, *28*, 49–58. [CrossRef]

48. Randis, R.; Darmadi, D.B.; Gapsari, F.; Sonief, A.A.; Akpan, E.D.; Ebenso, E.E. The Potential of Nanocomposite-Based Coatings for Corrosion Protection of Metals: A Review. *J. Mol. Liq.* **2023**, *390*, 123067. [CrossRef]
49. Przystupa, K. Research on the Durability and Reliability of Industrial Layered Coatings on Metal Substrate Due to Abrasive Wear. *Materials* **2023**, *16*, 1779. [CrossRef]
50. Mustapa, S.A.S.; Ramli Sulong, N.H. Performance of Palm Oil Clinker as a Bio-Filler with Hybrid Fillers in Intumescent Fire Protective Coatings for Steel. *JSM* **2017**, *46*, 2489–2496. [CrossRef]
51. Pei, X.-Q.; Bennewitz, R.; Kasper, C.; Tlatlik, H.; Bentz, D.; Becker-Willinger, C. Tribological Synergy of Filler Components in Multifunctional Polyimide Coatings. *Adv. Eng. Mater.* **2017**, *19*, 1600363. [CrossRef]

surfaces

Review

Industrial Piping System: Design and Corrosion Protection

David Coverdale Rangel Velasco [1,2,*], Victor Paes Dias Gonçalves [1], Michel Picanço Oliveira [3], Noan Tonini Simonassi [1], Felipe Perissé Duarte Lopes [1] and Carlos Maurício Fontes Vieira [1]

1 Advanced Materials Laboratory (LAMAV), State University of the Northern Rio de Janeiro (UENF), Av. Alberto Lamego, 2000, Campos dos Goytacazes 28013-602, Brazil; victor_paesdias@yahoo.com.br (V.P.D.G.); noansimonassi@uenf.br (N.T.S.); perisse@uenf.br (F.P.D.L.); vieira@uenf.br (C.M.F.V.)
2 Department of Mechanical Engineering (CCBEMCF), Fluminense Federal Institute of Education, Science and Technology (IFFluminense), Estr. dos Búzios, Cabo Frio 28293-660, Brazil
3 Wood and Forestry Science Department (DCFM), Federal University of Espírito Santo (UFES), Av. Governador Lindemberg, 316, Jerônimo Monteiro 29550-000, Brazil; michel.oliveira@ufes.br
* Correspondence: david.velasco@iff.edu.br

Abstract: Piping system failures in process industries pose significant financial, environmental, and social risks, with inadequate design and corrosion being major contributors. This review synthesizes the academic and normative literature on pipeline design and anti-corrosive protection strategies, providing a comprehensive examination of pipeline layout determination, material selection, and methods for mitigating corrosion. A particular focus is placed on organic coating as a pivotal strategy for corrosion reduction, with in-depth insights into their selection and evaluation criteria. By highlighting best practices and advancements in design and protection strategies, this review aims to enhance the overall integrity and safety of piping systems. The findings are intended to support industry professionals in implementing more effective measures to prevent pipeline failures and improve system reliability, while also presenting recent advances and current demands.

Keywords: industrial piping design; anti-corrosion protection; coatings

1. Introduction

Piping systems have existed for millennia, being fundamental to the functioning of society and considered the most important infrastructure in today's world [1–3]. They are defined as a set of tubes and their accessories are used to transport liquids, gases, and suspended solids. Every industry relies on piping systems, and their importance is particularly pronounced in process industries, where they account for an average of 20 to 25% of installation costs, 45 to 50% of equipment assembly costs, and 20% of the total project cost [4].

To maximize process profitability, piping systems are generally designed to last over 15 years, with this duration often extending in developed countries. For instance, in Calgary (Canada), a significant portion of the sewage infrastructure is old: 5.24% of the pipes are over 65 years old, and 3.4% have been in place for more than 100 years. There is a trend towards not only designing longer-lasting systems but also extending the life of existing ones, raising concerns about the safety and reliability of aging pipelines [5,6].

The tendency for reliability to decrease over the lifespan of pipelines is evident in increased downtime and accidents [7]. Studies such as Kidam and Hurme [8], shown in Table 1, indicate that most accidents in process industries occur within piping systems.

Table 1. Equipment with the highest frequency of accidents in chemical industries [8].

Process Equipment	Kidam and Hurme [8]	Duguid [9]	Vílchez et al. [10]	Instone [11]	Marsh Inc. [12]	Mean
Piping system	25	33	16	14	33	24
Reactors	14	9	14	5	10	10
Storage tank	14	20	2	14	17	13
Pressure vessels	10	9	18	3	5	9
Heat transfer eq.	8	11	6	19	4	10
Separation eq.	7	-	-	5	3	5
Others + unknown	22	18	44	40	28	30

Such behavior is undesirable, especially in process industries where greater reliability is required than in other sectors. In this type of industry, there is usually continuous operation, with no daily stops for inspection and/or maintenance; continuous process chains, where the stoppage of a section can results in the shutdown of the entire plant; and the transportation of hazardous products, which increases socio-environmental impacts in the event of leaks or accidents [13].

Thus, new methodologies that manage and optimize risks, project costs, maintenance, and availability have been developed [14,15]. Among these, studies on the influence of certain factors on pipeline reliability have intensified [16–18].

However, although the use of new tools such as finite elements, the Internet of Things (IoT), Cyber–Physical Systems (CPSs), machine learning, etc., are gaining ground, inadequate design remains the main cause of pipeline accidents [19–22]. Table 2 shows the largest contributors and sub-contributors to pipeline accidents. In this table, inadequate layout stands out as the main cause of accidents, accounting for almost 50% of them.

Table 2. Major contributors and sub-contributors in pipeline accidents [8].

Sub-Contributors / Main Contributors	N° Main Contributors	Human/Organizational	Contamination	Heat Transfer	Flow Related	Reaction	Layout	Fab. Const. & Inst.	Corrosion	Construction Material	Mechanical Failure	Static Electricity	External Factor	Vibration	Erosion	Utility Related
Human/organizational	**12**		3	3	2	2	3	1	1		1				1	
Contamination	**5**	3		1		2			2	1	1					
Heat transfer	**7**	3						4		1		2				
Flow related	**9**	6	2			2	1		1						1	1
Layout	**19**	8	7	2	5	3		1	2					1	1	
Fab. const. & inst.	**17**	4		4	1		1			2	5			6		
Corrosion	**9**	2			2		1	2		2			4		2	
Construction material	**13**	3			3	1		4	7		1			1	1	
Utility related	**1**				1			1								
Sub-contributor (total)	**92**	**29**	**12**	**10**	**14**	**10**	**6**	**13**	**13**	**6**	**8**	**2**	**4**	**8**	**6**	**1**

There are no cases where the following contributors are main: reaction, mechanical failure, static electricity, external factor, vibration, and erosion.

Another significant contributor is corrosion. Its effects are generally mitigated during repairs, as predictive maintenance can minimize the occurrence of accidents. However, studies such as those by Kidam and Hurme [8] and Biezma et al. [13] have observed that corrosion is associated with 1 in 4 accidents.

The latest report by the European Gas Pipeline Incident Data Group (EGIG) identifies corrosion as a leading cause of accidents, as well as external interference (Figure 1). Unlike external interference, however, corrosion does not show a downward trend, and its impact on accidents is likely to become even more significant in the future [23]. Consequently, most studies on pipeline reliability and failures currently focus on or relate to corrosion, not only due to the risk of accidents but also because it reduces the time between maintenance intervals, as well as the useful life and reliability of the system [24,25].

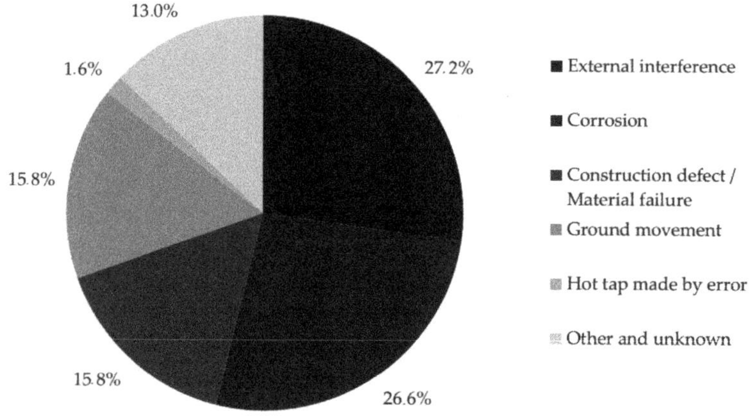

Figure 1. Distribution of pipeline failures by cause [23].

In this sense, this work aims to present a narrative review, covering the main recommendations for layout preparation, accessories used in pipelines, calculations for pipeline design, and corrosion protection mechanisms (with an emphasis on organic coatings). This review gathers information from the various types of literature to enable the realization of safer and more profitable projects.

2. Recommendations for Industrial Layout and Projects

In addition to articles, this section primarily references piping design standards from the American Society of Mechanical Engineers (ASME), the main international developer of codes and standards associated with practices in this area. However, standards from the following organizations were also used as references: American National Standards Institute (ANSI), American Society for Testing and Materials (ASTM), American Petroleum Institute (API), American Water Works Association (AWWA), National Fire Protection Association (NFPA), and Petróleo Brasileiro SA (Petrobras), among others.

2.1. Layout

To design an efficient industrial plant layout, it is necessary to understand the installation site, the plant's objectives, and the operation of the equipment. The layout should be designed to reduce piping, structures, energy consumption, and machine power, while maintaining safety and facilitating operations and maintenance, as well as allowing for future expansion in piping arrangements. In this sense, it is generally recommended to install piping above ground level [26].

The layout begins with a conceptual study of the terrain, analyzing the preliminary locations of the main units, equipment, roads, buildings, safety fences, location of flares, heaters, process units, direction of prevailing winds, pipe racks, and sleepers. Subsequently, two orthogonal directions are determined as references, termed North–South and East–West, which will guide all the constructions within the industrial facility. In the next stage, information such as the basic activities of the industry, the calculation of the area required for these activities, and the block diagram of material circulation are collected. This information is obtained during the process design, which forms the basic design for the installation's operation [26–28].

Barker [27] presents a logical design diagram (Figure 2). It shows that, before the development of the plot plan, project input data and project design data must be established. Project input data can be classified into three categories: project design data, supplier data, and internally generated engineering data. Additionally, engineering companies and clients typically use Process Industry Practice (PIP) plant layout design specifications that must also be met.

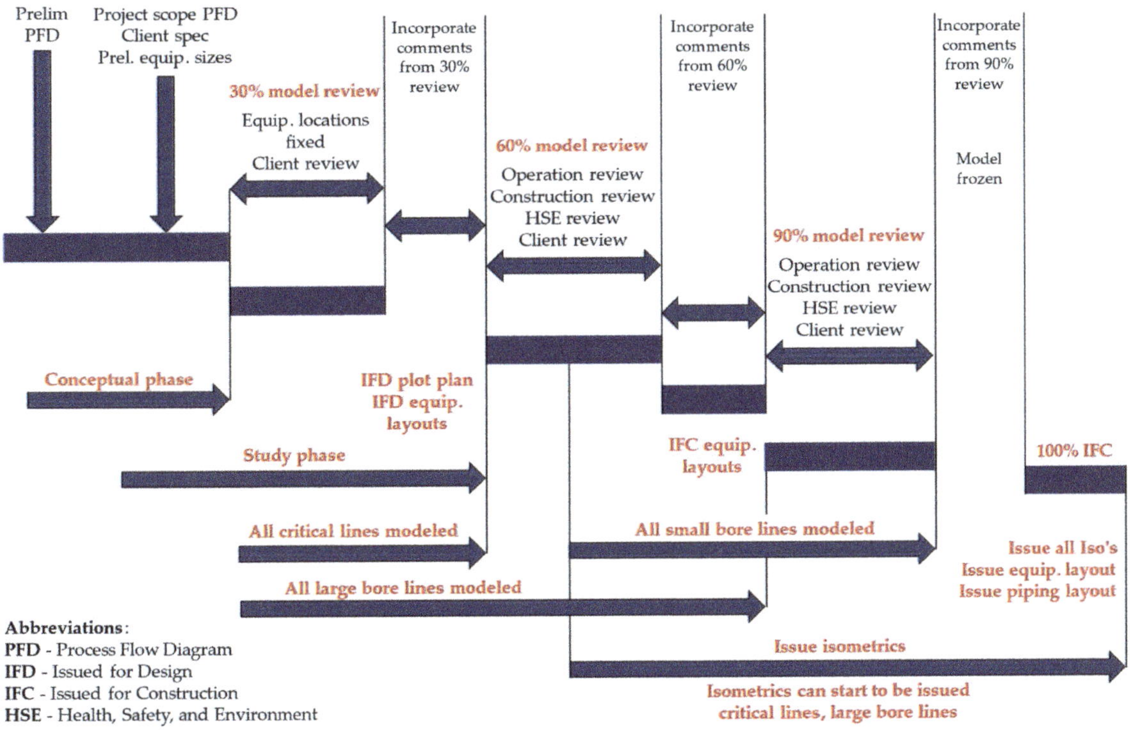

Figure 2. Logical design diagram [27].

The layout of an industrial facility is carried out through the development of several drawings, such as the following:

- Block Flow Diagram (BFD);
- Process Flow Diagram (PFD);
- Piping and Instrumentation Diagrams (P&IDs or PIDs);
- Equipment layout drawings;
- Plot plans;
- Isometric drawings;
- Control instrumentation drawings.

2.1.1. Classification of Risk Areas

The classification of risk areas is essential, as there are atmospheres prone to explosions due to the type of flammable substance present in the environment and there is a probability that these substances will be released into the external environment. This is intended to allow electrical devices to be operated safely in the environment, as well as to signal existing risks [29,30].

Some measures must be taken in places where there may be an explosive atmosphere, such as eliminating the likelihood of an explosive gas atmosphere occurring around the ignition source, developing an explosion-proof (containment) design, limiting the range of explosion flames and explosion pressures to a safe level. If there is a high likelihood of an explosive atmosphere occurring, the use of electrical appliances should be limited to those that have an extremely low likelihood of creating an ignition source [29,31].

According to the National Fire Protection Association (NFPA) 70 [32], classifications should be made according to the properties of the flammable gas, vapor produced by flammable liquid, vapors produced by combustible liquid, combustible dust, or fibers/flyings that may be present, and the likelihood that a flammable or combustible concentration or quantity is present. In addition to defining the type of hazardous substance and quantities sufficient to cause explosions, this publication classifies them according to Table 3. Each condition can be classified according to the type of material and whether it is flammable or combustible (Class), the frequency where this condition occurs (Division), and the risk (Zone).

Table 3. Classification of risk areas [32].

Classification	Description
Class I	Flammable vapors and gases present
Class II	Combustible dust may be found
Class III	Presence of flammable fibers or particles
Division 1	Flammable concentration exists under normal or frequent operating conditions
Division 2	Flammable concentration exists under abnormal operating conditions
Zone 0 (if Class I) Zone 20 (if Class II ou III)	Location where the hazard is present for long periods of time
Zone 1 (if Class I) Zone 21 (if Class II ou III)	Location where flammable concentrations are likely to exist under normal operating conditions
Zone 2 (if Classe I) Zone 22 (if Classe II ou III)	Location where flammable concentrations are unlikely to occur and if they do, they will exist for short periods of time

In this context, specific standards such as IEC 60079-10-1 provide essential guidelines for the proper selection and construction of structures [33].

2.1.2. Piping

The route of the piping passages is previously established in the plant layout study. To facilitate maintenance and contribute to project cost savings, the arrangement of non-underground piping is usually chosen in an industrial installation. The general rules for this arrangement are to form parallel groups of the same elevation, alternating larger diameters with smaller diameters. This way, the number of supports can be reduced, making the project more cost effective. Whenever possible, the piping should follow orthogonal directions and have different elevations for different directions, allowing for crossings and branches [34].

For ease of operation, maintenance, and safety, usually the piping is arranged on supports, classified according Figure 3. It is recommended that the passageways should be in the central part of the land, adjacent to the process areas, utility areas, and fluid storage areas. Spaces are also reserved in pipeline bridges for the passage of instrumentation ducts and electrical cables [26].

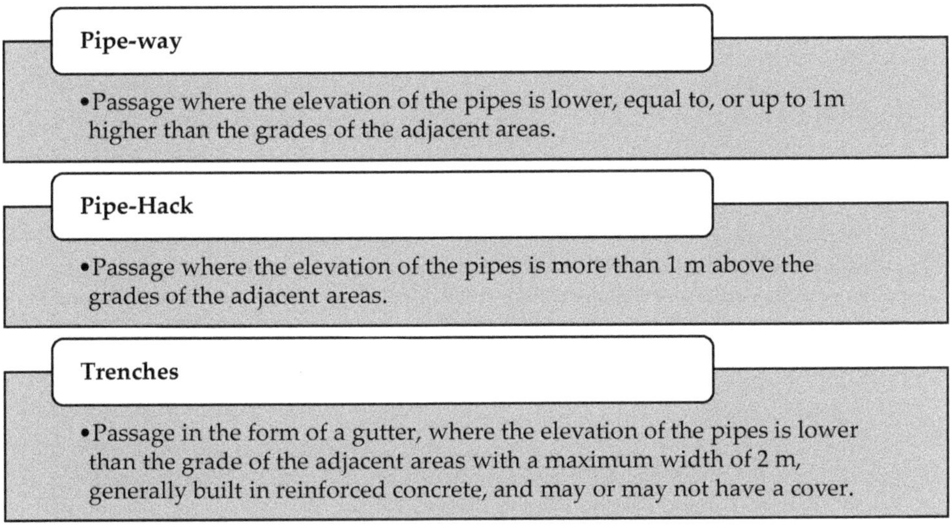

Figure 3. Classification of piping passage structures according Petrobras [26].

ASME B31 classifies pipelines according to the degree of danger related to the fluid being carried, with category "D" having a lower risk and thus less rigorous design, examination, and testing requirements, while category "M" has a high degree of danger and requires more stringent requirements. Calculations are made considering the design conditions such as pressures, temperatures, and loads to which the pipelines may be subjected [34,35]. Thus, ASME B31 is currently subdivided into seven standards that are presented in Table 4.

Table 4. Standards for piping design based on flow conditions [36–42].

Standard	Description
ASME B31.1	Power Piping
ASME B31.3	Process Piping
ASME B31.4	Pipeline Transportation Systems for Liquid Hydrocarbon and Other Liquids
ASME B31.5	Refrigeration Piping and Heat Transfer Components
ASME B31.8	Gas Transmission and Distribution Piping Systems
ASME B31.9	Building Services Piping
ASME B31.12	Hydrogen Piping and Pipelines

Additionally, ASME B31G is utilized to determine the remaining strength of corroded pipes, which differs from the standards presented in Table 4 [43].

2.1.3. Drains, Purgers, and Vents

Drains serve the function of completely emptying the pipes. They can be described as valved branches for draining the pipes and are placed at all low points of the piping system, whereas vents allow air to enter to prevent a vacuum and are placed at high points. Low points refer to sections of the pipe that have a lower elevation than the adjacent sections, while high points are higher than the surrounding sections. Drains are also installed in check valves when they are placed in vertical pipes to allow for emptying. Depending on the piping system, gate valves for liquids, plug valves for gases, or ball valves for both gases and corrosive services can be used. Drains and vents must be flanged when used for high-risk services such as pipelines transporting toxic fluids or hydrogen [44].

Traps are intended to separate or eliminate condensate that forms in steam pipelines and heating equipment, preventing steam from escaping. Like drains, these devices must be located at low points in the pipeline. Their use is recommended before block valves, control valves, check valves, closed ends of blind flanges, steam engine inlets, and in long sections of the pipeline [4,45].

2.1.4. Connections

According to Stewart [34], the selection of tubular connections depends on factors such as pipe diameter, pipe material, pressure and temperature ratings, and service requirements. In the oil industry, the most common connections are threaded connections, welded connections, bell spigot, groove, and clamp. Accessories include elbows and 45° and 90° return bends, eccentric and concentric reducers, olets, weldolets, elbolets, latrolets, sweeplets, and flanges.

Flanges are essential components in a piping system, used to connect pipes, valves, and vessels. They are generally classified according to their pressure classes, which dictates their strength and suitability for different applications. Flanges facilitate the assembly and disassembly of piping systems, making them crucial for maintenance and operational flexibility. They provide secure connections and are designed to handle various pressures and temperatures, depending on the system requirements [46].

Table 5 presents the main standards utilized for flanges, outlining the specifications and requirements for their application in different piping systems.

Table 5. Main flanges standards [47–58].

Standard	Description
ASME B16.1	Gray Iron Pipe Flanges and Flanged Fittings
ASME B16.5	Pipe Flanges and Flanged Fittings
ASME B16.24	Cast Copper Alloy Pipe Flanges, Flanged Fittings, and Valves
ASME B16.36	Orifice Flanges
ASME B16.42	Ductile Iron Pipe Flanges and Flanged Fittings
ASME B16.47	Large Diameter Steel Flanges
ASME B16.48	Line Blanks
AWWA C115	Flanged Ductile-Iron Pipe with Ductile-Iron or Gray-Iron Threaded Flanges
AWWA C207	Steel Pipe Flanges for Waterworks Service
MSS SP-44	Steel Pipeline Flanges
MSS SP-51	Class 150LW Corrosion Resistant Flanges and Cast Flanged Fittings
MSS SP-65	High-Pressure Chemical Industry Flanges and Threaded Stubs for Use with Lens Gaskets

The arrangement of eccentric reductions must be carefully considered regarding their positioning. For instance, when used in horizontal lines, they should maintain the same elevation at the bottom of the pipes. However, when used at the inlets of suction pumps, the positioning should be leveled from above if the suction line is vertical to prevent the formation of air pockets. If the suction line is vertical and the flow is downward, the reductions should be leveled downward to avoid fluid stagnation, which can lead to drainage issues [4].

2.1.5. Valves

Valves are essential components in piping systems and are classified based on their function into blocking, regulating, and control valves. The design should minimize the use of valves due to their high-pressure losses and cost. They should be easily accessible for operation and maintenance. Blocking valves are recommended at the boundaries of each process unit, while check valves should be placed close to the equipment's interconnection with the lines. For control valves, blocking valves and a bypass line with a regulating valve should be installed to allow repairs without interrupting the flow. Valves should not have stems facing downward to avoid debris accumulation in the valve bonnet. In pump suction lines, it is preferable for the stems to be horizontal or inclined to avoid air pockets [4,26].

When selecting valves, factors such as their function, material suitability, pressure and temperature ratings, plant service life, connection ends, operation, weight, availability, maintenance, and cost must be considered [59]. Several standards provide requirements for these valves, and the main standards currently in use are presented in Table 6.

Table 6. Main valve standards [60–80].

Standard	Description
ASME B16.10	Face-to-Face and End-to-End Dimensions of Valves
ASME B16.34	Valves-Flanged, Threaded, and Welding End
API 526	Flanged Steel Pressure-relief Valves
API 594	Check Valves: Flanged, Lug, Wafer, and Butt-welding
API 599	Metal Plug Valves—Flanged, Threaded, and Welding Ends
API 600	Steel Gate Valves—Flanged and Butt-welding Ends, Bolted Bonnets
API 602	Gate, Globe, and Check Valves
API 603	Corrosion-resistant, Bolted Bonnet Gate Valves—Flanged and Butt-welding Ends
API 608	Metal Ball Valves—Flanged, Threaded, and Welding Ends
API 609	Butterfly Valves: Double-flanged, Lug- and Wafer-type, and Butt-welding Ends
AWWA C500	METAL-SEATED GATE VALVES FOR WATER SUPPLY SERVICE
AWWA C504	RUBBER-SEATED BUTTERFLY VALVES
MSS SP-42	Corrosion-Resistant Gate, Globe, Angle, and Check Valves with Flanged and Butt Weld Ends
MSS SP-70	Gray Iron Gate Valves, Flanged and Threaded Ends
MSS SP-71	Gray Iron Swing Check Valves, Flanged and Threaded Ends
MSS SP-72	Ball Valves with Flanged or Butt-Welding Ends for General Service
MSS SP-80	Bronze Gate, Globe, Angle, and Check Valves
MSS SP-81	Stainless-Steel or Stainless-Steel-Lined, Bonnetless, Knife Gate Valves with Flanged Ends
MSS SP-85	Gray Iron Globe and Angle Valves, Flanged and Threaded Ends
MSS SP-88	Diaphragm Valves
MSS SP-105	Instrument Valves for Code Applications

2.1.6. Supports

Pipes are usually supported by supports or hangers designed by engineers. There are various standards for the design of pipe supports. This standard serves as a guide for the selection, design, and production of pipe supports and defines their different types [81]. Pipe supports must be sized considering their location, type of pipe and fluid flow, line temperature, height, width, sizes and numbers, anchors, and loads [27,34].

The type of support and its interaction with the pipe are crucial, as they can wear away coatings due to friction and/or accumulate water, making them more vulnerable to corrosion. When the support is fixed and attached to the pipe, it becomes an anchoring point, reducing the pipe's flexibility and increasing the stresses caused by thermal expansion. Therefore, several studies are currently being conducted to investigate this accessory and the mechanisms involved [82–84].

2.2. Piping Project

The design of a pipeline can be divided into several stages, including material selection, nominal diameter choice, wall thickness calculation, loads acting on the pipeline, structural resistance, span calculations between supports, pressure loss calculations of the conveyed fluid, and flexibility.

2.2.1. Material Selection

The pipeline material must meet requirements such as mechanical resistance, chemical resistance, and thermal resistance. Inadequate material selection can lead to corrosion, line rupture, cracks, safety risks, or excessive costs. Therefore, material selection requires the consideration of data such as operating pressure and temperature, type of conveyed fluid (analyzing corrosion and contamination aspects), material cost, required safety degree, external overloads, and resistance to fluid flow in the pipeline [4].

Standardization organizations, such as ASTM, provide documents specifying material requirements, including physical, mechanical, or chemical properties, safety, quality, and performance criteria, as well as testing methods to verify compliance with each requirement. According to ASTM, a range of materials is used for pipe manufacturing; however, carbon steel pipes are the most frequently used due to their cost and mechanical properties [85,86].

Standards, such as those indicated in Table 4, specify the permissible stresses for the materials used, considering working conditions and design temperature. This information is essential for the pipe selection stage. However, the ongoing development and optimization of materials are crucial [87,88].

Some piping will need to have some thermal insulation that is categorized into groups based on their temperature ranges. Table 7 presents the respective groups and materials used, along with their applicability and possible materials.

Table 7. Main materials used as thermal insulation according to temperature and applications [89].

Insulation Classification	Application Examples	Material Examples
Low Temperature (up to 90 °C)	Refrigerators, Cold and Hot Water Systems, Storage Tanks	Cork, Expanded Polystyrene, 85% Magnesia, Mineral Fibers, Polyurethane, and Wood
Medium Temperature (90 °C to 325 °C)	Heating and Steam Raising Equipment, Steam Lines, Flue Ducts	Asbestos, Calcium Silicate, 85% Magnesia, and Mineral Fibers
High Temperature (Above 325 °C)	Super-heated Steam System, Oven Dryer, and Furnaces	Asbestos, Calcium Silicate, Mineral Fiber, Mica and Vermiculite-based Insulation, Fireclay- or Silica-based Insulation, and Ceramic Fiber

Thermal insulation materials can also be classified as organic and inorganic. Organic materials are based on hydrocarbon polymers, such as Thermocol (expanded polystyrene) and Polyurethane Foam (PUF). Inorganic insulation, on the other hand, is based on siliceous, aluminous, and calcium materials in fibrous or particulate forms, such as mineral wool and calcium silicate, among others [90].

Pipe manufacturing is generally standardized by norms such as ASME B.36.10M and ASME B.36.19M, with custom-made pipes being expensive to produce [91,92]. Thus, selecting the appropriate pipe involves initially analyzing the nominal diameter and schedule to be chosen from commercial options. The calculation of the nominal diameter is directly related to the pressure drop in the pipe. Larger diameters result in higher pipe costs but lower pressure drops, which implies reduced machine power and energy consumption. Therefore, the internal diameter of pipes is typically determined based on tabulated economic velocities, which vary by country and material used [93].

Wall thickness calculation is generally performed according to Equation (1), if it meets the usage conditions expressed in Equations (2) and (3) [37]. Additionally, it should be noted that the standards governing pipe manufacturing include manufacturing tolerances that must be considered when ordering the pipes [91,92].

$$t = \frac{P \cdot D}{2(S_h \cdot E + P \cdot Y)} + C_r \tag{1}$$

$$P/(S_h \cdot E) < 0.385 \tag{2}$$

$$t < D/6 \tag{3}$$

where

P: Internal design pressure (kgf/cm^2);
D: External diameter of the pipe (mm);
S_h: Allowable stress of the material at the design temperature (kgf/cm^2);
E: Welding efficiency [values provided in ASME B31] (%);
t: Wall thickness (mm);
Y: Reduction coefficient (%);
C_r: Margins for corrosion, erosion, as well as thread and chamfer openings (mm).

2.2.2. Thermal Insulation

The purpose of thermal insulation is to reduce heat loss along a pipeline, make the surface safer to touch, and prevent condensation on cold surfaces to avoid resulting corrosion. When designing thermal insulation materials, factors such as location, surface dimensions, temperature conditions, type of internal or external insulation, required finish, and service requirements (e.g., compression resistance, fire resistance, vibration resistance) must be considered [94].

From a personal protection perspective, it is desirable for the external surface of the pipeline to be at a temperature close to that of the human body, as required by various standards. If this requirement is not met, modifications around the pipeline should be made to increase protection for those nearby [95,96]. However, new insulation materials are being developed that, while allowing the surface to reach high temperatures, have significantly lower thermal exchange capacities. These materials offer an alternative to conventional solutions and have the potential to reduce water accumulation, which can lead to corrosion and compromise insulation efficiency [97–99].

For the use of thermal insulation in industrial pipelines, studies comparing insulation costs with the costs related to energy dissipation along the pipeline are necessary to reduce costs and achieve project efficiency. Factors such as the pipeline's location, ambient temperature, air humidity, and wind speed are also analyzed. Practical tables provide established values for thermal insulation thicknesses for different materials based on pipe diameters and operating temperatures [95,96]. Additionally, other methods can be applied according to project specificities [100,101].

2.2.3. Flexibility

Pipeline flexibility is crucial due to temperature variations, which cause thermal expansion in the material. These expansions result in increased longitudinal sections at high temperatures and contraction at low temperatures, which can induce stresses according to system restrictions. Failure to adequately account for flexibility can lead to catastrophic failures in industrial installations, such as pipeline section collapse, fatigue, leaks, distortions in valves, equipment, and pipelines. Therefore, it is recommended that pipelines be designed to enhance flexibility, reduce disproportionalities, and maintain symmetry in their dimensioning [93,102]. ASME B31 specifies three main requirements for pipeline flexibility [35]:

- The calculated stress range at any point due to system displacements must not exceed permissible limits;
- The calculated reaction forces must not be detrimental to supports or connected equipment;
- The calculated movement of the pipeline must be within any prescribed limits.

When a layout does not meet design requirements, adjustments are generally made because, while expansion joints can be used, their application should be avoided due to their limited reliability and potential as weak points in the design [103]. Although flexibility calculations can be performed using equations, they are typically conducted with the aid of spreadsheet and modeling software, which allows for meeting regulatory requirements more efficiently and quickly adjusting calculations as needed. Additionally, specialized software provides more precise calculations [93,104].

3. Corrosion Protection Mechanisms

Steel is the main material used in industrial pipelines, with 90% of them being made of low carbon steel, given its mechanical properties, weldability, ease of forming, and low cost [4]. However, unlike materials such as aluminum and stainless steel, which form an oxide layer that tends to remain adhered to the surface, minimizing corrosion, carbon steel generates a layer that comes off more easily, making it susceptible to corrosion in various environments [105–107]. ISO 12944-2 [108] classifies the degree of corrosivity according to the thickness and mass lost after the first year of exposure to a given environment, with the classifications for low carbon steel being presented in Table 8.

Table 8. Corrosivity classification for low carbon steel in atmospheric environments [108].

Degree of Corrosivity		Thickness Loss * (μm)	Mass Loss * (g/m^2)
C1	Very low	≤1.3	≤10
C2	Low	1.3 to 25	10 to 200
C3	Medium	25 to 50	200 to 400
C4	High	50 to 80	400 to 650
C5	Very high	80 to 200	650 to 1500
CX	Extreme	200 to 700	1500 to 5500

* Values relative to the first year.

Corrosion can occur through chemical or electrochemical action. The main factors that contribute to the corrosive process include the type of soil or metals in contact, the fluids transported, the environmental conditions, the external loads, and the existence of stray currents [109–111]. Figure 4 illustrates an example of the aggressive nature of corrosion in metallic systems. In this case, a process pipeline of a Floating Production Storage and

Offloading (FPSO), exposed to the marine environment, experienced galvanic corrosion, and differential aeration corrosion, resulting in significant material degradation. After only five years of operation, the pipeline had deteriorated to the condition shown in the image, emphasizing the rapid progression of these corrosion mechanisms and their detrimental impact on the structural integrity of the component.

Figure 4. Degradation of an FPSO pipeline due to galvanic and differential aeration corrosion after five years of operation.

In this regard, the adoption of combined approaches for corrosion control is essential, especially in units like FPSOs, as repairs to this equipment take time, particularly when the acquisition of specific materials, such as certified ones, is required. These strategies help minimize degradation, reduce downtime, and extend the service life of the structure, ensuring greater operational efficiency and cost savings [112,113].

The practical methods used to reduce the corrosion rate in metallic materials are shown in Figure 5, where the environmental conditions in which these methods are normally used are represented by the letters: A (atmosphere), W (submerged in water), and G (underground).

Process Modification

- Structure design (A, W, G)
- Surface conditions (A, W, G)
- Cathodic protection (W, G)

Modification of Corrosive Environment

- Deaeration of water or neutral solution (W)
- Purification or reduction in air humidity (A)
- Addition of corrosion inhibitors (W/A and G in special cases)

Metal Modification

- Increased purity (A, W, G)
- Addition of elements: alloying (A, W, G)
- Heat treatment (A, W, G)

Protective Coatings

- Coatings with reaction products: chemical or electrochemical treatment of the metal surface (A and W)
- Organic coatings: paints, resins or polymers, etc. (A, W, and G)
- Inorganic coatings: enamels, cements (A, W, and G)
- Metallic coatings (A, W, and G)
- Temporary protectors (A)

Figure 5. Methods of reducing corrosion rates [114].

Furthermore, minimizing mechanical effects can mitigate the effects of corrosion. Surface wear (abrasion, erosion, cavitation, fretting, etc.) can not only accelerate corrosion in low carbon steels, but also in materials that form an adherent oxide layer. These effects act synergistically, meaning that the resulting degradation is greater than the sum of the individual effects due to one agent enhancing the action of the other [115–118].

In addition to the effects resulting from surface wear, the application of stress, whether cyclical or continuous, can also act synergistically with corrosion (stress corrosion/fatigue). The application of stress can generate and propagate cracks that favor corrosion (which act as an anodic region), while corrosion can reduce the amount of material, increasing stress and/or acting as a stress concentrator [119–122].

Similarly, microbiologically influenced corrosion (MIC) is another critical factor, as it results from the metabolic activity of microorganisms that interact synergistically with electrochemical and chemical processes, accelerating material degradation. This phenomenon can lead to severe forms of localized corrosion, such as pitting and under-deposit corrosion, especially in environments with stagnant flow and the presence of water [123,124].

Aiming to take advantage of the characteristics of low carbon steel, combat corrosion, and minimizing costs, the main form of pipeline protection is through the application of coatings [25].

3.1. Piping Coatings

Piping protection through coating makes it possible to minimize corrosion in several ways, for example, by acting as a barrier between the flowing fluid and the pipe, stabilizing the oxide layer, promoting cathodic protection, etc. [125–129].

Figure 6 shows the publications evolution of papers and patent families on anticorrosive coatings for industrial pipelines. Both types of publications are growing, with a greater number of patents and greater growth in the publication of articles, especially in recent years. In the period from 2011 to 2023, the number of patents grew by 80%, while the number of publications increased by 490%, indicating a significant increase in academic research in this segment.

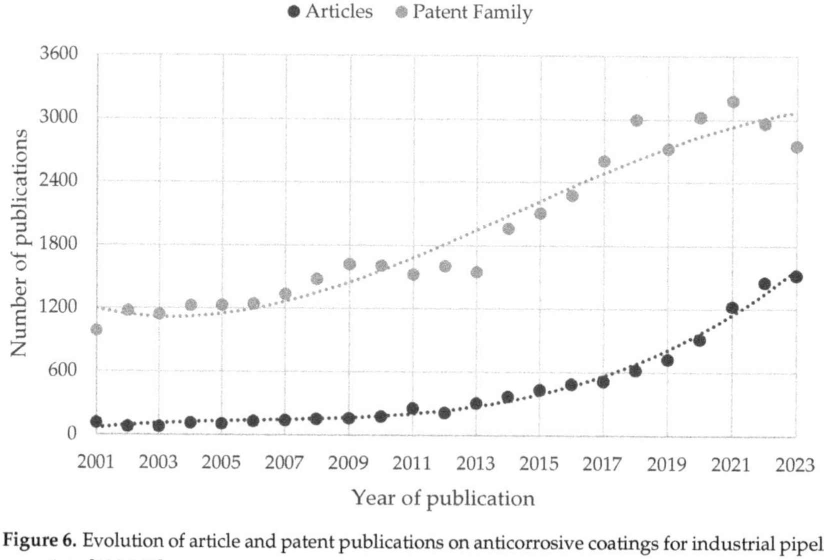

Figure 6. Evolution of article and patent publications on anticorrosive coatings for industrial pipelines over time [130,131].

Coatings can be classified as metallic and non-metallic, are subclassified as organic and inorganic, and are generally applied after the pipe has been manufactured. Among the most used metallic coatings, galvanization and bimetallic (cladded) sheets stand out. Galvanization is the most common metallic coating and is often used in conjunction with other protection methods. This involves the application of metallic coating, and its main advantages include low cost and corrosion rate, as well as the possibility of protecting the pipe even when the coating has its continuity compromised [132–134].

Table 9 shows the corrosivity levels of zinc, and when compared with low carbon steel (Table 8), a loss of material that is considered low for carbon steel would be an extreme loss for zinc, given its lower corrosion rate.

Table 9. Corrosivity classification for zinc in atmospheric environments [108].

Degree of Corrosivity		Thickness Loss * (μm)	Mass Loss * (g/m^2)
C1	Very low	≤0.1	≤0.7
C2	Low	0.1 to 0.7	0.7 to 5
C3	Medium	0.7 to 2.1	5 to 15
C4	High	2.1 to 4.2	15 to 30
C5	Very high	4.2 to 8.4	30 to 60
CX	Extreme	8.4 to 25	60 to 180

* Values relative to the first year.

Another type of metal coating consists of manufacturing pipes with clad layer. That is, bimetallic layers that can be obtained by various methods such as co-rolling, explosion bonding, and welding. Its advantage is that it allows the use of two metals, usually carbon steel and stainless steel, to combine their characteristics, being a good solution for services with high corrosion rates and chemical incompatibility of the pipe material with the fluid being flowed [87,135,136].

"Organic–metal" coatings are conductive or semi-conductive organic coatings, such as doped polymers and polymer matrix composites [137–139]. However, although their application was proposed about 40 years ago, there is still a challenge in understanding the behavior of these materials, which is one of the biggest challenges in using this solution. Nevertheless, there are already robust solutions in the literature and on the market, such as zinc-rich coatings that are marketed as an alternative to galvanization and as a way of repairing damage that the latter may have received during transportation [140–142].

The main inorganic coatings applied to pipelines are concrete, glass, and porcelain [4]. These materials have high resistance to corrosion and temperature, with glass and porcelain being expensive, fragile, and difficult to handle, but they are recommended especially when absolute purity of the conveyed fluid is desired. Concrete, on the other hand, is a cheap coating, and can be applied at higher temperatures, reaching up to 1800 °C depending on its composition. However, an anticorrosive coating needs to be continuous; otherwise, corrosion will occur in a concentrated manner at the exposed site. Due to the impossibility of guaranteeing the non-appearance of cracks, concrete is not recommended for services involving high levels of corrosion and responsibility [25,143].

Another inorganic material is geopolymer, which consists of the union of precursor elements, rich in aluminosilicates, and an activating alkaline solution. Its application as an anticorrosive coating is still in the early stages of development, but studies and patents indicate it as a promising material [143–145].

Organic coatings can be divided into two main groups: thermoplastic/rigid and elastomeric. Elastomeric coatings are notable for their ability to inhibit corrosion while offering excellent anti-abrasive protection. Thermoplastic/rigid coatings stand out not only for their high corrosion resistance, but also for their low cost and coefficient of friction. It should also be added that both coatings have low density [4,146].

Criteria for Selection and Development of Organic Coatings

Inorganic coatings are generally zinc-based and have undergone extensive standardization and studies. Thus, the development of new materials has focused on organic coatings, the selection of which is more complex [130,131]. This section focuses on anticorrosive coatings, addressing the main materials used and their desirable properties.

The performance of coatings is linked to their adhesion, barrier properties, ability to inhibit corrosion, as well as their thermal, chemical, and mechanical resistance [146]. However, their effectiveness depends on prior preparation of the substrate on which it will be coated. Such preparation is generally carried out with the aim of providing greater roughness and cleanliness to the surface [86]. Therefore, it is common to specify minimum surface preparation conditions. These specifications generally consist of a minimum roughness value and degree of cleaning, as well as a maximum time between surface cleaning and application of the coating [147,148]. Currently, the main reference used to measure the degree of cleaning of metal substrates is ISO 8501-1 [149].

Another factor that must be observed before qualifying the coating is associated with control over the quality of the coating application. In general, in addition to requiring the qualification of the professional who will apply the coating, the standards provide for control over the conditions of the environment in which the material will be applied, the thickness of the coating, and the integrity of the applied material. These variables have a significant influence on the performance of the materials, having specific criteria that vary depending on the material and thickness used [150,151].

The main polymers used in organic coatings are Acrylics, Epoxy, Polyurethanes, and Vinyls [146]. Of these, acrylic and vinyl coatings stand out for being resistant to abrasion, impacts, and deformations, as well as for their sustainability [152,153]. Epoxies stand out for their corrosion resistance to various media, less contraction during curing, thermal stability, mechanical strength, and durability [154,155]. Finally, polyurethane coatings stand out for their mechanical properties, corrosion resistance in various environments, and resistance to ultraviolet radiation, being one of the main materials used in external coatings [156,157].

In addition to the materials mentioned earlier, there are multilayer coatings that combine the characteristics of different materials to maximize performance and reduce costs [158,159]. A notable example is polyethylene coating, which offers excellent mechanical and thermal properties, as well as weather resistance. This type of coating is often applied both during pipe manufacturing and in the field, typically over an epoxy layer [160]. This combination of polyethylene and epoxy stands out as one of the most effective, being recognized in EGIG reports as one of the highest-performing options [23].

It is still possible to combine polymers and make modifications to their manufacturing processes, which makes it possible to modify their properties for specific applications [157,161]. In addition to matrices, it is common to use particulates to generate visual and performance changes in organic coatings [162]. In general, these are named as pigments classified as organic and inorganic, with the latter being subclassified as shown in Figure 7.

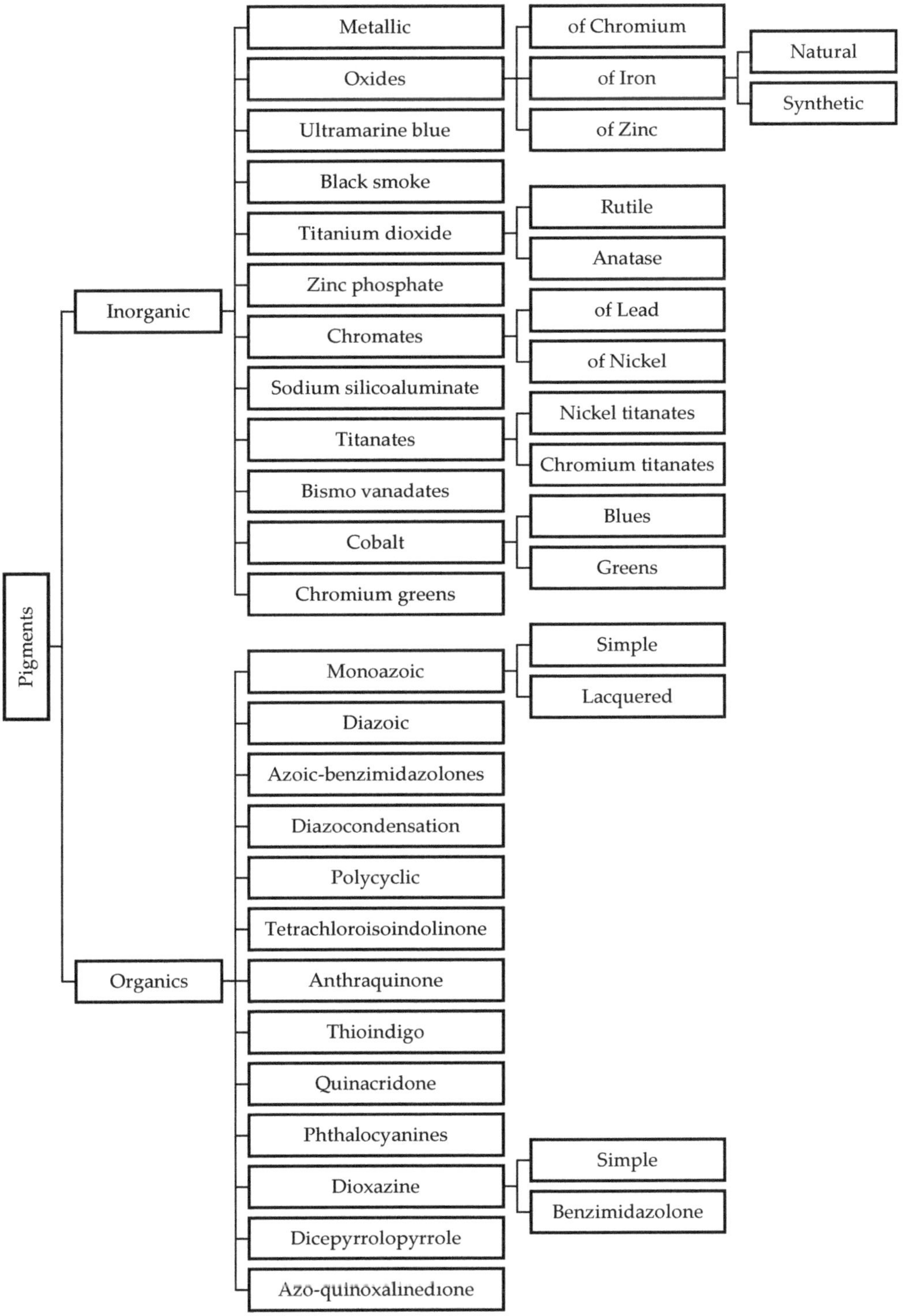

Figure 7. Main types of pigment [163].

Currently, the development of more sustainable (water-based, bio-based, and using waste) and intelligent coatings (with hyperbranching, antimicrobial action, self-repair, corrosion sensitivity, self-cleaning, and anti-fouling action) has intensified [146,164]. Thus, we can observe the use of increasingly varied dispersed phases with strategic functions, such as nanomaterials, microcapsules, waste, etc. [165–167].

A representative example of this evolution is the advancement in controlling microbiologically influenced corrosion (MIC), which has minimized the use of heavy metals and biocides in pursuit of more sustainable strategies [168–170].

To optimize the properties of a polymeric coating, additives can be added to a coating. The main additives are shown in Table 10, including solvents and plasticizers, which are the most used. However, their use must be evaluated according to the purpose of the material's application, as it may result in undesirable changes in other properties [171].

Table 10. Main coating additives and their applications [163,172–174].

Additive	Function
Antifoaming agents	Reduces foam formation during product processing and application
Anti-skin	Minimizes the formation of skins during storage and/or curing of coatings
Anti-sedimenting	Prevents the formation of hard, non-homogenizable sediment at the bottom of the packaging
Bactericide	Prevents the development of bacteria that cause putrefaction of liquid-phase plastic paints
Catalysts	Accelerates the curing process of a material
Coalescent	Reduces the minimum film formation temperature of aqueous products
Deaerators	Reduces the presence of voids in the coating
Deflocculators	Reduces the support capacity of particulates
Dispersant	Facilitates the dispersion of pigments and fillers, preventing their re-agglomeration
Thickener	Increases the viscosity of the coating
Fungicide	Prevents the development of fungi in the dry film
Insecticide	Ensures the elimination of insects in contact with the dry film
Wetting agent	Reduces the interfacial tension between the solid phase and the liquid phase
Plasticizer	Gives flexibility to the dry film
Preservatives	Acts to minimize the socio-environmental impacts of the coating and/or acts to protect the coating from various external factors
Rheological	Provide safe and practical means of regulating the fluidity of a resin
Drying agent	Reduce the drying, hardening, and/or curing time of the coating
Solvents	Solubilize and dissolve other materials
Surfactants	Improve the coating by controlling surface tension, stabilizing interfaces, and optimizing physical properties such as fluidity and spreading
Surface	Correct surface defects of the coating
Wetting	Accelerate the wetting process and reduce interfacial tension

Table 11 presents 20 of the main tests for evaluating the performance of pipeline coatings and their respective reference standards. This table is based on standards such as ISO 21809-2:2014 from the International Organization for Standardization (ISO), SP-2217:2017 from Petroleum Development Oman (PDO), N-2843 C and N-2912 A from Petróleo Brasileiro S.A. (PETROBRAS), SP 0394 from the National Association of Corrosion Engineers (NACE), and M501 from Norsk Sokkels Konkurranseposisjon (NORSOK), among others.

Table 11. Main tests for evaluating the performance of pipe coatings [150,175–194].

Test	Standards
Cross section porosity	NACE SP 0394
Interface porosity	NACE SP 0394
Interface contamination	NACE SP 0394
Penetration resistance	ASTM G17
Flexibility	ASTM D522/D522M I NACE SP 0394
Variation in gas pressure	PDO SP-2217
Pull off	ASTM D4541
Cathodic disbondment	ASTM G8/ASTM G95
Strained coating, Cathodic disbondment	ISO 21809-2
Electrochemical Impedance Spectroscopy	ASTM G3
Impact resistance	ASTM G14 I NACE SP 0394
Abrasion resistance	ASTM D4060
Water vapor permeability	ASTM D1653
Water absorption	ASTM D870
Water resistance	ASTM D2247 I ISO 2812-2
Accelerated weathering	ASTM D4587
Chemical resistance	ASTM D6943
Salt spray	ASTM B117
Autoclave	NACE TM0185 & ASTM D714
Differential Scanning Calorimetry	ASTM E1356

Of these, the adhesion test stands out because it is often used alongside other tests to determine if there has been a loss of adhesion when the coating is exposed to certain conditions. However, it is essential to standardize not only the test parameters but also the equipment and plates used. Fragata et al. [195] observed significant variations in results when the same coating was evaluated in different laboratories and noted that variations in plate thickness could lead to significant differences in the results obtained.

Finally, it is important to emphasize that the tests and conditions used for coating qualification can vary. Therefore, while Table 4 is a good reference for development purposes, a complementary analysis is recommended for commercial purposes to meet the demands of potential buyers. For instance, both SP-2217:2017 from PDO and N-2912 A from Petrobras include chemical resistance tests, but they analyze different fluids or concentrations. The former assesses resistance to HCl (pH 2.5–3); HF (pH 2.5–3); H_2SO_4 (pH 2); NaCl and H_2SO_4 (pH 3); NaCl; and Toluene, while the latter evaluates resistance to NaOH (30%); H_2SO_4 (40%); xylene; coke naphtha; and fuel ethanol [148,150].

Additionally, besides the tests used for qualification requirements, other tests are desirable for characterizing the coating. Table 12 presents 10 of the main properties and the corresponding standards.

Table 12. Main complementary tests for pipe coatings [151,196–204].

Test	Standards
Specific mass	ASTM D1475
Dry film roughness	PETROBRAS N-2843 C
Decay	ASTM D4400
Consistency	ASTM D562
Pot-Life	ABNT NBR 15742
Hardness	ASTM D2240
Solids by mass	ASTM D2369
Solids by volume	ISO 3233-1
Linear thermal expansion	ASTM E228
Drying time/painting interval	ASTM D1640

4. Challenges and Future Perspectives

Figure 8 shows the variation in the causes of pipeline failures in Europe, the segment with the highest incidence of pipeline accidents. It is observed that failures have become increasingly less frequent, with a notable reduction in those caused by external interference. Regarding corrosion, despite the significant growth in the length of pipeline systems (400% between 1970 and 2010), pipelines built since the end of the last century have adopted more efficient protection mechanisms, resulting in the probability of failure not increasing with the age of the pipeline [23].

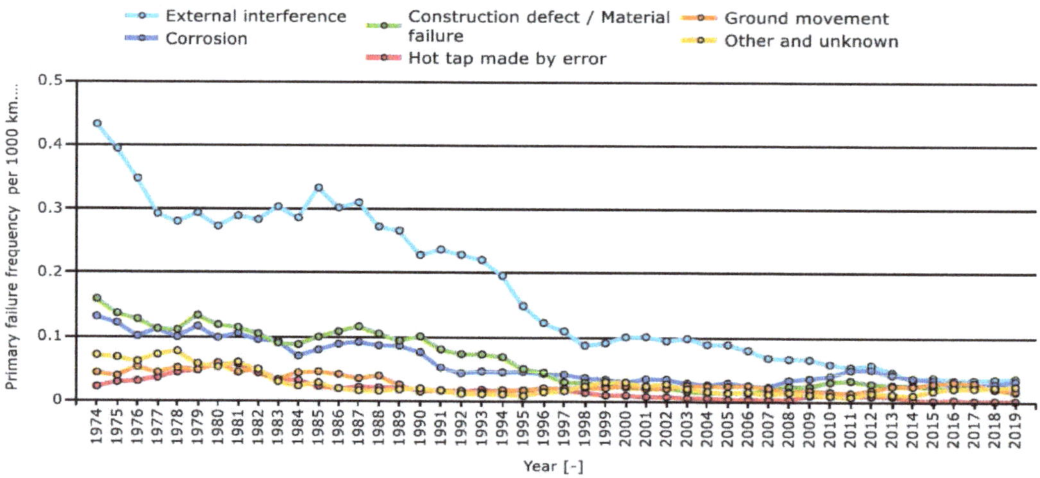

Figure 8. Primary failure frequency per 1000 km by cause (five-year moving average) [23].

However, the growing complexity of technological systems in industries has increased efficiency, but also the risk of accidents. In surveys detailing human errors, a significant incidence of accidents due to human error is observed during stages such as design, maintenance, and operation, especially in less developed countries, due to the lower enforcement of safety standards. Thus, the trend points to the continuous need to improve business procedures and their enforcement. Although new technologies have been developed to assist such processes, the human factor can also present resistance to their adoption [205–208].

Additionally, incorporating a life cycle cost analysis for pipelines is a methodology that can promote the adoption of more efficient and safer measures during the design phase. This approach helps counterbalance the tendency to prioritize short-term costs at the expense of long-term benefits and safety and aligns with what is observed in other types of projects [209,210].

An example of a challenge arising from current technological advances is the transportation of green hydrogen. Although green hydrogen is a promising alternative for the energy transition, its handling and transportation present significant challenges. One of the main issues is hydrogen embrittlement, a phenomenon in which hydrogen penetrates metallic materials, such as steel, causing degradation of their mechanical properties and making them more susceptible to cracks and fractures [211–214].

Moreover, beyond the challenges of designing a pipeline system to transport green hydrogen, plans are being made to repurpose structures originally intended for transporting natural gas to carry hydrogen, as well as mixtures of natural gas and hydrogen [215,216]. While this process remains costly, it is anticipated that with technological advancements and economies of scale, economic viability will significantly improve [217,218].

An example of technological development is the application of advanced techniques to characterize and inspect materials. These methodologies enable not only the production and selection of materials with optimized properties but also contribute to maximizing operational efficiency and minimizing failures, resulting in greater safety and reliability of industrial systems [219–221].

5. Conclusions

The field of pipeline engineering is both extensive and technically demanding, with its complexity increasing in parallel with technological advancements. This review has aimed to underscore the most critical factors in ensuring the security, reliability, and profitability of pipeline systems, with a particular focus on design considerations and anticorrosive protection. Additionally, it has addressed the challenges and future directions that are pertinent to this evolving field.

In terms of project execution, a structured sequence of essential steps was outlined, covering the classification of risk areas, the strategic distribution of pipelines and their accessories, and the crucial calculations required for selecting appropriate materials. With respect to anticorrosive protection, the review explored the fundamental corrosion mechanisms that affect pipelines, along with a range of mitigation strategies. Special emphasis was placed on the selection of coatings, a primary method for safeguarding industrial pipelines, and the criteria and methodologies for evaluating the performance of polymeric materials were discussed in depth.

The analysis of current trends reveals a positive trajectory, with a notable decrease in the frequency of pipeline failures, suggesting that projects are safer and more resilient. Effective management of corrosion, even in aging pipelines, has further reduced the risk of failures over time. However, it is important to note that this progress is not uniformly distributed across the globe. In developing regions, the limited adoption of advanced technologies and best practices continues to result in a higher incidence of pipeline-related incidents, highlighting the need for increased attention to human factors within industrial operations.

At last, while significant strides have been made in the field of pipeline engineering, there remains a pressing need to develop more efficient and innovative solutions to address emerging challenges. The transportation of green hydrogen, for instance, presents complex issues, such as hydrogen embrittlement, that require comprehensive research and development to ensure safe and sustainable implementation. Future efforts must continue to focus on enhancing the safety, reliability, and efficiency of pipeline systems, particularly in the context of the global energy transition and reuse of existing structures for other applications.

Author Contributions: Conceptualization, D.C.R.V. and F.P.D.L.; validation, M.P.O. and N.T.S.; resources, C.M.F.V.; writing—original draft preparation, D.C.R.V.; writing—review and editing, V.P.D.G. and F.P.D.L.; project administration, C.M.F.V. All authors have read and agreed to the published version of the manuscript.

Funding: CAPES, CNPQ (process n° 302976/2022-1), FAPERJ (process n° E-26/200.658/2022, E-26/203.874/2022, E-26/200.847/2021, and E-26/201.628/2021), and FAPES (process n° 932/2023).

Acknowledgments: The authors would like to thank CAPES, CNPQ, FAPERJ, FAPES, UFES, and UENF for their financial support and for fostering research.

Conflicts of Interest: The authors declare no conflicts of interest.

Abbreviations

The following abbreviations are used in this manuscript:

ANSI	American National Standards Institute
API	American Petroleum Institute
ASME	American Society of Mechanical Engineers
ASTM	American Society for Testing and Materials
AWWA	American Water Works Association
BFD	Block Flow Diagram
CPS	Cyber-Physical Systems
EGIG	European Gas Pipeline Incident Data Group
FPSO	Floating Production Storage and Offloading
HSE	Health, Safety, and Environment
IFC	Issued for Construction
IFD	Issued for Design
IoT	Internet of Things
ISO	International Organization for Standardization
MIC	Microbiologically Influenced Corrosion
MSS	Manufacturers Standardization Society
NACE	National Association of Corrosion Engineers
NORSOK	Norsk Sokkels Konkurranseposisjon
NFPA	National Fire Protection Association
P&ID	Piping and Instrumentation Diagrams
Petrobras	Petróleo Brasileiro SA
P&ID	Piping and Instrumentation Diagrams
PDO	Petroleum Development Oman
PFD	Process Flow Diagram
PID	Piping and Instrumentation Diagrams
PIP	Process Industry Practice
PUF	Polyurethane Foam

References

1. Angelakis, A.N.; Passchier, C.W.; Valipour, M.; Krasilnikoff, J.A.; Tzanakakis, V.A.; Ahmed, A.T.; Baba, A.; Kumar, R.; Bilgic, E.; Capodaglio, A.G.; et al. Evolution of Tunneling Hydro-Technology: From Ancient Times to Present and Future. *Hydrology* **2023**, *10*, 190. [CrossRef]
2. Campbell, J.W.P. The Development of Water Pipes: A Brief Introduction from Ancient Times until the Industrial Revolution. In *Proceedings of the Eighth Annual Conference of the Construction History Society*; Queens' College: Cambridge, UK, 2021.
3. Mazumder, R.K.; Salman, A.M.; Li, Y.; Yu, X. Reliability Analysis of Water Distribution Systems Using Physical Probabilistic Pipe Failure Method. *J. Water Resour. Plan. Manag.* **2019**, *145*, 04018097. [CrossRef]
4. Telles, P.C.d.S. *Tubulacoes Industriais: Materiais, Projeto e Montagem*, 10th ed.; LTC: Rio de Janeiro, Brazil, 2018; ISBN 978-85-216-1289-6.
5. Snider, B.; McBean, E.A. Improving Urban Water Security through Pipe-Break Prediction Models: Machine Learning or Survival Analysis. *J. Environ. Eng.* **2020**, *146*, 04019129. [CrossRef]
6. Balekelayi, N.; Tesfamariam, S. Statistical Inference of Sewer Pipe Deterioration Using Bayesian Geoadditive Regression Model. *J. Infrastruct. Syst.* **2019**, *25*, 04019021. [CrossRef]
7. Lee, J.; Park, C.Y.; Baek, S.; Han, S.H.; Yun, S. Risk-Based Prioritization of Sewer Pipe Inspection from Infrastructure Asset Management Perspective. *Sustainability* **2021**, *13*, 7213. [CrossRef]
8. Kidam, K.; Hurme, M. Analysis of Equipment Failures as Contributors to Chemical Process Accidents. *Process Saf. Environ. Prot.* **2013**, *91*, 61–78. [CrossRef]
9. Duguid, I. Take This Safety Database to Heart. *Chem. Eng.* **2001**, *108*, 80–84.
10. Vílchez, J.A.; Sevilla, S.; Montiel, H.; Casal, J. Historical Analysis of Accidents in Chemical Plants and in the Transportation of Hazardous Materials. *J. Loss Prev. Process Ind.* **1995**, *8*, 87–96. [CrossRef]
11. Instone, B.T. Losses in the Hydrocarbon Process Industries. In Proceedings of the 6th Int. Symposium Loss Prevention and Safety Promotion in the Chemical Industries, Oslo, Norway, 19–22 June 1989; pp. 118–119.

12. Marsh Inc. *Thirty-Year Review of One Hundred of the Largest Property Damage Losses in the Hydrocarbon-Chemical Industries*; Marsh Inc.: New York, USA, 1987.
13. Biezma, M.V.; Andrés, M.A.; Agudo, D.; Briz, E. Most Fatal Oil & Gas Pipeline Accidents through History: A Lessons Learned Approach. *Eng. Fail. Anal.* **2020**, *110*, 104446. [CrossRef]
14. Fan, L.; Su, H.; Wang, W.; Zio, E.; Zhang, L.; Yang, Z.; Peng, S.; Yu, W.; Zuo, L.; Zhang, J. A Systematic Method for the Optimization of Gas Supply Reliability in Natural Gas Pipeline Network Based on Bayesian Networks and Deep Reinforcement Learning. *Reliab. Eng. Syst. Saf.* **2022**, *225*, 108613. [CrossRef]
15. Wang, W.; Zhang, Y.; Li, Y.; Hu, Q.; Liu, C.; Liu, C. Vulnerability Analysis Method Based on Risk Assessment for Gas Transmission Capabilities of Natural Gas Pipeline Networks. *Reliab. Eng. Syst. Saf.* **2022**, *218*, 108150. [CrossRef]
16. Moskovitch, Y.; Jagadish, H.V. Reliability at Multiple Stages in a Data Analysis Pipeline. *Commun. ACM* **2022**, *65*, 118–128. [CrossRef]
17. Nahal, M.; Chateauneuf, A.; Sahraoui, Y. Reliability Analysis of Irregular Zones in Pipelines under Both Effects of Corrosion and Residual Stress. *Eng. Fail. Anal.* **2019**, *98*, 177–188. [CrossRef]
18. Yu, W.; Huang, W.; Wen, K.; Zhang, J.; Liu, H.; Wang, K.; Gong, J.; Qu, C. Subset Simulation-Based Reliability Analysis of the Corroding Natural Gas Pipeline. *Reliab. Eng. Syst. Saf.* **2021**, *213*, 107661. [CrossRef]
19. Al-Sabaeei, A.M.; Alhussian, H.; Abdulkadir, S.J.; Jagadeesh, A. Prediction of Oil and Gas Pipeline Failures through Machine Learning Approaches: A Systematic Review. *Energy Rep.* **2023**, *10*, 1313–1338. [CrossRef]
20. Li, R.; Chen, B.-Q.; Guedes Soares, C. Design Equation of Buckle Propagation Pressure for Pipe-in-Pipe Systems. *J. Mar. Sci. Eng.* **2023**, *11*, 622. [CrossRef]
21. Li, W.; Ye, Z.; Wang, Y.; Yang, H.; Yang, S.; Gong, Z.; Wang, L. Development of a Distributed MR-IoT Method for Operations and Maintenance of Underground Pipeline Network. *Tunn. Undergr. Space Technol.* **2023**, *133*, 104935. [CrossRef]
22. Wong, B.; McCann, J.A. Failure Detection Methods for Pipeline Networks: From Acoustic Sensing to Cyber-Physical Systems. *Sensors* **2021**, *21*, 4959. [CrossRef] [PubMed]
23. *EGIG 11th Report of the European Gas Pipeline Incident Data Group*; EGIG: Groningen, The Netherlands, 2020.
24. Soomro, A.A.; Mokhtar, A.A.; Kurnia, J.C.; Lashari, N.; Sarwar, U.; Jameel, S.M.; Inayat, M.; Oladosu, T.L. A Review on Bayesian Modeling Approach to Quantify Failure Risk Assessment of Oil and Gas Pipelines due to Corrosion. *Int. J. Press. Vessel. Pip.* **2022**, *200*, 104841. [CrossRef]
25. Farh, H.M.H.; Ben Seghier, M.E.A.; Zayed, T. A Comprehensive Review of Corrosion Protection and Control Techniques for Metallic Pipelines. *Eng. Fail. Anal.* **2022**, *143*, 106885. [CrossRef]
26. *PETROBRAS N-0057 H: Projeto Mecânico de Tubulações Industriais*; LTC: Rio de Janeiro, Brazil, 2017.
27. Barker, G.B. *The Engineer's Guide to Plant Layout and Piping Design for the Oil and Gas Industries*; Gulf Professional Publishing: Woburn, MA, USA, 2017.
28. Couper, J.R.; Penney, W.R.; Fair, J.R.; Walas, S.M. *Chemical Process Equipment: Selection and Design*, 3rd ed.; Butterworth-Heinemann: Woburn, MA, USA, 2010.
29. Kumar Singh, A. Overview of Hazardous Locations. In *Explosion-Proof Equipment in Hazardous Area*; Springer Nature Singapore: Singapore, 2023; pp. 1–22.
30. Oliver, E.; Rudrapada, G.; Sarada, S. Recommendations for Design Basis Approach to Fire Protection and Safety Systems for US Offshore Wind Industry. In *Offshore Technology Conference*; OTC: Richardson, TX, USA, 2023.
31. Hasan, R.; Brownlee, R.; Mendler, R.; D'Souza, M. Expected Sources of Energies in Hazardous Spectrum and Their Control: Ensuring the Safe Use of Mobile Devices in the Work Environment. *IEEE Ind. Appl. Mag.* **2024**, *30*, 42–51. [CrossRef]
32. *NFPA NFPA 70*; National Electrical Code (NEC). NFPA: Quincy, MA, USA, 2023.
33. *IEC 60079-10-1*; Explosive Atmospheres—Part 10-1: Classification of Areas—Explosive Gas Atmospheres. IEC: Geneva, Switzerland, 2020.
34. Stewart, M. *Surface Production Operations*; Professional Publishing: Woburn, MA, USA, 2021.
35. *ASME B31*; Pressure Piping. ASME: New York, NY, USA, 2024.
36. *ASME B31.1*; Power Piping. ASME: New York, NY, USA, 2022.
37. *ASME B31.3*; Process Piping. ASME: New York, NY, USA, 2023.
38. *ASME B31.4*; Pipeline Transportation Systems for Liquid Hydrocarbon and Other Liquids. ASME: New York, NY, USA, 2022.
39. *ASME B31.5*; Refrigeration Piping and Heat Transfer Components. ASME: New York, NY, USA, 2023.
40. *ASME B31.8*; Gas Transmission and Distribution Piping Systems. ASME: New York, NY, USA, 2022.
41. *ASME B31.9*; Building Services Piping. ASME: New York, NY, USA, 2020.
42. *ASME B31.12*; Hydrogen Piping and Pipelines. ASME: New York, NY, USA, 2024.
43. *ASME B31G*; Manual for Determining the Remaining Strength of Corroded Pipelines. ASME: New York, NY, USA, 2023.
44. *Petrobras N-0108 E*; Vents and Drains for Piping and Equipment. LTC: Rio de Janeiro, Brazil, 2014.
45. *PETROBRAS N-0116 E*; Sistemas de Purga de Vapor em Tubulações e Equipamentos. LTC: Rio de Janeiro, Brazil, 2013.

46. SMITH, P. *The Fundamentals of Piping Design*; Elsevier: Amesterdam, The Netherlands, 2023.
47. *ASME B16.1*; Gray Iron Pipe Flanges and Flanged Fittings: Classes 25, 125, And 250. ASME: New York, NY, USA, 2020.
48. *ASME B16.5*; Pipe Flanges and Flanged Fittings: NPS 1/2 Through NPS 24, Metric/Inch Standard. ASME: New York, NY, USA, 2020.
49. *ASME B16.24*; Cast Copper Alloy Pipe Flanges, Flanged Fittings, And Valves: Classes 150, 300, 600, 900, 1500, And 2500. ASME: New York, NY, USA, 2021.
50. *ASME B16.36*; Orifice Flanges. ASME: New York, NY, USA, 2020.
51. *ASME B16.42*; Ductile Iron Pipe Flanges and Flanged Fittings: Classes 150 and 300. ASME: New York, NY, USA, 2021.
52. *ASME B16.47*; Large Diameter Steel Flanges: NPS 26 Through NPS 60, Metric/Inch Standard. ASME: New York, NY, USA, 2020.
53. *ASME B16.48*; Line Blanks. AWWA: Denver, CO, USA, 2020.
54. *AWWA C115*; Flanged Ductile-Iron Pipe with Ductile-Iron or Gray-Iron Threaded Flanges. AWWA: Denver, CO, USA, 2020.
55. *AWWA C207*; Steel Pipe Flanges for Waterworks Service, Sizes 4 in. Through 144 in. (100 MM Through 3600 MM). AWWA: Denver, CO, USA, 2023.
56. *MSS SP-44*; Steel Pipeline Flanges. MSS: Alexandria, VA, USA, 2019.
57. *MSS SP-51*; Class 150LW Corrosion Resistant Flanges and Cast Flanged Fittings. MSS: Alexandria, VA, USA, 2021.
58. *MSS SP-65*; High Pressure Chemical Industry Flanges and Threaded Stubs for Use with Lens Gaskets. MSS: Alexandria, VA, USA, 2024.
59. Aziz, A.S.B.A. *The Basics of Chemical Plant Operations*; Universiti College TATI: Kemaman, Malaysia, 2023.
60. *ASME B16.10*; Face-to-Face and End-to-End Dimensions of Valves. AWWA: Denver, CO, USA, 2023.
61. *ASME B16.34*; Valves-Flanged, Threaded, and Welding End. AWWA: Denver, CO, USA, 2021.
62. *API STD 526*; Flanged Steel Pressure-Relief Valves. API: Washington, DC, USA, 2023.
63. *API STD 594*; Check Valves: Flanged, Lug, Wafer, and Butt-Welding. API: Washington, DC, USA, 2022.
64. *API STD 599*; Metal Plug Valves—Flanged, Threaded, and Welding Ends. API: Washington, DC, USA, 2020.
65. *API STD 600*; Steel Gate Valves—Flanged and Butt-Welding Ends, Bolted Bonnets. API: Washington, DC, USA, 2024.
66. *API STD 602*; Gate, Globe, and Check Valves for Sizes DN 100 (NPS 4) and Smaller for the Petroleum and Natural Gas Industries. API: Washington, DC, USA, 2024.
67. *API STD 603*; Corrosion-Resistant, Bolted Bonnet Gate Valves—Flanged and Butt-Welding Ends. API: Washington, DC, USA, 2018.
68. *API STD 608*; Metal Ball Valves—Flanged, Threaded, and Welding Ends. API: Washington, DC, USA, 2020.
69. *API STD 609*; Butterfly Valves: Double-Flanged, Lug- and Wafer-Type, and Butt-Welding Ends. API: Washington, DC, USA, 2021.
70. *AWWA C500*; Metal-Seated Gate Valves for Water Supply Service. AWWA: Denver, CO, USA, 2019.
71. *AWWA C504*; Rubber-Seated Butterfly Valves. AWWA: Denver, CO, USA, 2023.
72. *MSS SP-42*; Corrosion-Resistant Gate, Globe, Angle, and Check Valves with Flanged and Butt Weld Ends (Classes 150, 300, & 600). MSS: Alexandria, VA, USA, 2013.
73. *MSS SP-70*; Gray Iron Gate Valves, Flanged and Threaded Ends. MSS: Alexandria, VA, USA, 2011.
74. *MSS SP-71*; Gray Iron Swing Check Valves, Flanged and Threaded Ends. MSS: Alexandria, VA, USA, 2018.
75. *MSS SP-72*; Ball Valves with Flanged or Butt-Welding Ends for General Service. MSS: Alexandria, VA, USA, 2010.
76. *MSS SP-80*; Bronze Gate, Globe, Angle, and Check Valves. MSS: Alexandria, VA, USA, 2019.
77. *MSS SP-81*; Stainless-Steel or Stainless-Steel-Lined, Bonnetless, Knife Gate Valves with Flanged Ends. MSS: Alexandria, VA, USA, 2021.
78. *MSS SP-85*; Gray Iron Globe & Angle Valves, Flanged and Threaded Ends. MSS: Alexandria, VA, USA, 2011.
79. *MSS SP-88*; Diaphragm Valves. MSS: Alexandria, VA, USA, 2021.
80. *MSS SP-105*; Instrument Valves for Code Applications. MSS: Alexandria, VA, USA, 2023.
81. *MSS SP-58*; Pipe Hangers and Supports–Materials, Design, Manufacture, Selection, Application, and Installation. MSS: Alexandria, VA, USA, 2018.
82. Abajuo, C.; Nwamaghinna, N.; Osinjolu, T.; Court, J.-P.; Matteoli, F.; Paillusseau, C. Replacement of Structural Support on a Brownfield FPSO Using Cold Work: A Case Study on Bonga FPSO's Pipe Support Replacement. In *Offshore Technology Conference*; OTC: Richardson, TX, USA, 2024.
83. Mensah, A.; Sriramula, S. Probabilistic Finite Element-Based Reliability of Corroded Pipelines with Interacting Corrosion Cluster Defects. *Int. J. Press. Vessel. Pip.* **2024**, *207*, 105086. [CrossRef]
84. Pialucha, T.; Pavlakovic, B.; Alleyne, D.; Cawley, P. Quantitative Measurement of Remnant Thickness in Corrosion under Pipe Supports. *Insight—Non-Destr. Test. Cond. Monit.* **2020**, *62*, 642–648. [CrossRef]
85. Marzuki, M.; Turmizi, T.; Bukhari, B.; Juhan, J.; Ibrahim, A.; Khatami, A. Mechanical Properties Analysis of ASME SA-106 Grade B Pipe Material Welded Joints in Combination of SMAW and GTAW Welding. *J. Polimesin* **2023**, *21*, 140–145.
86. Velasco, D.C.R.; Lopes, F.P.D.; Simonassi, N.T.; Vieira, C.M.F. Description of the Surface Preparation of Metallic Substrates by Rotating Blaster. *Braz. J. Dev.* **2022**, *8*, 24260–24271. [CrossRef]

87. Li, L.; Niu, X.; Han, B.; Song, L.; Li, X. Microstructure and Properties of Laser Cladding Coating at the End of L415/316L Bimetal Composite Pipe. *Int. J. Press. Vessel. Pip.* **2022**, *195*, 104568. [CrossRef]
88. Wang, H.; Wang, T.; Yang, S.; Gao, J.; Yu, Y.; Tao, H. Ductile Burst Behavior of High Pressure X100 Steel Pipe Considering Hydrogen Damage. *Int. J. Hydrogen Energy* **2024**, *58*, 362–379. [CrossRef]
89. Bokde, R.D.; Bankar, V.H. Paper on Design and Analysis of Bending Furnace Using Light Diesel Oil as an Alternative Fuel. *GRD J. Eng.* **2019**, *1*, 64–77.
90. Atal, S.; Dwivedi, D.; Tripathi, N.M. Corrosion under Insulation (CUI) in Oil and Gas Industries. In *Functional Materials for the Oil and Gas Industry*; CRC Press: Boca Raton, FL, USA, 2023; pp. 127–138.
91. *ASME B36.10M*; Welded and Seamless Wrought Steel Pipe. ASME: New York, NY, USA, 2018.
92. *ASME B36.19M*; Stainless Steel Pipe. ASME: New York, NY, USA, 2018.
93. Azevedo, V.M.; Velasco, D.C.R.; Gusmão, L.C.B. Projeto de uma linha de vapor entre caldeira e turbina. *Rev. FOCO* **2024**, *17*, e5708. [CrossRef]
94. Bahadori, A. *Hazardous Area Classification in Petroleum and Chemical Plants: A Guide to Mitigating Risk*, 1st ed.; CRC Press: Boca Raton, FL, USA, 2013.
95. *Petrobras N-0894*; Design of Cold Insulation. LTC: Rio de Janeiro, Brazil, 2022.
96. *Petrobras N-0550*; Projeto de Isolamento Térmico a Alta Temperatura. LTC: Rio de Janeiro, Brazil, 2024.
97. Cosentino, L.; Fernandes, J.; Mateus, R. A Review of Natural Bio-Based Insulation Materials. *Energies* **2023**, *16*, 4676. [CrossRef]
98. Malachy Udowo, V.; Yan, M.; Liu, F.; Ikeuba, A.I. Corrosion under Insulation of Pipeline Steel in Groundwater Containing Sulphate-Reducing Bacteria. *Corros. Eng. Sci. Technol.* **2023**, *58*, 549–560. [CrossRef]
99. Wu, J.; Ghosh, S.; Gonuguntla, M.; Liu, F.; Jin, Y. A Review of Research Methods for Corrosion under Insulation. *Corros. Rev.* **2023**, *41*, 263–276. [CrossRef]
100. Wang, Y.; Wang, C.; Gao, S.; Zheng, X.; Darkwa, J. The Impact of Thermal Insulation on Cooling Energy Consumption and Optimal Insulation Thickness for Underground Tunnel. *Sustain. Energy Technol. Assess.* **2021**, *47*, 101495. [CrossRef]
101. Zhang, L.; Zou, Y.; Yang, Y.; Chen, X.; Dai, Y.; Zhou, C.; Xu, H. Design and Optimization of Thermal Insulation Structure for High-Temperature Pipeline inside the Lower Tank Wall. *Ann. Nucl. Energy* **2023**, *192*, 109988. [CrossRef]
102. Hwang, S.-Y.; Kim, M.-S.; Lee, J.-H. Thermal Stress Analysis of Process Piping System Installed on LNG Vessel Subject to Hull Design Loads. *J. Mar. Sci. Eng.* **2020**, *8*, 926. [CrossRef]
103. Becth, C., IV. *Process Piping: The Complete Guide to the ASME B31.3*, 4th ed.; ASME: New York, NY, USA, 2021.
104. Zahid, U.; Khan, S.Z.; Khan, M.A.; Bukhari, H.J.; Ahmed, I.; Khan, K.A. A Methodology for Flexibility Analysis of Process Piping. *Proc. Inst. Mech. Eng. Part E J. Process Mech. Eng.* **2018**, *232*, 751–761. [CrossRef]
105. Paz-Martínez-Viademonte, M.; Abrahami, S.T.; Hack, T.; Burchardt, M.; Terryn, H. A Review on Anodizing of Aerospace Aluminum Alloys for Corrosion Protection. *Coatings* **2020**, *10*, 1106. [CrossRef]
106. Shaik, N.B.; Pedapati, S.R.; B A Dzubir, F.A. Remaining Useful Life Prediction of a Piping System Using Artificial Neural Networks: A Case Study. *Ain Shams Eng. J.* **2022**, *13*, 101535. [CrossRef]
107. Velasco, D.C.R.; Oliveira, D.L.R.; Lopes, F.P.D.; Souza, D.; Vieira, C.M.F. Influence of the Use of Anti-Bubble Additives on the Permeability and Porosity of Anticorrosive Coatings. In *TMS Annual Meeting & Exhibition*; Springer Nature: Cham, Switzerland, 2024; pp. 109–116.
108. *ISO 12944-2*; Paints and Varnishes—Corrosion Protection of Steel Structures by Protective Paint Systems: Part 2—Classification of Environments. ISO: Geneva, Switzerland, 2017.
109. Al-Janabi, Y.T. An Overview of Corrosion in Oil and Gas Industry. In *Corrosion Inhibitors in the Oil and Gas Industry*; Wiley: Hoboken, NJ, USA, 2020; pp. 1–39.
110. Hussein Farh, H.M.; Ben Seghier, M.E.A.; Taiwo, R.; Zayed, T. Analysis and Ranking of Corrosion Causes for Water Pipelines: A Critical Review. *NPJ Clean Water* **2023**, *6*, 65. [CrossRef]
111. Oshin, T.A.; Fakinle, B.S.; Oyewole, O. The Role of Microbes in the Inhibition of the Atmospheric Corrosion of Steel Caused by Air Pollutants. *Corros. Rev.* **2023**, *41*, 635–647. [CrossRef]
112. Pereira, A.A.; Neves, A.C.; Ladeira, D.; Caprace, J.-D. Corrosion Prediction of FPSOs Hull Using Machine Learning. *Mar. Struct.* **2024**, *97*, 103652. [CrossRef]
113. Oluwatayo, M.; Atebe, O.; Okafor, F.; Nebo, O.; Etuk, O.; Nwaonu, S.; Akanni, J.; Onukwugha, P. Bonga FPSO: Corrosion Management in the First Deep Water FPSO in the Gulf of Guinea—A Methodical Approach to Inspections and Fabric Maintenance. In *Offshore Technology Conference*; OTC: Richardson, TX, USA, 2024.
114. Gentil, V. *Corrosão*, 7th ed.; LTC: Rio de Janeiro, Brazil, 2022; ISBN 978-85-216-1804-1.
115. Cao, B.S.; Wu, C.L.; Wang, L.; Zhang, S.; Zhang, C.H.; Sun, X.Y. Effect of Residual Stress and Phase Constituents on Corrosion-Cavitation Erosion Behavior of 304 Stainless Steel by Iso-Material Manufacturing of Laser Surface Melting. *J. Mater. Res. Technol.* **2023**, *26*, 6532–6551. [CrossRef]

116. Ibrahimi, B.E.; Nardeli, J.V.; Guo, L. An Overview of Corrosion. Sustainable Corrosion Inhibitors I: Fundamentals, Methodologies, and Industrial Applications. *Methodol. Ind. Appl.* **2021**, *1*, 1–19.
117. Maher, M.; Iraola-Arregui, I.; Ben Youcef, H.; Rhouta, B.; Trabadelo, V. The Synergistic Effect of Wear-Corrosion in Stainless Steels: A Review. *Mater. Today Proc.* **2022**, *51*, 1975–1990. [CrossRef]
118. Ralls, A.M.; Menezes, P.L. Revealing the Fretting Corrosion Degradation Mechanisms of Friction Stir Processed Cold Spray Additively Manufactured Stainless Steel. *Int. J. Adv. Manuf. Technol.* **2024**, *130*, 2855–2876. [CrossRef]
119. Behvar, A.; Aghayar, Y.; Avateffazeli, M.; Tridello, A.; Benelli, A.; Paolino, D.S.; Mohammadi, M.; Haghshenas, M. Synergistic Impact of Corrosion Pitting on the Rotating Bending Fatigue of Additively Manufactured 316L Stainless Steel: Integrated Experimental and Modeling Analyses. *Int. J. Fatigue* **2024**, *188*, 108491. [CrossRef]
120. Fathyunes, L.; Mohtadi-Bonab, M.A. A Review on the Corrosion and Fatigue Failure of Gas Turbines. *Metals* **2023**, *13*, 701. [CrossRef]
121. Matlakhov, A.N. *Corrosão e Proteção dos Materiais;* Matlakhova, L.A., Ed.; Paco Editorial: Jundiaí, Brazil, 2021.
122. Tian, H.; Cui, Z.; Ma, H.; Zhao, P.; Yan, M.; Wang, X.; Cui, H. Corrosion Evolution and Stress Corrosion Cracking Behavior of a Low Carbon Bainite Steel in the Marine Environments: Effect of the Marine Zones. *Corros. Sci.* **2022**, *206*, 110490. [CrossRef]
123. Knisz, J.; Eckert, R.; Gieg, L.M.; Koerdt, A.; Lee, J.S.; Silva, E.R.; Skovhus, T.L.; An Stepec, B.A.; Wade, S.A. Microbiologically Influenced Corrosion—More than Just Microorganisms. *FEMS Microbiol. Rev.* **2023**, *47*, fuad041. [CrossRef]
124. Abilio, A.d.A.; Wolodko, J.D.; Eckert, R.B.; Skovhus, T.L. Development of an Expert System for Assessing Failures in Oil and Gas Pipelines Due to Microbiologically Influenced Corrosion (MIC). *Eng. Fail. Anal.* **2024**, *163*, 108426. [CrossRef]
125. Hussain, A.K.; Seetharamaiah, N.; Pichumani, M.; Chakra, C.S. Research Progress in Organic Zinc Rich Primer Coatings for Cathodic Protection of Metals—A Comprehensive Review. *Prog. Org. Coat.* **2021**, *153*, 106040. [CrossRef]
126. Nazari, M.H.; Zhang, Y.; Mahmoodi, A.; Xu, G.; Yu, J.; Wu, J.; Shi, X. Nanocomposite Organic Coatings for Corrosion Protection of Metals: A Review of Recent Advances. *Prog. Org. Coat.* **2022**, *162*, 106573. [CrossRef]
127. Sharun, V.; Rajasekaran, M.; Kumar, S.S.; Tripathi, V.; Sharma, R.; Puthilibai, G.; Sudhakar, M.; Negash, K. Study on Developments in Protection Coating Techniques for Steel. *Adv. Mater. Sci. Eng.* **2022**, *2022*, 2843043. [CrossRef]
128. Velasco, D.C.R.; Lopes, F.P.D.; Simonassi, N.T.; Vieira, C.M.F. Influência Da Incorporação de Resíduos Na Permeabilidade Ao Vapor de Água de Compósitos de Matriz Epoxídica/Influence of the Incorporation of Waste on the Water Vapor Permeability of Epoxide Matrix Composites. *Braz. J. Dev.* **2022**, *8*, 24272–24282. [CrossRef]
129. Sheydaei, M. The Use of Plant Extracts as Green Corrosion Inhibitors: A Review. *Surfaces* **2024**, *7*, 380–403. [CrossRef]
130. WIPO PatentScope. Available online: https://patentscope.wipo.int/ (accessed on 1 February 2025).
131. Elsevier ScienceDirect. Available online: https://www.sciencedirect.com/ (accessed on 1 February 2025).
132. Chen, T.; Shang, T.; Jiang, G.; Teng, H.; Liu, W.; He, X.; Zhou, X.; Liu, W.; Cheng, X.; Li, X.; et al. Effect of SO2 and NH3 on the Corrosion Failure of Zn-Al-Mg Coatings. *Surf. Coat. Technol.* **2024**, *493*, 131244. [CrossRef]
133. Fayomi, O.S.I.; Akande, I.G.; Sode, A.A. Corrosion Prevention of Metals via Electroless Nickel Coating: A Review. *J. Phys. Conf. Ser.* **2019**, *1378*, 022063. [CrossRef]
134. Verma, C.; Alameri, A.; Barsoum, I.; Alfantazi, A. Review on Corrosion-Related Aspects of Metallic Alloys Additive Manufactured with Laser Powder Bed-Fusion (LPBF) Technology. *Prog. Addit. Manuf.* **2024**, 1–29. [CrossRef]
135. Chakravarti, B.; Baker, B.A.; Maitra, D. Method for Making Clad Metal Pipe. US10112254B2, 2015.
136. Zhang, W.; Xin, S.; Huang, Q.; Jiao, H. Study on the Thermal Control Performance of Mg-Li Alloy Micro-Arc Oxidation Coating in High-Temperature Environments. *Surfaces* **2024**, *7*, 969–978. [CrossRef]
137. Bazli, L.; Yusuf, M.; Farahani, A.; Kiamarzi, M.; Seyedhosseini, Z.; Nezhadmansari, M.; Aliasghari, M.; Iranpoor, M. Application of Composite Conducting Polymers for Improving the Corrosion Behavior of Various Substrates: A Review. *J. Compos. Compd.* **2020**, *2*, 228–240. [CrossRef]
138. Bubnova, O.; Khan, Z.U.; Wang, H.; Braun, S.; Evans, D.R.; Fabretto, M.; Hojati-Talemi, P.; Dagnelund, D.; Arlin, J.-B.; Geerts, Y.H.; et al. Semi-Metallic Polymers. *Nat. Mater.* **2014**, *13*, 190–194. [CrossRef]
139. Xie, J.; Ewing, S.; Boyn, J.-N.; Filatov, A.S.; Cheng, B.; Ma, T.; Grocke, G.L.; Zhao, N.; Itani, R.; Sun, X.; et al. Intrinsic Glassy-Metallic Transport in an Amorphous Coordination Polymer. *Nature* **2022**, *611*, 479–484. [CrossRef]
140. Bai, M.; Liu, T.; Liu, B.; Li, Y.; Yu, H.; Zhao, Y.; Yang, C.; Song, L.; Liu, W. Preparation and Properties of Polyurethane Cold Galvanizing Coatings with Phosphoric Acid Modified Zinc Powder. *Surf. Coat. Technol.* **2024**, *489*, 131128. [CrossRef]
141. Lin, X.-Y.; Yu, Y.-H.; Tse, M.-M.; Cheng, S.-H.; Liu, Y.-H.; Lu, K.-L. Semiconductive (Cu–S)n Metal–Organic Framework Incorporated Polyimide Nanocomposite Coatings Forming an Oxide Barrier for Durable Anticorrosion Effects. *Prog. Org. Coat.* **2023**, *180*, 107562. [CrossRef]
142. Riaz, U.; Nwaoha, C.; Ashraf, S.M. Recent Advances in Corrosion Protective Composite Coatings Based on Conducting Polymers and Natural Resource Derived Polymers. *Prog. Org. Coat.* **2014**, *77*, 743–756. [CrossRef]
143. Jiang, C.; Wang, A.; Bao, X.; Ni, T.; Ling, J. A Review on Geopolymer in Potential Coating Application: Materials, Preparation and Basic Properties. *J. Build. Eng.* **2020**, *32*, 101734. [CrossRef]

144. Azevedo, A.R.G.; Marvila, M.T.; Vieira, C.M.F. Película de Reparo e Proteção a Base de Tinta Geopolimérica Para Aplicação Em Substratos de Concreto. BR 10 2020 008058 0 A2 2020, 3 November 2021.
145. Junior Linhares, J.A.; de Azevedo, A.R.G.; Marvila, M.T.; Teixeira, S.R.; Fediuk, R.; Vieira, C.M.F. Influence of Processing Parameters Variation on the Development of Geopolymeric Ceramic Blocks with Calcined Kaolinite Clay. *Case Stud. Constr. Mater.* **2022**, *16*, e00897. [CrossRef]
146. Thomas, D.; Philip, E.; Sindhu, R.; Ulaeto, S.B.; Pugazhendhi, A.; Awasthi, M.K. Developments in Smart Organic Coatings for Anticorrosion Applications: A Review. *Biomass Convers. Biorefinery* **2022**, *12*, 4683–4699. [CrossRef]
147. *NORSOK M-501*; Surface Preparation and Protective Coating. NORSOK: Lysaker, Norway, 2022.
148. *PETROBRAS N-2912 A*; Novolac Epoxy Paint. LTC: Rio de Janeiro, Brazil, 2016.
149. *ISO 8501-1*; Preparation of Steel Substrates before Application of Paints and Related Products—Visual Assessment of Surface Cleanliness—Part 1: Rust Grades and Preparation Grades of Uncoated Steel Substrates and of Steel Substrates after Overall Removal of Previous Coatings. ISO: Geneva, Switzerland, 2007.
150. *PDO SP-2217*; Specification of Internal FBE and Liquid Coating Systems For Carbon Steel Piping and Line Pipes. Petroleum Development Oman: Muscat, Oman, 2017.
151. *PETROBRAS N-2843 C*; Internal Coating of Pipes. LTC: Rio de Janeiro, Brazil, 2020.
152. Faccini, M.; Bautista, L.; Soldi, L.; Escobar, A.M.; Altavilla, M.; Calvet, M.; Domènech, A.; Domínguez, E. Environmentally Friendly Anticorrosive Polymeric Coatings. *Appl. Sci.* **2021**, *11*, 3446. [CrossRef]
153. McInerney, M.K.; Marsh, C.P.; Hock, V.F.; Beitelman, A.D.; Skaja, A.; Carlson, T.A. *Use of Coatings on Hydraulic Steel Structures: Part 1-Overview and Field Survey*; U.S. Army Corps of Engineers: Fort Belvoir, VA, USA, 2016.
154. Baig, M.M.A.; Samad, M.A. Epoxy\Epoxy Composite\Epoxy Hybrid Composite Coatings for Tribological Applications—A Review. *Polymers* **2021**, *13*, 179. [CrossRef] [PubMed]
155. Wang, Y.; Ou, B.; Zhu, P.; Niu, B.; Guo, Y.; Zhi, Q. High Mechanical Strength Aluminum Foam Epoxy Resin Composite Material with Superhydrophobic, Anticorrosive and Wear-Resistant Surface. *Surf. Interfaces* **2022**, *29*, 101747. [CrossRef]
156. Ma, Y.; Ye, Y.; Wan, H.; Chen, L.; Zhou, H.; Chen, J. Chemical Modification of Graphene Oxide to Reinforce the Corrosion Protection Performance of UV-Curable Polyurethane Acrylate Coating. *Prog. Org. Coat.* **2020**, *141*, 105547. [CrossRef]
157. Rabbani, S.; Bakhshandeh, E.; Jafari, R.; Momen, G. Superhydrophobic and Icephobic Polyurethane Coatings: Fundamentals, Progress, Challenges and Opportunities. *Prog. Org. Coat.* **2022**, *165*, 106715. [CrossRef]
158. Qu, L.; Wang, Q.; Xu, S.; Wang, N.; Shi, Z. Chloride Corrosion Resistance of Double-Layer Anticorrosive Coating in Simulated Concrete Pore Solution. *Constr. Build. Mater.* **2021**, *295*, 123682. [CrossRef]
159. Zhang, Z.; Liu, Y.; Yan, H.; Hu, C.; Huang, Y. Ultralong-Term Durable Anticorrosive Coatings by Integration of Double-Layered Transfer Self-Healing Ability, Fe Ion-Responsive Ability, and Active/Passive Functional Partitioning. *ACS Appl. Mater. Interfaces* **2024**, *16*, 1564–1577. [CrossRef]
160. Ghosal, S.; Dey, R.; Duari, B. Evaluation of Coatings Suitability for Buried SS316L Pipelines. *Prog. Color Color. Coat.* **2023**, *16*, 361–375.
161. Mohamed, M.G.; Ahmed, N.M.; Mohamed, W.S.; Mabrouk, M.R. Novel Water-Based Coatings of Acrylic-Polyurethane Reinforced with Mixed Metal Pigment for Oil and Gas Pipelines Protection. *Prog. Org. Coat.* **2020**, *149*, 105941. [CrossRef]
162. Buyondo, K.A.; Kasedde, H.; Kirabira, J.B. A Comprehensive Review on Kaolin as Pigment for Paint and Coating: Recent Trends of Chemical-Based Paints, Their Environmental Impacts and Regulation. *Case Stud. Chem. Environ. Eng.* **2022**, *6*, 100244. [CrossRef]
163. Fazenda, J.M.R. *Tintas e Vernizes: Ciência e Tecnologia, ABRAFATI*, 4th ed.; Blucher: São Paulo, Brazil, 2009.
164. Cui, G.; Bi, Z.; Wang, S.; Liu, J.; Xing, X.; Li, Z.; Wang, B. A Comprehensive Review on Smart Anti-Corrosive Coatings. *Prog. Org. Coat.* **2020**, *148*, 105821. [CrossRef]
165. Chowaniec, A.; Sadowski, Ł.; Żak, A. The Chemical and Microstructural Analysis of the Adhesive Properties of Epoxy Resin Coatings Modified Using Waste Glass Powder. *Appl. Surf. Sci.* **2020**, *504*, 144373. [CrossRef]
166. Olivieri, F.; Castaldo, R.; Cocca, M.; Gentile, G.; Lavorgna, M. Mesoporous Silica Nanoparticles as Carriers of Active Agents for Smart Anticorrosive Organic Coatings: A Critical Review. *Nanoscale* **2021**, *13*, 9091–9111. [CrossRef]
167. Ouarga, A.; Lebaz, N.; Tarhini, M.; Noukrati, H.; Barroug, A.; Elaissari, A.; Ben Youcef, H. Towards Smart Self-Healing Coatings: Advances in Micro/Nano-Encapsulation Processes as Carriers for Anti-Corrosion Coatings Development. *J. Mol. Liq.* **2022**, *354*, 118862. [CrossRef]
168. Kalajahi, S.T.; Misra, A.; Koerdt, A. Nanotechnology to Mitigate Microbiologically Influenced Corrosion (MIC). *Front. Nanotechnol.* **2024**, *6*, 1340352. [CrossRef]
169. Abdolahi, A.; Hamzah, E.; Ibrahim, Z.; Hashim, S. Application of Environmentally-Friendly Coatings Toward Inhibiting the Microbially Influenced Corrosion (MIC) of Steel: A Review. *Polym. Rev.* **2014**, *54*, 702–745. [CrossRef]

170. Balakrishnan, A.; Govindaraj, S.; Dhaipule, N.G.K.; Thirumalaisamy, N.; Anne, R.S.; Sublime, N.; Philip, J. Enhancing Microbiologically Influenced Corrosion Protection of Carbon Steels with Silanized Epoxy-Biocide Hybrid Coatings. *Environ. Sci. Pollut. Res.* **2024**, *31*, 13302–13326. [CrossRef]
171. Teixeira, S.C.; Silva, R.R.A.; de Oliveira, T.V.; Stringheta, P.C.; Pinto, M.R.M.R.; Soares, N.d.F.F. Glycerol and Triethyl Citrate Plasticizer Effects on Molecular, Thermal, Mechanical, and Barrier Properties of Cellulose Acetate Films. *Food Biosci.* **2021**, *42*, 101202. [CrossRef]
172. Colomer, I.; Chamberlain, A.E.R.; Haughey, M.B.; Donohoe, T.J. Hexafluoroisopropanol as a Highly Versatile Solvent. *Nat. Rev. Chem.* **2017**, *1*, 88. [CrossRef]
173. Du, K.; Xie, H.; Hu, G.; Peng, Z.; Cao, Y.; Yu, F. Enhancing the Thermal and Upper Voltage Performance of Ni-Rich Cathode Material by a Homogeneous and Facile Coating Method: Spray-Drying Coating with Nano-Al_2O_3. *ACS Appl. Mater. Interfaces* **2016**, *8*, 17713–17720. [CrossRef]
174. Valencia, C.; Valencia, Y.; Tovar, C.D.G. Synthesis and Application of a Cationic Polyamine as Yankee Dryer Coating Agent for the Tissue Paper-Making Process. *Polymers* **2020**, *12*, 173. [CrossRef]
175. *NACE SP 0394*; Application, Performance, and Quality Control of Plant-Applied Single-Layer Fusion-Bonded Epoxy External Pipe Coating. NACE: Houston, TX, USA, 2013.
176. *ASTM G17*; Standard Test Method for Penetration Resistance of Pipeline Coatings (Blunt Rod). ASTM: West Conshohocken, PA, USA, 2020.
177. *ASTM D522/D522M*; Standard Test Methods for Mandrel Bend Test of Attached Organic Coatings. ASTM: West Conshohocken, PA, USA, 2021.
178. *ASTM D4541*; Standard Test Method for Pull-Off Strength of Coatings Using Portable Adhesion Testers. ASTM: West Conshohocken, PA, USA, 2022.
179. *ASTM G8*; Standard Test Methods for Cathodic Disbonding of Pipeline Coatings. ASTM: West Conshohocken, PA, USA, 2019.
180. *ASTM G95*; Standard Test Method for Cathodic Disbondment Test of Pipeline Coatings (Attached Cell Method). ASTM: West Conshohocken, PA, USA, 2021.
181. *ISO 21809-2*; Petroleum and Natural Gas Industries—External Coatings for Buried or Submerged Pipelines Used in Pipeline Transportation Systems—Part 2: Single Layer Fusion-Bonded Epoxy Coatings. ISO: Geneva, Switzerland, 2014.
182. *ASTM G3*; Standard Practice for Conventions Applicable to Electrochemical Measurements in Corrosion Testing. ASTM: West Conshohocken, PA, USA, 2019.
183. *ASTM G14*; Standard Test Method for Impact Resistance of Pipeline Coatings (Falling Weight Test). ASTM: West Conshohocken, PA, USA, 2018.
184. *ASTM D4060*; Standard Test Method for Abrasion Resistance of Organic Coatings by the Taber Abraser. ASTM: West Conshohocken, PA, USA, 2019.
185. *ASTM D1653*; Standard Test Methods for Water Vapor Transmission of Organic Coating Films. ASTM: West Conshohocken, PA, USA, 2021.
186. *ASTM D870*; Standard Practice for Testing Water Resistance of Coatings Using Water Immersion. ASTM: West Conshohocken, PA, USA, 2020.
187. *ASTM D2247*; Standard Practice for Testing Water Resistance of Coatings in 100 % Relative Humidity. ASTM: West Conshohocken, PA, USA, 2020.
188. *ISO 2812-2*; Paints and Varnishes—Determination of Resistance to Liquids—Part 2: Water Immersion Method. ISO: Geneva, Switzerland, 2018.
189. *ASTM D4587*; Standard Practice for Fluorescent UV-Condensation Exposures of Paint and Related Coatings. ASTM: West Conshohocken, PA, USA, 2019.
190. *ASTM D6943*; Standard Practice for Immersion Testing of Industrial Protective Coatings and Linings. ASTM: West Conshohocken, PA, USA, 2019.
191. *ASTM B117*; Standard Practice for Operating Salt Spray (Fog) Apparatus. ASTM: West Conshohocken, PA, USA, 2019.
192. *NACE TM0185*; Evaluation of Internal Plastic Coatings for Corrosion Control of Tubular Goods by Autoclave Testing. NACE: Houston, TX, USA, TX, 2006.
193. *ASTM D714*; Test Method for Evaluating Degree of Blistering of Paints. ASTM: West Conshohocken, PA, USA, 2017.
194. *ASTM E1356*; Standard Test Method for Assignment of the Glass Transition Temperatures by Differential Scanning Calorimetry. ASTM: West Conshohocken, PA, USA, 2016.
195. Fragata, F.; Ordine, A.P.; Amorim, C.C.; Ferraz, O. Ensaios de aderência de esquemas de pintura pelo método de tração (pull-off)—Avaliação de fatores que influenciam os resultados. *Corrosão Protecção Mater.* **2008**, *27*, 122–129.
196. *ASTM D1475*; Standard Test Method for Density of Liquid Coatings, Inks, and Related Products. ASTM: West Conshohocken, PA, USA, 2020.

197. *ASTM D4400*; Standard Test Method for Sag Resistance of Paints Using a Multinotch Applicator. ASTM: West Conshohocken, PA, USA, 2018.
198. *ASTM D562*; Standard Test Method for Consistency of Paints Measuring Krebs Unit (KU) Viscosity Using a Stormer-Type Viscometer. ASTM: West Conshohocken, PA, USA, 2018.
199. *ABNT NBR 15742*; Pot-Life Determination of Paints and Varnishesvarnis. ABNT: Rio de Janeiro, Brazil, 2015.
200. *ASTM D2240*; Standard Test Method for Rubber Property—Durometer Hardness. ASTM: West Conshohocken, PA, USA, 2021.
201. *ASTM D2369*; Standard Test Method for Volatile Content of Coatings. ASTM: West Conshohocken, PA, USA, 2020.
202. *ISO 3233-1*; Paints and Varnishes—Determination of Percentage Volume of Non-Volatile Matter—Part 1: Method Using a Coated Test Panel to Determine Non-Volatile Matter and to Determine Dry-Film Density by the Archimedes' Principle. ISO: Geneva, Switzerland, 2019.
203. *ASTM E228*; Standard Test Method for Linear Thermal Expansion of Solid Materials With a Push-Rod Dilatometer. ASTM: West Conshohocken, PA, USA, 2022.
204. *ASTM D1640/D1640M*; Standard Test Methods for Drying, Curing, or Film Formation of Organic Coatings. ASTM: West Conshohocken, PA, USA, 2022.
205. Gonyora, M.; Ventura-Medina, E. Investigating the Relationship between Human and Organisational Factors, Maintenance, and Accidents. The Case of Chemical Process Industry in South Africa. *Saf. Sci.* **2024**, *176*, 106530. [CrossRef]
206. Jung, S.; Woo, J.; Kang, C. Analysis of Severe Industrial Accidents Caused by Hazardous Chemicals in South Korea from January 2008 to June 2018. *Saf. Sci.* **2020**, *124*, 104580. [CrossRef]
207. Trávníček, P.; Tichá, Z.; Kotek, L. Statistical and Text Analysis of Major Accidents to the Environment in European Union. *Process Saf. Progress.* **2021**, *40*, 367–374. [CrossRef]
208. Waqar, A.; Othman, I.; González-Lezcano, R.A. Challenges to the Implementation of BIM for the Risk Management of Oil and Gas Construction Projects: Structural Equation Modeling Approach. *Sustainability* **2023**, *15*, 8019. [CrossRef]
209. Obradović, D.; Marenjak, S.; Šperac, M. Estimating Maintenance Costs of Sewer System. *Buildings* **2023**, *13*, 500. [CrossRef]
210. Robaina, R.G.; Bastos, E.F.; Sousa, M.A.F.d.; Velasco, D.C.R.; Gallo, D.P. Projeto de Uma Rebobinadeira de Fita Poliéster Utilizada Na Fabricação de Dutos Flexíveis/Design of a Polyester Tape Rewinder Used in the Manufacturing of Flexible Pipelines. *Braz. J. Dev.* **2021**, *7*, 101676–101688. [CrossRef]
211. Badakhsh, A.; Bhagavathy, S.M. Caveats of Green Hydrogen for Decarbonisation of Heating in Buildings. *Appl. Energy* **2024**, *353*, 122152. [CrossRef]
212. Li, H.; Cao, X.; Liu, Y.; Shao, Y.; Nan, Z.; Teng, L.; Peng, W.; Bian, J. Safety of Hydrogen Storage and Transportation: An Overview on Mechanisms, Techniques, and Challenges. *Energy Rep.* **2022**, *8*, 6258–6269. [CrossRef]
213. Risco-Bravo, A.; Varela, C.; Bartels, J.; Zondervan, E. From Green Hydrogen to Electricity: A Review on Recent Advances, Challenges, and Opportunities on Power-to-Hydrogen-to-Power Systems. *Renew. Sustain. Energy Rev.* **2024**, *189*, 113930. [CrossRef]
214. Saborit, E.; García-Rosales Vazquez, E.; Storch de Gracia Calvo, M.D.; Rodado Nieto, G.M.; Martínez Fondón, P.; Abánades, A. Alternatives for Transport, Storage in Port and Bunkering Systems for Offshore Energy to Green Hydrogen. *Energies* **2023**, *16*, 7467. [CrossRef]
215. Jia, G.; Lei, M.; Li, M.; Xu, W.; Li, R.; Lu, Y.; Cai, M. Hydrogen Embrittlement in Hydrogen-Blended Natural Gas Transportation Systems: A Review. *Int. J. Hydrogen Energy* **2023**, *48*, 32137–32157. [CrossRef]
216. Zvirko, O.; Nykyforchyn, H.; Krechkovska, H.; Tsyrulnyk, O.; Hredil, M.; Venhryniuk, O.; Tsybailo, I. Evaluating Hydrogen Embrittlement Susceptibility of Operated Natural Gas Pipeline Steel Intended for Hydrogen Service. *Eng. Fail. Anal.* **2024**, *163*, 108472. [CrossRef]
217. Ghomashchi, R. Green Energy Revolution and Substitution of Hydrocarbons with Hydrogen: Distribution Network Infrastructure Materials. *Energies* **2023**, *16*, 8020. [CrossRef]
218. Ma, N.; Zhao, W.; Wang, W.; Li, X.; Zhou, H. Large Scale of Green Hydrogen Storage: Opportunities and Challenges. *Int. J. Hydrogen Energy* **2024**, *50*, 379–396. [CrossRef]
219. Zhang, Z.; Khong, J.C.; Koe, B.; Luo, S.; Huang, S.; Qin, L.; Cipiccia, S.; Batey, D.; Bodey, A.J.; Rau, C.; et al. Multiscale Characterization of the 3D Network Structure of Metal Carbides in a Ni Superalloy by Synchrotron X-Ray Microtomography and Ptychography. *Scr. Mater.* **2021**, *193*, 71–76. [CrossRef]
220. Rifai, D.; Abdalla, A.; Razali, R.; Ali, K.; Faraj, M. An Eddy Current Testing Platform System for Pipe Defect Inspection Based on an Optimized Eddy Current Technique Probe Design. *Sensors* **2017**, *17*, 579. [CrossRef]
221. Peruń, G. Advances in Non-Destructive Testing Methods. *Materials* **2024**, *17*, 554. [CrossRef]

surfaces

MDPI

Article

Biopolymer-Based Microencapsulation of Bioactive Compounds: Evaluation of the Impact of Encapsulated Compound Characteristics on Process Efficiency

Sarah Hamid [1], Hamza Moussa [2], Mohamed Malik Mahdjoub [2], Ismail Berrabah [3,4], Nadjet Djihad [1], Amel Attia [1], Naima Fadloun Oukil [1], Mustapha Mounir Bouhenna [5], Hichem Tahraoui [6,7] and Abdeltif Amrane [7,*]

1 Laboratoire de Biotechnologie Végétale et Ethnobotanique (LBVEB), Faculté des Sciences de la Nature et de la Vie, Université de Bejaia, Bejaia 06000, Algeria; sarah.hamid@univ-bejaia.dz (S.H.); nadjet.djihad@univ-bejaia.dz (N.D.); amel.attia@univ-bejaia.dz (A.A.); naima.fadloun@univ-bejaia.dz (N.F.O.)

2 Département des Sciences Biologiques, Faculté des Sciences de la Nature et de la Vie et des Sciences de la Terre, Université de Bouira, Bouira 10000, Algeria; h.moussa@univ-bouira.dz (H.M.); m.mahdjoub@univ-bouira.dz (M.M.M.)

3 Laboratoire des Matériaux Polymères Avancés (LMPA), Faculté de Technologie, Université de Bejaia, Bejaia 06000, Algeria; ismail.berrabah@univ-bejaia.dz

4 Laboratory of Analysis and Conformity Control of Industrial Products (LACCIP), TeleghmaPlast, Mila 43000, Algeria

5 Centre de Recherche Scientifique et Technique en Analyses Physico-Chimiques (CRAPC), BP384, Bou-Ismail, Tipaza 42004, Algeria; mustapha.bouhenna@crapc.dz

6 Laboratory of Biomaterials and Transport Phenomena, University of Medea, Medea 26000, Algeria; hichemm.tahraouii@gmail.com

7 Ecole Nationale Supérieure de Chimie de Rennes, University of Rennes, CNRS, ISCR—UMR6226, F-35000 Rennes, France

* Correspondence: abdeltif.amrane@univ-rennes.fr

Abstract: Complex coacervation using proteins and polysaccharides enables efficient microencapsulation with high thermal stability, facilitating continuous core component release and yielding coacervates with superior properties for diverse applications. This study investigates the use of casein and pectin for microencapsulating *Ocimum basilicum* L. essential oil (EO) and phenolic extract (PE). Microencapsulation yield and efficiency were 85.3% and 89.8% for EO microcapsules (EO-MC) and 53.1% and 70.0% for PE microcapsules (PE-MC). Optical microscopy revealed spherical microcapsules; EO-MC had smooth surfaces, while PE-MC had porous surfaces. Thermal analysis showed stability, with both types exhibiting two stages of weight loss. XRD analysis indicated increased crystallinity in EO-MC and high crystallinity in PE-MC due to phenolic interactions. FTIR spectroscopy confirmed molecular interactions, including hydrogen bonding between phenolic compounds and the biopolymer matrix and amide bonds between the carboxyl groups of pectin and the amino groups of casein, ensuring the successful encapsulation of the bioactive compounds. These findings highlight the potential of casein and pectin for microencapsulating extracts, particularly EOs, for food industry applications.

Keywords: encapsulated compounds; biopolymer; freeze-drying; thermal behavior; crystallinity

1. Introduction

The genus *Ocimum*, Lamiaceae, commonly known as basil, showcases remarkable diversity and is predominantly cultivated in Mediterranean regions [1–4]. Among its various species, *Ocimum basilicum* L., or sweet basil, stands out as a popular culinary

https://doi.org/10.3390/surfaces8010015

herb available in fresh and dried forms, owing to its multifaceted pharmacological effects [5,6]. Basil, with its extensive range of applications, demonstrates potent antimicrobial, antifungal, antioxidant, and anti-inflammatory properties, making it a promising candidate for therapeutic interventions in both human and animal health [1,7–9]. *Ocimum basilicum* L. comprises numerous active compounds within its essential oil (EO), including phenols, alcohols, esters, and oxides, contributing to its distinct aroma and therapeutic potential [10,11]. Furthermore, the EO of sweet basil finds applications in pharmaceuticals for treating various ailments such as coughs, headaches, and gastrointestinal issues [12,13], while also being extensively utilized in the food and perfume industries due to its flavoring and aromatic properties [14–16]. Extracts derived from *Ocimum basilicum* L. leaves encompass a wide array of bioactive components, including rosmarinic acid, a phenolic compound renowned for its antioxidant qualities and therapeutic benefits against oxidative stress-related disorders and cancer proliferation [17–19]. Additionally, rosmarinic acid demonstrates hepatoprotective, lung-protective, and cardioprotective effects, attributed to its antioxidant and anti-inflammatory activities [20–23]. All these findings underscore the immense therapeutic potential of basil extracts and highlight the importance of further research to harness their benefits effectively [24,25].

Polysaccharides, particularly pectin, are pivotal renewable resources with diverse applications in the food and pharmaceutical industries. Pectin, renowned for its abundance, biodegradability, and versatility, acts as a crucial component within the plant cell wall, facilitating structural integrity and cell adhesion [26]. Beyond its role in plant physiology, pectin serves as a multifunctional ingredient in food production, contributing to texture and stability while serving as a gelling agent, thickener, emulsifier, and sugar substitute [27,28]. In pharmaceuticals, pectin functions as a valuable drug carrier, offering diverse applications in gel beads, matrices, and film-coated tablets [29,30].

Conversely, casein, a major milk protein, showcases remarkable properties such as exceptional surface activity and insolubility in aqueous mediums at low pH levels, making it invaluable across various applications [29,31]. Its unique structure endows it with excellent barrier properties against oxygen and non-polar molecules, rendering it suitable for the production of protein-based edible capsules [32]. Furthermore, its compatibility with other encapsulation materials and its potential to stabilize acidified dairy drinks highlight its versatility and importance in industrial applications [30]. The interest in studying the interactions between pectin and casein has surged, particularly in enhancing their physicochemical properties and exploring novel applications in the food and pharmaceutical sectors. Research has elucidated various factors influencing their interactions, including pH, temperature, ionic strength, and the presence of oil droplets, particularly in the development of acidic milk drinks [31].

Complex coacervation, a phenomenon driven by electrostatic interactions between oppositely charged biopolymers, finds wide applications in the food and pharmaceutical industries [26,33]. Influenced by factors like pH, temperature, and biopolymer ratios, it enables the encapsulation of bioactive compounds, ensuring protection and controlled release [34,35]. Despite being miscible in solvents, coacervates form through solute interactions, offering potential in food processing and encapsulation [36]. The process, crucial for creating coacervates, is utilized in various food applications and offers innovative solutions for enhancing food functionality and shelf-life through the improved stability of bioactive compounds [37–39]. This technology, driven by electrostatic attraction between proteins and polysaccharides, holds promise in protecting bioactive compounds and improving their stability during processing and storage, thereby offering innovative solutions for enhancing food functionality and shelf-life [40]. Several researchers delved into the microencapsulation of basil extracts. Ozdemir et al. [41] explored the microencapsulation efficiency of basil essential oil through various combinations of gum arabic, whey protein isolate, and maltodextrin. Similarly, Paul et al. [42] and Fadel et al. [43] examined the impact of coating materials such as maltodextrin, sodium caseinate, carboxymethyl cellulose, sodium alginate, and chitosan on the microencapsulation of basil oil using spray drying techniques. Expanding on this, chitosan and sodium alginate biopolymers were utilized by Thuekeaw et al. [44] to microencapsulate basil essential oil, employing a hot air oven for the drying process. Meanwhile, Tomé and da Silva [45] obtained basil ethanolic extract microcapsules using sodium alginate as the wall material, dried via air drying.

While numerous studies have explored the microencapsulation techniques, there is a paucity of investigation about understanding the potential of complex coacervation for encapsulating *Ocimum basilicum* L. essential oils and phenolic extract efficiently. This novel study aims to fill this gap by investigating the efficacy of complex coacervation employing casein from bovine milk and pectin from citrus peels as wall materials. The choice of freeze-drying as the drying method underscores the commitment to preserving the delicate bioactive compounds inherent in basil extracts. This innovative approach not only addresses a significant research gap but also holds promise for unlocking novel applications in various industries. What sets this study apart is its innovative approach to combining complex coacervation with these specific biopolymers for the encapsulation of basil oil and phenolic extract. By employing sophisticated physicochemical characterizations such as thermogravimetric and differential scanning calorimetry analysis, X-ray diffraction, and FTIR spectroscopy, this research delves into elucidating the moisture content, thermal behavior, crystallinity, and chemical structure of the resulting microcapsules. Additionally, the meticulous evaluation of particle size and distribution adds depth to this study's findings, contributing valuable insights into the potential applications of these microencapsulated basil extracts in various sectors.

2. Materials and Methods

2.1. Material and Reagents

Ocimum basilicum L. aerial parts were collected (Lakhdaria province of Bouira, Algeria), and air-dried for a week. An electrical grinder (SAYONA) was used to grind the plant into a fine powder, which was then sieved through a sieve shaker with a porosity of 100 μm. The powder was stored in an amber flask until use. Ethanol (C_2H_6O), hexane (C_6H_{14}), methanol (CH_3OH), casein from bovine milk ((Purity: 87–94% protein basis, high molecular weight phosphoprotein), (C7078, Lot ♯BCCD1666)), pectin (Ps) from *citrus* peel ((Purity: galacturonic acid ≥74.0% dried basis, poly-D-galacturonic acid methyl ester, molecular weight may range from 10 kDa to 300 kDa), (P9135, Lot ♯SLCB2455)), trichloroacetic acid ($C_2HCl_3O_2$), sodium sulfate (Na_2SO_4), sodium hydroxide (NaOH), and hydrochloric acid (HCl) were purchased from Sigma-Aldrich (Brøndby, Denmark). Sodium phosphate monobasic ($NaH_2PO_4.H_2O$) and sodium phosphate dibasic ($Na_2HPO_4.12H_2O$) were purchased from VWR Chemicals (Fontenay-sous-Bois, France).

2.2. Extraction of Bioactive Compounds

The bioactive compounds extracted from basil powder in this study employed unconventional extraction methods. Basil essential oil (EO) was obtained through microwave-assisted extraction using a Samsung ME6124ST (Samsung, Seremban, Malaysia) (2450 MHz, 1000 W) oven with interior dimensions of 373 mm × 233 mm × 393 mm (W × H × D) [46]. Following the protocol outlined by Bellik et al. [47], 20 g of basil powder was mixed with 200 mL of distilled water in a flat-bottom flask, and then subjected to microwave irradiation at 500 W for 15 min. Subsequently, the EO was separated from the water and stored in an amber flask at 4 °C.

The phenolic extract of basil (PE) was acquired via ultrasound-assisted extraction [48]. An amount of 10 g of basil powder was combined with 100 mL of ethanol (70% v/v) and set to ultrasound treatment in a bath operating at 40 kHz and 40 °C for 30 min, as described by Pedrosa et al. [49]. The resulting extract was obtained through centrifugation at 4025× g for 15 min using a Sigma 3–16 L centrifuge (Osterode, Germany), followed by filtration [50]. After solvent evaporation, the phenolic extract was dried using lyophilization (Christ Alpha 1-2 LD plus, Osterode am Harz, Germany) and stored at 4 °C.

2.3. Preparation of Microcapsules by Complex Coacervation and Freeze-Drying

The microcapsules containing basil EO and PE were prepared based on the methodologies outlined by Medeiros et al. [51] and Guazelli et al. [52], with adaptations. The essential oil was added directly while the PE extract was hydrated in deionized water. Biopolymer solutions were prepared by dispersing pectin and casein in distilled water, with a solid content of 10% w/v (5% casein and 5% pectin), and left to disperse fully at 4 °C overnight.

To create the microcapsules, the extracts were added to the casein solution (pH 8) at proportions of (1:10) and (5:10) substance-to-polymer for basil EO and PE, respectively. The mixture was homogenized at 362× g for 10 min using an Ultra-Turrax (OS40-IBX instrument, IKA-Werke GmbH & Co. KG, Staufen im Breisgau, Germany). The pectin solution was then slowly added while the mixture was agitated. Subsequently, the mixture was stirred continuously and the pH was gradually reduced to 3.5 using a solution of TCA (1.0 M).

To ensure successful complex coacervation and microcapsule precipitation, the mixture was left overnight at 4 °C. After filtration, the mixture was frozen at −25 °C in a Liebherr Comfort-NoFrost freezer (Ochsenhausen, Germany) before undergoing lyophilization for 48 h using the Christ Alpha 1-2 LD plus freeze dryer (0.1 mbar, −60 °C).

2.4. Determination of the Encapsulation Yield and Efficiency

The yield (Y%) of the freeze-dried microcapsules was determined using Equation (1) as described previously by Mebarki et al. [53], with modifications. Encapsulation efficiency was assessed by dispersing 500 mg of basil EO microcapsules (EOMC) and phenolic extract microcapsules (PEMC) in 2 mL of hexane and shaking the mixture for 10 min. After filtration (0.22 µm syringe filter), the amount of basil EO and PE on the microcapsule surfaces was quantified using a spectrophotometer (UV–visible, Optizan Pop, Daejeon, Republic of Korea) at 280 nm and 330 nm, respectively.

Alternatively, 1.5 mL of 2.0 M HCl was added to the microcapsules and stirred (161× g, 37 °C) for 5 min. Subsequently, 2 mL of ethanol was added, and the mixture was centrifuged at (2516× g, 25 °C) for 5 min using a Sigma centrifuge (Osterode, Germany). The amounts of basil EO and PE in the supernatant were determined spectroscopically at 280 nm and 330 nm, respectively, using calibration curves (EO: y = 8.7153x + 0.0132, R^2 = 0.99/PE: y = 0.5168x + 0.0307, R^2 = 0.9958). Encapsulation efficiency (EE %) was calculated using Equation (2) [53–55].

$$Y(\%) = \frac{\text{Dried microcapsules (g)}}{\text{Initial amounnt of substance (g)} + \text{Polymers (g)}} * 100 \quad (1)$$

$$EE(\%) = \frac{\text{Total amount of substance} - \text{Surface amount substance}}{\text{Total substance amount}} * 100 \quad (2)$$

2.5. Morphology and Size Characteristics

The morphology and size of the microcapsules were assessed through optical microscopy (Axiom 2000, OPTICA, Ponteranica, Italy) and scanning electron microscopy (Thermo Fisher Scientific (formerly FEI Company), FEI Quanta 250, Hillsboro, Oregon,

USA), following the procedures outlined by Benammar et al. [56]. Additionally, the size distribution of the dried microcapsules was analyzed using specialized software.

2.6. Moisture Content

Moisture content in the freeze-dried microcapsules was performed according to the protocol outlined by Bakry et al. [57]. Basil EOMC and PEMC samples were weighed and placed in glass Petri dishes, then heated in an oven at 105 °C until a constant weight was achieved, indicating complete evaporation of moisture from the microcapsules.

2.7. Thermal Behavior, Crystallinity, and Structure of Microcapsules

The thermal properties of casein (CN), pectin (PN), EOMC, and PEMC were evaluated according to the procedure described by Chaib et al. [58]. Thermogravimetric analysis (TG-DSC) was performed using a SDT Q600 analyzer (TA Instruments, New Castle, Delaware, USA) under a nitrogen atmosphere (N_2). Each sample (20 mg) was heated from 25 °C to 650 °C at a rate of 10 °C/min. The microcapsules' crystalline structures and their phase compositions were examined using X-ray diffractometry with a panalytical-Empyrean X-ray diffractometer (Almelo, Netherlands). The instrument was equipped with a copper tube and utilized Cu-Kα radiation with a wavelength of 1.5418 Å at 45 kV and 40 mA. The analysis involved examining samples across a diffraction angular range of 5–70° with a step of 2θ = 0.0262°. The peak position (2θ), full width at half maximum (FWHM), crystallite size (nm), and lattice strain (%) of the samples were obtained using X-PertHighScore software (version 3.0e (3.0.5)) with the Scherrer equation (Equation (3)) [52]. Also, the degree of crystallinity (Xc) of the pectin/casein complex was calculated by Equation (4) [26].

$$D = \frac{K.\lambda}{\beta.cos\theta} \tag{3}$$

$$Xc = \frac{Area\ of\ crystalline\ peaks}{Area\ of\ all\ peaks\ (Crystalline + Amorphous)} \tag{4}$$

The infrared spectra were collected using an attenuated total reflectance (ATR) cell in an infrared spectrum (FTIR) instrument (JASCO Corporation, Jasco-ATR PRO 450-FT/IR4200, Tokyo, Japan) in a wavenumber range of 4000 to 400 cm^{-1} and a resolution of 1 cm^{-1} [59,60].

2.8. Statistical Analysis

All experiments were conducted in triplicate (n = 3), and the results are expressed as mean ± standard deviation (SD). Statistical analysis was performed using two-way ANOVA, followed by Sidak's multiple comparisons test to assess differences among groups. Graphs and statistical evaluations (assuming a normal distribution) were generated using Origin (version 2019b (9.6.5.169)) and GraphPad software (9.5.0 (730)). A p-value of less than 0.05 was considered statistically significant.

3. Results

3.1. Encapsulation Yield and Efficiency

Microencapsulation yield and efficiency are critical parameters in the encapsulation process, determining the success and cost-effectiveness of the technique. Yield refers to the proportion of core material successfully encapsulated within the microcapsules, typically expressed as a percentage of the initial amount of the core material [61]. A high yield indicates efficient encapsulation, minimizing wastage and ensuring the maximum utilization of materials. On the other hand, efficiency encompasses various factors such as encapsulation rate, uniformity of capsule size, and stability of the encapsulated material.

Achieving high efficiency implies not only a high yield but also consistent and reliable encapsulation performance, leading to desired product characteristics such as prolonged release profiles, improved stability, and enhanced bioavailability [62,63].

In this study, the microencapsulation of EO and PE yielded distinct results in terms of both yield and efficiency (Table 1). The yield of EO microencapsulation (EO-MC) was notably higher at 85.3% compared to PE microencapsulation (PE-MC) at 53.1%. Similarly, the efficiency of EO-MC was substantially higher at 89.8 ± 1.4% compared to 70.0 ± 1.01% for PE-MC.

Table 1. Characteristics of EO and PE microcapsules.

Sample	Yield (%)	Efficiency (%)	Moisture (%)	Size (μm)
EO-MC	85.3 ± 1.8 [a]	89.8 ± 1.4 [a]	4.08 ± 0.47 [a]	9.70 ± 2.82 [a]
PE-MC	22.297 ± 0.5 [b]	70.0 ± 1.01 [b]	4.41 ± 0.25 [a]	6.95 ± 1.94 [a]

Values with different letters indicate statistically significant differences ($p < 0.05$).

3.2. Microcapsule Morphology and Particle Size Distribution

The microcapsules derived from both EO-MC and PE-MC exhibited a characteristic spherical shape, as evident from the results obtained through SEM (Figure 1a,b), indicative of the typical outcome of microcapsules produced via complex coacervation.

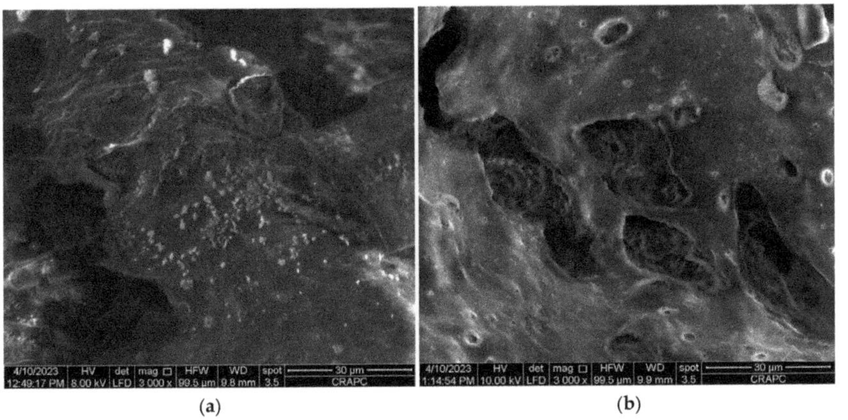

Figure 1. Microcapsules observation under scanning electron microscopy: (**a**) EO-MC; (**b**) PE-MC.

The SEM analysis of basil PE-MC unveiled a notably porous surface texture (Figure 1b), contrasting with the smoother surface of EO-MC. These figures further illustrate the successful microencapsulation process utilizing casein and pectin as encapsulating agents through coacervation, with clear evidence of a protective layer enveloping the core, indicative of the presence of the extracts.

On the other hand, particle size holds significant importance, particularly concerning the utilization of food ingredients, as excessively large particles can yield an unpleasant mouthfeel. Conversely, there are cases where large particles are pursued, such as when aiming for visible particle presence in products. Furthermore, larger particles have the potential to facilitate an extended release of active compounds [61,64]. Both EO-MC and PE-MC displayed comparable sizes on average, with EO-MC showing a slightly larger average size of 9.70 ± 2.82 μm (Table 1, Figure 2a). On the other hand, PE-MC exhibiting a slightly smaller average particle size of 6.95 ± 1.94 μm (Figure 2b).

(a)

(b)

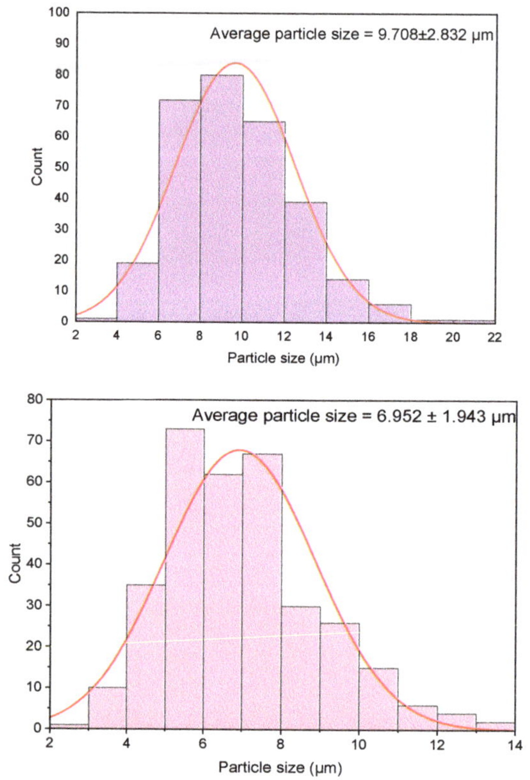

Figure 2. Particle size average and distribution of the microcapsules: (**a**) EO-MC; (**b**) PE-MC, the red line in both figures represents the normal distribution of particle size.

3.3. Moisture Determination Results

The moisture content of microcapsules obtained through complex coacervation was determined to be 4.08 ± 0.47% for EO-MC and 4.41 ± 0.25% for PE-MC (Table 1). These results align closely with the minimum moisture content specifications required in the food industry for dried powders, which typically range between 3% and 4% [65].

3.4. Thermal Behavior

The thermal behavior of wall materials (casein and pectin) and the obtained EO-MC and PE-MC were evaluated using thermogravimetric analysis (TGA) and differential scanning calorimetry (DSC). The TGA curves exhibited three distinct stages of thermal degradation, characterized by weight loss relative to temperature. The wall material curves were previously presented in Hamid et al. [55]. Briefly, two distinct thermal events were observed for pectin (50–100 °C and 200–250 °C), attributed to thermal salvation and degradation of pectin chains. The weight losses for these stages were 11% and 62%, respectively, with a total weight loss of 73%. Casein exhibited two thermal stages between 50–250 °C and 280–370 °C, resulting in a total weight loss of 76%, with maximum loss at 280 °C.

The thermal stability of the microcapsules EO-MC and PE-MC is illustrated in Figure 3a,b. EO-MC presented two stages of weight loss with a total weight loss of 69.68%. The first stage occurred from 20 °C to 135 °C, with a weight loss percentage of 4.24%, and the second from 125 °C to 650 °C with 65.62%. The PE-MC showed similar behavior, with two major stages of weight loss: 8.97% from 25 °C to 190 °C and 70.17% from 200 °C to 650 °C, with a total weight loss of 79.14%.

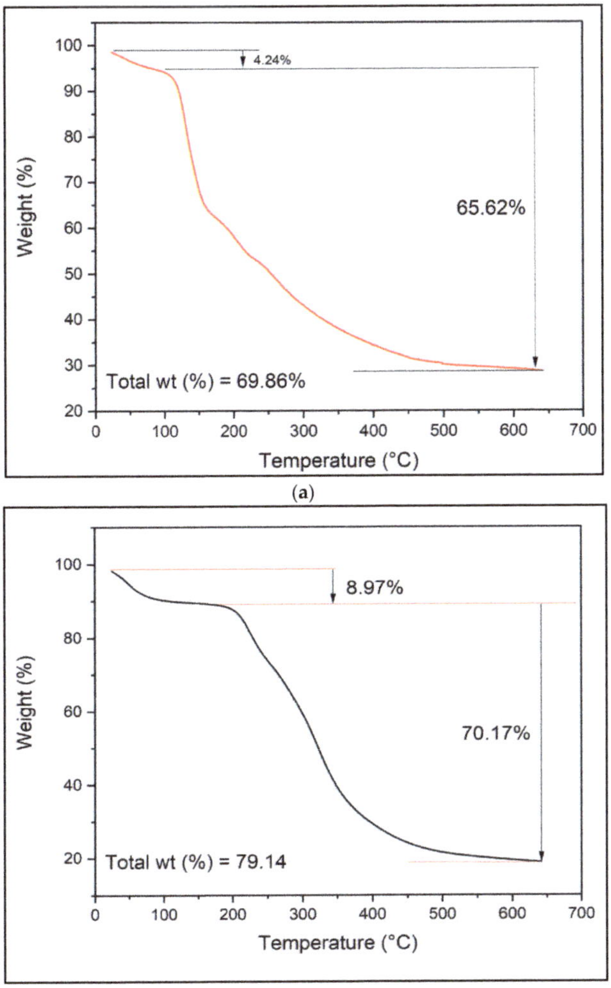

Figure 3. Thermogravimetric analysis (TGA) of EO-MC (**a**) and PE-MC (**b**) (for interpretation of the references to color in this legend, the reader is referred to the web version of this article).

3.5. Crystallinity (X-Ray Diffraction)

The crystalline behavior and phase composition of the pectin/casein complex, basil essential oil microcapsules (EO-MC), and phenolic extract microcapsules (PE-MC) were investigated through X-ray diffraction (XRD) analysis.

Figure 4a,b illustrates the XRD patterns, while Table 2 summarizes the crystallographic parameters for these materials. In Figure 4a, the XRD pattern of the pectin/casein complex reveals two broad peaks spanning the range of 9.9° to 19.9°, with an additional small peak observed at 2θ = 21.75°. These features are indicative of the presence of amorphous structures, attributed to the casein component [66]. The XRD pattern of EO-MC exhibits three new peaks at 31.8°, 45.5°, and 56.6°, corresponding to characteristic peaks of pectin [67], suggesting the presence of pectin crystals, likely formed due to essential oil-induced re-crystallization. Additionally, the peak at 21.75° shifts to 22.2° and becomes more intense, indicating changes in casein's crystallinity within the polymer complex.

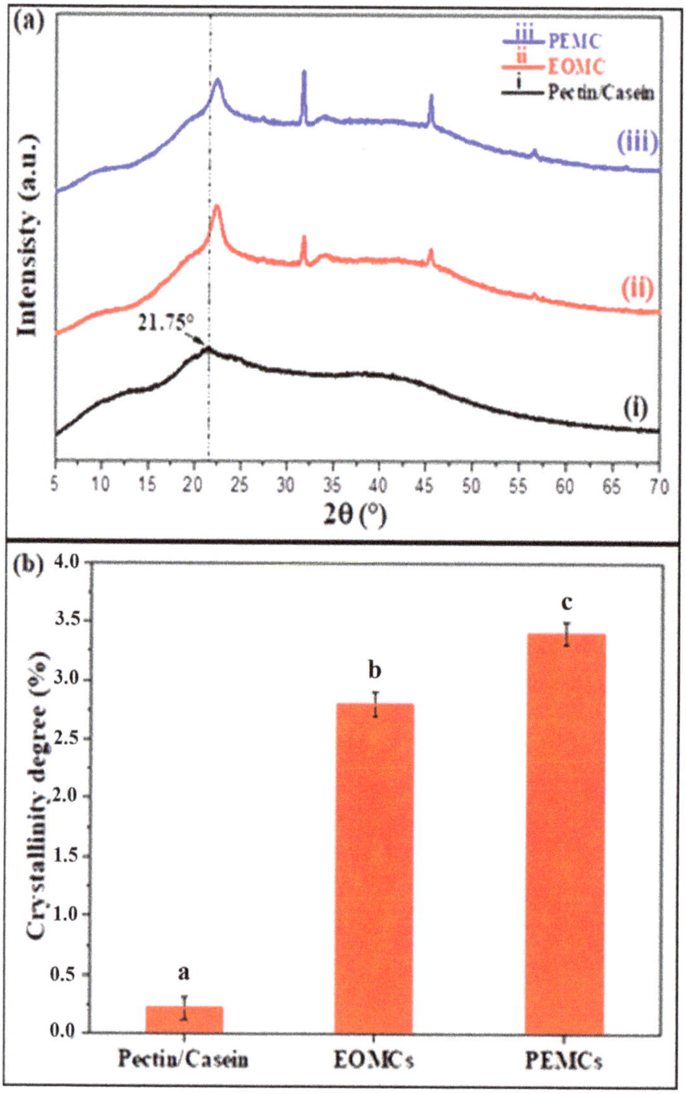

Figure 4. XRD patterns (**a**) of pectin/casein, EO-MC, and PE-MC. Effect of different microcapsule formulations on the degree of crystallinity of the pectin/casein complex (**b**); values with different letters indicate statistically significant differences ($p < 0.05$).

Table 2 shows the decreased full-width at half maximum (FWHM), increased crystallite size, and reduced lattice strain for this casein peak, supporting enhanced crystallinity in the pectin/casein complex due to the microcapsule formulation.

Table 2. Crystallographic parameters of samples obtained by XRD pattern.

Sample	2θ (°)	FWHM (°)	Crystallite Size (nm)	Strain (%)
Pectin + Casein	21.747	0.461	14.6	1.047
EO-MC	22.204	0.358	23.1	0.796
PE-MC	22.297	0.614	13.4	1.359

FWHM: full width at half maximum.

The XRD pattern of basil PE-MC shows an increase in the intensity of characteristic pectin peaks, suggesting the enhanced crystallinity of pectin chains, likely due to physical interactions between the phenolic substances and the pectin. However, there is a slight decrease in the intensity of the casein characteristic peak, potentially caused by reduced pectin/casein interface interactions as a result of increased pectin crystallinity promoted by the phenolic molecules [68]. This is supported by an increase in the value of FWHM and the lattice strain, along with a decrease in crystallite size, corresponding to the characteristic peak of casein (as indicated in Table 1).

Figure 4b illustrates the impact of various microcapsule formulations on the degree of crystallinity of the pectin/casein complex, with values marked by different letters indicating statistically significant differences ($p < 0.05$). The data reveal that the unmodified pectin/casein complex exhibits a low crystallinity, characteristic of an amorphous structure, while the introduction of specific microcapsule formulations, such as EOMCs and PEMCs, substantially increases the degree of crystallinity. This enhancement suggests that the encapsulation process induces a more ordered molecular arrangement within the biopolymer matrix, likely due to interactions between the encapsulated components and the pectin/casein network. Notably, the formulation differences between EO-MCs and PE-MCs further imply that subtle variations in the encapsulation protocol or the nature of the additives can significantly influence the structural ordering of the complex, which may, in turn, affect the stability and release properties of the encapsulated bioactive compounds. This observation underscores the pivotal role of these bioactive compounds and their synergistic interactions in modulating the crystalline architecture of the complex [69]. Notably, such structural modifications are anticipated to influence the controlled release profile of the encapsulated drug. Enhanced crystallinity reinforces the structural integrity of the pectin/casein matrix, thereby reducing molecular diffusion, impeding the release kinetics of the active ingredient, and improving the overall stability of the microcapsules [70].

3.6. Infrared Spectroscopy

Figure 5a,b illustrates the FTIR spectrum of microencapsulated samples (EO-MC and PE-MC) alongside a comparison of the wave numbers from the FTIR spectra corresponding to the basil compounds and wall biopolymers.

The FTIR analysis validated the successful encapsulation process by detecting the presence of basil EO and phenolic extract in the casein/pectin microcapsules. This analysis revealed similarities among the samples, with variations observed in peak width and intensities, likely due to differences in the type and quantities of compounds used to formulate the microcapsules. The basil EO displayed distinctive peaks, revealing its chemical composition. These included broad peaks in the ranges of 3000–2800 cm^{-1} and 3750–3500 cm^{-1}, indicating C–H stretching vibrations and O–H/N–H vibrations, respectively [58]. A strong C–O stretching vibration band was observed between 1200 and 1100 cm^{-1}.

Additionally, peaks from 1600 to 1000 cm^{-1} confirmed the oil's aromatic nature, with C=C stretching of a benzene ring, O–H bending, and C–O stretching of an aromatic group [69]. In the case of basil PE, bands identified within the wavenumber range of 1800 to

750 cm^{-1} are associated with polyphenols found in plants, such as C–OH, C–O–C, and C–C stretching vibrations [71]. The absorption bands of the aromatic ring stretching vibrations of C–H were observed at 2900 cm^{-1} [72]. The band detected between 3700 and 3300 cm^{-1} is attributed to hydrogen bonding and the stretching of the O–H bond in carbohydrates and carboxylic acids [32,73]. The functional groups of the wall biopolymers (casein and pectin) were previously detailed in our study [55]. FT-IR analysis confirmed the generation of amide bonds due to the interaction between the free carboxyl groups of pectin and the amino groups of casein in the complex.

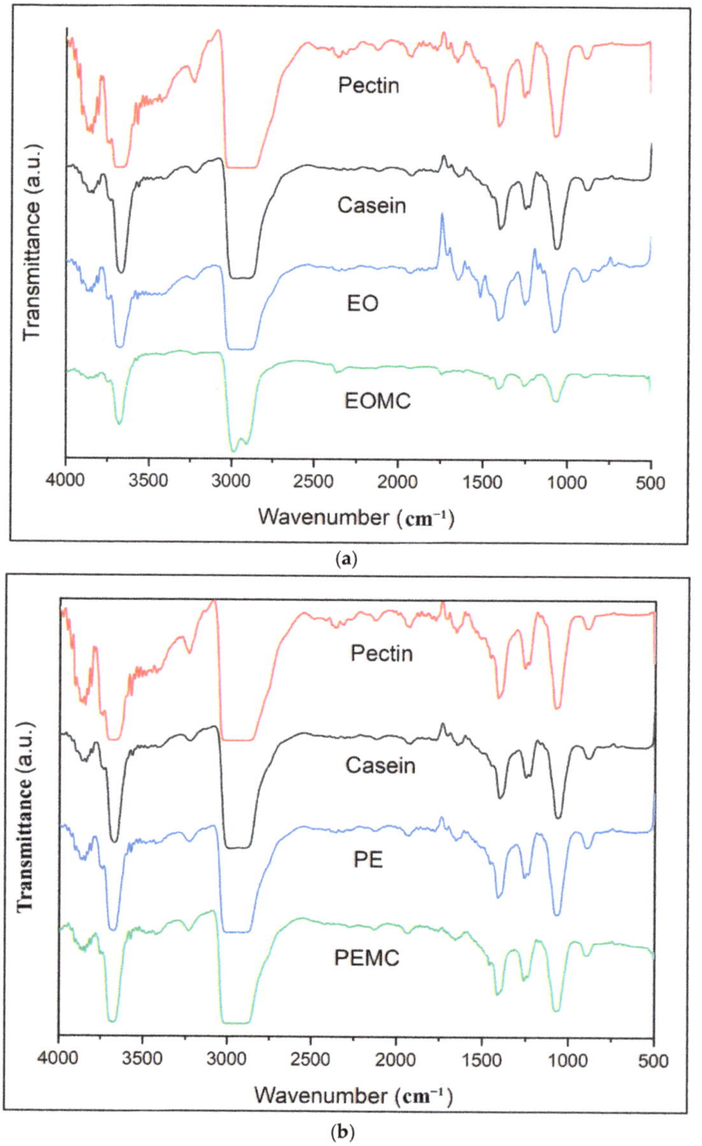

Figure 5. FTIR spectra of pectin and casein, basil essential oil and phenolic extract, and the obtained microcapsules (**a**) EO and (**b**) PE.

4. Discussion

Basil extracts exhibited significant biological activity, attributed to their rich composition of phenolic and volatile compounds. The basil phenolic extract (PE) demonstrated potent antioxidant properties, while the basil essential oil (EO) exhibited remarkable antimicrobial efficacy. These findings highlight the promising bioactivity of basil extracts, suggesting their potential applications in health and culinary contexts (Erez et al., 2024 [1], Chu 2022 [6], Pedrosa et al., 2021 [49], Moussa et al., 2024 [74], Darrag et al., 2022 [75], Fernandes et al. [76], Fayoumi et al. [77], and Rezzoug et al. [78]. This underscores the utility of basil extracts, not only as functional ingredients in food but also in therapeutic applications, further supporting the relevance of encapsulation techniques to enhance their stability and bioactivity.

In terms of microencapsulation, the results indicated that the encapsulation of the essential oil (EO) was more efficient than the phenolic extract (PE). The higher yield and efficiency of EO microcapsules (EO-MC) suggest a more successful encapsulation process, likely due to the inherent chemical properties and solubility of essential oils compared to phenolic compounds. Essential oils are typically composed of smaller, volatile compounds with low molecular weights, which make them more readily encapsulated in contrast to larger, more complex phenolic compounds. Furthermore, the hydrophobic nature of essential oils facilitates their encapsulation in hydrophobic matrices such as those formed by pectin and casein in complex coacervation [27]. In line with this, García-Saldaña et al. [79] encapsulated broccoli seed extract rich in sulforaphane using complex coacervation with two biopolymer combinations (gelatin/pectin and gelatin/gum Arabic), reporting higher yield values of 81% and 85%, respectively. However, their encapsulation efficiency was lower than that observed in this study, with values of 18% and 12%, respectively. Similarly, De Souza et al. [61] achieved encapsulation yield and efficiency values close to those of our study, with values of 65% and 70%, respectively, when encapsulating cinnamon bark extract in gelatin and κ-carrageenan. In contrast, Chaib et al. [58] encapsulated *Thymus vulgaris* essential oil in gelatin B and gum Arabic, reporting an efficiency of 98%, which aligns closely with our results, although the yield (66%) was slightly lower than ours.

These differences are consistent with the typical outcomes observed in microencapsulation studies using complex coacervation. The porous surface observed on the PE-MC microcapsules, in contrast to the smoother surface of EO-MC, can be explained by the physical and chemical characteristics of the encapsulated compounds. Phenolic compounds, being larger and more complex than compounds present in essential oils, may influence the formation and structure of the microcapsules during coacervation. The uneven distribution or aggregation of phenolic compounds within the microcapsule core likely contributes to the creation of void spaces or pores on the surface during drying. Additionally, the solubility and interaction differences between the encapsulating biopolymers (casein and pectin) and the core compounds (essential oil and phenolic extract) may lead to variations in the surface morphology. Essential oils, composed of smaller and more homogeneous molecules, may result in a smoother surface due to more uniform interactions with the encapsulating agents, whereas phenolic extracts, with their larger and more heterogeneous composition, tend to create a more porous surface. This observation is supported by previous studies that have examined the morphology of microcapsules encapsulating essential oils and phenolic extracts from various plant sources [80–84].

Notably, the particle sizes observed in this study were considerably smaller compared to those reported in the literature. For example, Silva et al. [85] encapsulated green tea phenolic extract using cashew gum/maltodextrin and reported an average particle size of 10 μm. Campini et al. [81] observed microcapsules of *Cinnamomum cassia* essential oil encapsulated in polylactic acid with an average particle size of 80 μm. In comparison, our study achieved smaller microcapsules, especially for essential oils, demonstrating the effectiveness of the current encapsulation technique and the biopolymer combination used. This result is also consistent with our previous study on the microencapsulation of *Citrus limon* EO, which yielded spherical-shaped microcapsules with an average size of 25 μm [37], highlighting the suitability of the chosen biopolymers for achieving smaller particle sizes.

The particle size of microcapsules plays a crucial role in determining their functionality and application potential. In this study, the microcapsules exhibited a particle size of less than 10 μm, which, while beneficial for enhancing surface area, solubility, and bioavailability, may also lead to increased interparticle interactions that could affect their flowability and dispersibility. To address these potential challenges, various strategies can be employed, including the incorporation of anti-caking agents, surface modifications, or controlled agglomeration during processing. Moreover, the specific application of these microcapsules will dictate whether their small size presents a limitation or an advantage. In pharmaceutical and food formulations, for example, finer particles can facilitate improved dissolution rates and controlled release properties, making them highly desirable. Therefore, despite potential concerns regarding interparticle interactions, the reduced particle size of the microcapsules remains advantageous for their intended applications.

Moisture content is a critical factor influencing the stability and shelf life of microcapsules. Complex coacervation generally results in microcapsules with lower moisture content due to water evaporation during the encapsulation process. In our study, moisture content was relatively low in both EO-MC and PE-MC microcapsules. This is in contrast to the findings of Silva et al. [86], who reported a moisture content of 5.96% in microcapsules of lycopene encapsulated in gelatin/pectin. Vonghirundecha et al. [87] reported a moisture content ranging from 5.10% to 6.65% for Moringa oleifera phenolic extract encapsulated in maltodextrin, which was notably higher than our values for PE-MC. Conversely, Wang et al. [88] observed a moisture content of 3.70% in microcapsules of tuna oil encapsulated in gelatin/sodium hexametaphosphate. Low moisture content in microcapsules is particularly beneficial in food applications, as it helps prevent microbial growth and reduces biochemical reactions, thereby enhancing the shelf life of products [28].

The thermogravimetric analysis (TGA) of microcapsules revealed distinct thermal events corresponding to various stages of degradation. The first stage of weight loss is attributed to the evaporation of residual water and the release of volatile compounds that are not encapsulated. This initial loss reflects heat absorption and is consistent with typical moisture loss or desorption phenomena observed at lower temperatures [51]. In the second stage, both the encapsulated compounds and the wall materials undergo thermal degradation, involving the breakdown of proteins and polysaccharides. This is indicative of the thermal decomposition of the biopolymers that make up the wall materials, leading to significant mass loss. This observation is consistent with the findings reported in the literature, such as those of Radünz et al. [69] and Tavares and Zapata Noreña [89], who describe similar degradation processes for proteins and polysaccharides under elevated temperatures. The thermal degradation patterns observed in this study align with previous research on the microencapsulation of volatile and phenolic compounds, which has demonstrated that encapsulation enhances the thermal stability of bioactive substances [51,90–92]. The microcapsules' ability to protect the encapsulated compounds from degradation at high temperatures supports the notion that encapsulation is an effective strategy for improving the thermal stability of sensitive bioactive agents.

An X-ray diffraction (XRD) analysis of the microcapsules revealed important crystallographic changes. The broad peaks observed in the pectin/casein complex suggest an amorphous structure, likely due to the high casein ratio and the preparation method, which inhibits crystallization [53,66]. The appearance of new peaks in the EO-MC pattern, particularly at 31.8°, 45.5°, and 56.6°, suggests pectin crystallization induced by the essential oil. The shift and increased intensity of the peak at 21.75° further indicate changes in casein's crystallinity, reflecting the impact of the encapsulation process on the polymer complex. The enhanced crystallinity of pectin in the PE-MC formulation is likely a result of interactions between phenolic substances and pectin, promoting a more ordered structure [53]. However, the decreased intensity of the casein peak in the PE-MC pattern may be due to reduced interactions between pectin and casein, caused by the increased crystallinity of pectin [68]. These crystallographic changes are crucial because they affect the microcapsules' structural integrity, potentially influencing the release profiles of the active ingredients. Enhanced crystallinity in the polymer matrix may provide a better barrier effect, slowing down the release and contributing to the stability of the microcapsules [93].

Finally, the results of FTIR spectroscopy confirmed the successful encapsulation of basil essential oil (EO) and phenolic extract (PE) within the pectin/casein microcapsules. The characteristic peaks corresponding to the molecular structures of the essential oil and phenolic compounds were observed in the spectra of the microcapsules, confirming their presence. Specifically, the basil EO exhibited distinctive peaks in the 3000–2800 cm^{-1} and 3750–3500 cm^{-1} ranges, attributed to C–H stretching vibrations and O–H/N–H vibrations, respectively. The strong C–O stretching vibration band was detected between 1200 and 1100 cm^{-1}, while peaks from 1600 to 1000 cm^{-1} confirmed the oil's aromatic nature. Similarly, the presence of phenolic compounds in the microcapsules was confirmed by absorption bands in the 1800–750 cm^{-1} range, corresponding to C–OH, C–O–C, and C–C stretching vibrations. Furthermore, modifications in the spectra of the microcapsules, particularly the reduction or shifting of peaks associated with the functional groups of pectin and casein, suggest molecular interactions between the encapsulated compounds and the biopolymer matrix. The disappearance or shift of peaks in the regions associated with pectin's free carboxyl groups (-COOH, ~1730 cm^{-1}) and casein's amine groups ($-NH_2$, ~1550 cm^{-1}) indicates the formation of amide bonds, confirming electrostatic interactions between pectin and casein in the coacervation process. Additionally, the hydrogen bonding interactions between phenolic compounds and the biopolymer network are supported by changes in the 3700–3300 cm^{-1} region, associated with O–H stretching vibrations. These results are consistent with previous studies that have demonstrated successful encapsulation processes for volatile and phenolic compounds [78,94]. The successful encapsulation process and the chemical stability of the bioactive compounds further support the potential applications of these microcapsules in pharmacology and nutraceuticals [88,94].

5. Conclusions

This study effectively evaluated the complex coacervation method using a combination of biopolymers (casein and pectin) for the microencapsulation of bioactive compounds derived from *O. basilicum* L. The resulting microparticles of basil essential oil and phenolic extract exhibited promising characteristics. EO-MC demonstrated superior microencapsulation yield (85.3%) and efficiency (89.8%) compared to PE-MC (53.1% and 70.0%) for yield and efficiency, respectively, while both types displayed spherical morphology with PE-MC exhibiting a notably porous surface texture. An analysis revealed smaller particle sizes for PE-MC (6.95 ± 1.94 µm) compared to EO-MC (9.70 ± 2.82 µm). The moisture content aligned with the food industry standards for dried powders with values less than 5% for both EO-MC and PE-MC. Thermogravimetric and differential scanning calorimetry analysis indicated two stages of weight loss for both EO-MC and PE-MC, reflecting the evaporation of residual water and volatile compounds followed by the decomposition of microencapsulated compounds and wall materials. X-ray diffraction analysis showed enhanced crystallinity in EO-MC due to essential oil-induced re-crystallization and increased pectin crystallinity in PE-MC attributed to interactions with phenolic substances, impacting drug release and microcapsule stability. The results of FTIR spectroscopy confirmed molecular interactions and the formation of new chemical bonds between encapsulated compounds and wall materials. Specifically, hydrogen bonds are likely formed between phenolic compounds and the biopolymer matrix, and amide bonds are generated through interactions between the carboxyl groups of pectin and the amino groups of casein, facilitating the encapsulation of the bioactive compounds. Overall, these findings underscore the potential of the complex coacervation method using these biopolymers for the effective microencapsulation of bioactive compounds from plant sources, especially essential oils, with implications for various applications in food, nutraceuticals, and pharmaceuticals.

Author Contributions: Conceptualization, S.H., H.M., M.M.M. and I.B.; Methodology, S.H., M.M.B. and H.T.; Software, H.M., I.B. and H.T.; Validation, S.H. and H.M.; Formal analysis, S.H. and M.M.B.; Investigation, S.H., I.B., N.D. and A.A. (Amel Attia); Resources, M.M.M., N.F.O., M.M.B. and A.A. (Abdeltif Amrane); Writing—review & editing, S.H., H.M., N.D., A.A. (Amel Attia), N.F.O., H.T. and A.A. (Abdeltif Amrane); Supervision, M.M.M., N.F.O. and A.A. (Abdeltif Amrane). All authors have read and agreed to the published version of the manuscript.

Funding: This research received no external funding.

Institutional Review Board Statement: Not applicable.

Data Availability Statement: All data generated or analyzed during this study are included in this published article.

Acknowledgments: This study was supported by the Ministry of Higher Education and Scientific Research and the Directorate-General for Scientific Research and Technological Development of Algeria.

Conflicts of Interest: Author Ismail Berrabah was employed by the company "EURL TELEGHMA PLAST". The remaining authors declare that the research was conducted in the absence of any commercial or financial re-lationships that could be construed as a potential conflict of interest.

Abbreviations

The following abbreviations are used in this manuscript:

XRD	X-ray diffraction
TGA	Thermogravimetric analysis
FTIR	Fourier transform infrared
EO	Basil essential oil
PE	Phenolic extract
PE-MC	PE microcapsules
EO-MC	EO microcapsules

References

1. Erez, E.; Bayramoglu, B. Evaluation of the Extracts of Purple Basil (*Ocimum basilicum* L.) as Natural pH-Indicator Dyes Anticipated to Be Utilised in Intelligent Food Packaging—An Optimisation Study. *Color. Technol.* **2024**, *140*, 937–951. [CrossRef]
2. Falowo, A.B.; Mukumbo, F.E.; Idamokoro, E.M.; Afolayan, A.J.; Muchenje, V. Phytochemical Constituents and Antioxidant Activity of Sweet Basil (*Ocimum basilicum* L.) Essential Oil on Ground Beef from Boran and Nguni Cattle. *Int. J. Food Sci.* **2019**, *2019*, 2628747. [CrossRef] [PubMed]
3. Licon, C.C.; Moro, A.; Librán, C.M.; Molina, A.M.; Zalacain, A.; Berruga, M.I.; Carmona, M. Volatile Transference and Antimicrobial Activity of Cheeses Made with Ewes' Milk Fortified with Essential Oils. *Foods* **2020**, *9*, 35. [CrossRef] [PubMed]
4. Prinsi, B.; Morgutti, S.; Negrini, N.; Faoro, F.; Espen, L. Insight into Composition of Bioactive Phenolic Compounds in Leaves and Flowers of Green and Purple Basil. *Plants* **2020**, *9*, 22. [CrossRef]

5. Yahoum, M.M.; Toumi, S.; Tahraoui, H.; Lefnaoui, S.; Kebir, M.; Amrane, A.; Assadi, A.A.; Zhang, J.; Mouni, L. Formulation and Evaluation of Xanthan Gum Microspheres for the Sustained Release of Metformin Hydrochloride. *Micromachines* **2023**, *14*, 609. [CrossRef]
6. Dhama, K.; Sharun, K.; Gugjoo, M.B.; Tiwari, R.; Alagawany, M.; Iqbal Yatoo, M.; Thakur, P.; Iqbal, H.M.N.; Chaicumpa, W.; Michalak, I.; et al. A Comprehensive Review on Chemical Profile and Pharmacological Activities of *Ocimum basilicum*. *Food Rev. Int.* **2023**, *39*, 119–147. [CrossRef]
7. Chu, H.T.; Vu, T.N.; Dinh, T.T.; Do, P.T.; Chu, H.H.; Tien, T.Q.; Tong, Q.C.; Nguyen, M.H.; Ha, Q.T.; Setzer, W.N. Effects of supplemental light spectra on the composition, production and antimicrobial activity of *Ocimum basilicum* L. essential oil. *Molecules* **2022**, *27*, 5599. [CrossRef]
8. Comite, E.; El-Nakhel, C.; Rouphael, Y.; Ventorino, V.; Pepe, O.; Borzacchiello, A.; Vinale, F.; Rigano, D.; Staropoli, A.; Lorito, M.; et al. Bioformulations with Beneficial Microbial Consortia a Bioactive Compound and Plant Biopolymers Modulate Sweet Basil Productivity Photosynthetic Activity and Metabolites. *Pathogens* **2021**, *10*, 870. [CrossRef]
9. Jordán, M.J.; Quílez, M.; Luna, M.C.; Bekhradi, F.; Sotomayor, J.A.; Sánchez-Gómez, P.; Gil, M.I. Influence of Water Stress and Storage Time on Preservation of the Fresh Volatile Profile of Three Basil Genotypes. *Food Chem.* **2017**, *221*, 169–177. [CrossRef]
10. Bobakulov, K.; Ozek, G.; Ozek, T.; Asilbekova, D.T.; Abdullaev, N.D.; Sagdullaev, S.S.; Başer, K.H.C. Essential Oils and Lipids from the Flowers of Two Varieties of *Ocimum basilicum* L. Cultivated in Uzbekistan. *J. Essent. Oil Res.* **2020**, *32*, 323–330. [CrossRef]
11. Moayedi, F.; Kordi, S.; Mehrabi, A.A. Evaluation of Yield, Chemical Composition and Yield of Essential Oil of Four Cultivars of Sweet Basil (*Ocimum basilicum* L.) Affected by Different Levels of Nitrogen. *Acta Agric. Slov.* **2020**, *115*, 171–181. [CrossRef]
12. Ben Hadj Tahar, D.; Triki, Z.; Guendouz, M.; Tahraoui, H.; Zamouche, M.; Kebir, M.; Zhang, J.; Amrane, A. Characterization and Thermal Evaluation of a Novel Bio-BasedNatural Insulation Material from Posidonia Oceanica Waste: A Sustainable Solution for Building Insulation. *ChemEngineering* **2024**, *8*, 18. [CrossRef]
13. Solouki, A.; Zare Mehrjerdi, M.; Azimi, R.; Aliniaeifard, S. Improving Basil (*Ocimum basilicum* L.) Essential Oil Yield Following down-Regulation of Photosynthetic Functionality by Short-Term Application of Abiotic Elicitors. *Biocatal. Agric. Biotechnol.* **2023**, *50*, 102675. [CrossRef]
14. Boudoukhani, M.; Yahoum, M.M.; Ezzroug, K.; Toumi, S.; Lefnaoui, S.; Moulai-Mostefa, N.; Sid, A.N.E.H.; Tahraoui, H.; Kebir, M.; Amrane, A. Formulation and Characterization of Double Emulsions W/O/W Stabilized by Two Natural Polymers with Two Manufacturing Processes (Comparative Study). *ChemEngineering* **2024**, *8*, 34. [CrossRef]
15. Ferradj, S.; Yahoum, M.M.; Rebiha, M.; Nabi, I.; Toumi, S.; Lefnaoui, S.; Hadj-Ziane-Zafour, A.; Touzout, N.; Tahraoui, H.; Mihoub, A. Nanocurcumin-Based Sugar-Free Formulation: Development and Impact on Diabetes and Oxidative Stress Reduction. *Nanomaterials* **2024**, *14*, 1105. [CrossRef]
16. Đorđević, N.; Mančić, S.; Karabegović, I.; Cvetković, D.; Stanojević, J.; Savić, D.; Danilović, B. The Effect of Distillation Methods on the Yield, Composition and Biological Activity of Basil (*Ocimum basilicum* L.) Essential Oil. *Adv. Technol.* **2022**, *11*, 16–25. [CrossRef]
17. Abdel-Tawwab, M.; Adeshina, I.; Emikpe, B.O.; Jenyo-Oni, A.; Ajani, E.K.; Tiamiyu, L.O. Effect of Dietary Clove Basil, Ocimum Gratissimum, Leaves Extract on Healing of Artificially Wounded African Catfish, Clarias Gariepinus (B.), Juveniles. *J. Appl. Aquac.* **2019**, *31*, 289–300. [CrossRef]
18. Dahchour, A. Anxiolytic and Antidepressive Potentials of Rosmarinic Acid: A Review with a Focus on Antioxidant and Anti-Inflammatory Effects. *Pharmacol. Res.* **2022**, *184*, 106421. [CrossRef]
19. Wang, L.; Wang, H.; Chen, J.; Qin, Z.; Yu, S.; Zhou, J. Coordinating Caffeic Acid and Salvianic Acid A Pathways for Efficient Production of Rosmarinic Acid in *Escherichia coli*. *Metab. Eng.* **2023**, *76*, 29–38. [CrossRef]
20. Aloisio, C.; Razola-d, C.; Longhi, M.R.; Andreatta, A.E.; Verardo, V. Optimization of the Extraction Conditions of Bioactive Compounds from *Ocimum basilicum* Leaves Using Ultrasound-Assisted Extraction via a Sonotrode. *Molecules* **2023**, *28*, 5286. [CrossRef]
21. Lin, S.Y.; Wang, Y.Y.; Chen, W.Y.; Liao, S.L.; Chou, S.T.; Yang, C.P.; Chen, C.J. Hepatoprotective Activities of Rosmarinic Acid against Extrahepatic Cholestasis in Rats. *Food Chem. Toxicol.* **2017**, *108*, 214–223. [CrossRef] [PubMed]
22. Quan, W.; Liu, H.X.; Zhang, W.; Lou, W.J.; Gong, Y.Z.; Yuan, C.; Shao, Q.; Wang, N.; Guo, C.; Liu, F. Cardioprotective Effect of Rosmarinic Acid against Myocardial Ischaemia/Reperfusion Injury via Suppression of the NF-κB Inflammatory Signalling Pathway and ROS Production in Mice. *Pharm. Biol.* **2021**, *59*, 220–229. [CrossRef] [PubMed]
23. Zhang, Z.K.; Zhou, Y.; Cao, J.; Liu, D.Y.; Wan, L.H. Rosmarinic Acid Ameliorates Septic-Associated Mortality and Lung Injury in Mice via GRP78/IRE1α/JNK Pathway. *J. Pharm. Pharmacol.* **2021**, *73*, 916–921. [CrossRef] [PubMed]
24. Chadni, M.; Isidore, E.; Lagalle, F.; Langlait, M.; Dosso, A.; Ioannou, I. Optimization of the Supercritical Extraction of Rosmarinic Acid from Clary Sage Residue and the Antioxidant Activity of the Extracts. *J. Supercrit. Fluids* **2023**, *193*, 105830. [CrossRef]

25. Rodrigues, L.B.; Oliveira Brito Pereira Bezerra Martins, A.; Cesário, F.R.A.S.; Ferreira e Castro, F.; de Albuquerque, T.R.; Martins Fernandes, M.N.; Fernandes da Silva, B.A.; Quintans Júnior, L.J.; da Costa, J.G.M.; Melo Coutinho, H.D.; et al. Anti-Inflammatory and Antiedematogenic Activity of the *Ocimum basilicum* Essential Oil and Its Main Compound Estragole: In Vivo Mouse Models. *Chem.-Biol. Interact.* **2016**, *257*, 14–25. [CrossRef]
26. González Moreno, A.; Guzman-Puyol, S.; Domínguez, E.; Benítez, J.J.; Segado, P.; Lauciello, S.; Ceseracciu, L.; Porras-Vázquez, J.M.; Leon-Reina, L.; Heredia, A.; et al. Pectin-Cellulose Nanocrystal Biocomposites: Tuning of Physical Properties and Biodegradability. *Int. J. Biol. Macromol.* **2021**, *180*, 709–717. [CrossRef]
27. Brito de Souza, V.; Thomazini, M.; Chaves, I.E.; Ferro-Furtado, R.; Favaro-Trindade, C.S. Microencapsulation by Complex Coacervation as a Tool to Protect Bioactive Compounds and to Reduce Astringency and Strong Flavor of Vegetable Extracts. *Food Hydrocoll.* **2020**, *98*, 105244. [CrossRef]
28. Lan, Y.; Ohm, J.B.; Chen, B.; Rao, J. Microencapsulation of Hemp Seed Oil by Pea Protein Isolate−sugar Beet Pectin Complex Coacervation: Influence of Coacervation pH and Wall/Core Ratio. *Food Hydrocoll.* **2021**, *113*, 106423. [CrossRef]
29. Singh, J.; Kaur, K.; Kumar, P. Optimizing Microencapsulation of α-Tocopherol with Pectin and Sodium Alginate. *J. Food Sci. Technol.* **2018**, *55*, 3625–3631. [CrossRef]
30. Sood, A.; Saini, C.S. Utilization of Peel of White Pomelo for the Development of Pectin Based Biodegradable Composite Films Blended with Casein and Egg Albumen. *Food Chem. Adv.* **2022**, *1*, 100054. [CrossRef]
31. Wusigale; Liang, L.; Luo, Y. Casein and Pectin: Structures, Interactions, and Applications. *Trends Food Sci. Technol.* **2020**, *97*, 391–403. [CrossRef]
32. Awasthi, R.; Kulkarni, G.T.; Ramana, M.V.; de Jesus Andreoli Pinto, T.; Kikuchi, I.S.; Molim Ghisleni, D.D.; de Souza Braga, M.; De Bank, P.; Dua, K. Dual Crosslinked Pectin–Alginate Network as Sustained Release Hydrophilic Matrix for Repaglinide. *Int. J. Biol. Macromol.* **2017**, *97*, 721–732. [CrossRef]
33. Yun, P.; Devahastin, S.; Chiewchan, N. Microstructures of Encapsulates and Their Relations with Encapsulation Efficiency and Controlled Release of Bioactive Constituents: A Review. *Compr. Rev. Food Sci. Food Saf.* **2021**, *20*, 1768–1799. [CrossRef] [PubMed]
34. Muhoza, B.; Qi, B.; Harindintwali, J.D.; Farag Koko, M.Y.; Zhang, S.; Li, Y. Combined Plant Protein Modification and Complex Coacervation as a Sustainable Strategy to Produce Coacervates Encapsulating Bioactives. *Food Hydrocoll.* **2022**, *124*, 107239. [CrossRef]
35. Warnakulasuriya, S.N.; Nickerson, M.T. Review on Plant Protein–Polysaccharide Complex Coacervation, and the Functionality and Applicability of Formed Complexes. *J. Sci. Food Agric.* **2018**, *98*, 5559–5571. [CrossRef]
36. Eghbal, N.; Choudhary, R. Complex Coacervation: Encapsulation and Controlled Release of Active Agents in Food Systems. *LWT* **2018**, *90*, 254–264. [CrossRef]
37. Djihad, N.; Naima, F.O.; Petronilho, S.; Hamid, S.; Bedjou, F.N.E.; Coimbra, M.A. Microencapsulation of Citrus Limon Essential Oil by Complex Coacervation and Release Behavior of Terpenic and Derived Volatile Compounds. *Food Hydrocoll.* **2024**, *152*, 109830. [CrossRef]
38. Guo, Q.; Li, S.; Du, G.; Chen, H.; Yan, X.; Chang, S.; Yue, T.; Yuan, Y. Formulation and Characterization of Microcapsules Encapsulating Carvacrol Using Complex Coacervation Crosslinked with Tannic Acid. *LWT* **2022**, *165*, 113683. [CrossRef]
39. Muhoza, B.; Xia, S.; Cai, J.; Zhang, X.; Duhoranimana, E.; Su, J. Gelatin and Pectin Complex Coacervates as Carriers for Cinnamaldehyde: Effect of Pectin Esterification Degree on Coacervate Formation, and Enhanced Thermal Stability. *Food Hydrocoll.* **2019**, *87*, 712–722. [CrossRef]
40. Lan, Y.; Ohm, J.B.; Chen, B.; Rao, J. Phase Behavior and Complex Coacervation of Concentrated Pea Protein Isolate-Beet Pectin Solution. *Food Chem.* **2020**, *307*, 125536. [CrossRef]
41. Ozdemir, N.; Bayrak, A.; Tat, T.; Altay, F.; Kiralan, M.; Kurt, A. Microencapsulation of Basil Essential Oil: Utilization of Gum Arabic/Whey Protein Isolate/Maltodextrin Combinations for Encapsulation Efficiency and in Vitro Release. *J. Food Meas. Charact.* **2021**, *15*, 1865–1876. [CrossRef]
42. Paul, V.; Arvind; Rai, D.C.; Pandhi, S.; Seth, A. Effect of Coating Materials for Microencapsulation of Basil Oil Using Spray Drying. *Med. Plants* **2020**, *12*, 251–257. [CrossRef]
43. Fadel, H.H.M.; El-Ghorab, A.H.; Hussein, A.M.S.; El-Massry, K.F.; Lotfy, S.N.; Sayed Ahmed, M.Y.; Soliman, T.N. Correlation between Chemical Composition and Radical Scavenging Activity of 10 Commercial Essential Oils: Impact of Microencapsulation on Functional Properties of Essential Oils. *Arab. J. Chem.* **2020**, *13*, 6815–6827. [CrossRef]
44. Thuekeaw, S.; Angkanaporn, K.; Chirachanchai, S.; Nuengjamnong, C. Dual pH Responsive via Double—Layered Microencapsulation for Controlled Release of Active Ingredients in Simulated Gastrointestinal Tract: A Model Case of Chitosan-Alginate Microcapsules Containing Basil Oil (*Ocimum basilicum* Linn.). *Polym. Degrad. Stab.* **2021**, *191*, 109660. [CrossRef]
45. Tomé, A.C.; da Silva, F.A. Alginate Based Encapsulation as a Tool for the Protection of Bioactive Compounds from Aromatic Herbs. *Food Hydrocoll. Health* **2022**, *2*, 100051. [CrossRef]

46. Moussa, H.; Dahmoune, F.; Mróz, M.; Remini, H.; Kadri, N.; Hamid, S.; Kusznierewicz, B. Efficient Optimization Approaches for Microwave Assisted Extraction of High-Quality Antioxidant Compounds from *Salvia officinalis* L.: UHPLC-HRMS Differential Analysis of Phenolic Profiles Obtained by Ultrasound and Microwave Extraction. *Sustain. Chem. Pharm.* **2023**, *35*, 101194. [CrossRef]
47. Bellik, F.Z.; Benkaci-Ali, F.; Alsafra, Z.; Eppe, G.; Tata, S.; Sabaou, N.; Zidani, R. Chemical Composition, Kinetic Study and Antimicrobial Activity of Essential Oils from *Cymbopogon schoenanthus* L. Spreng Extracted by Conventional and Microwave-Assisted Techniques Using Cryogenic Grinding. *Ind. Crops Prod.* **2019**, *139*, 111505. [CrossRef]
48. Moussa, H.; Dahmoune, F.; Hentabli, M.; Remini, H.; Mouni, L. Optimization of Ultrasound-Assisted Extraction of Phenolic-Saponin Content from *Carthamus caeruleus* L. Rhizome and Predictive Model Based on Support Vector Regression Optimized by Dragonfly Algorithm. *Chemom. Intell. Lab. Syst.* **2022**, *222*, 104493. [CrossRef]
49. Pedrosa, M.C.; Ueda, J.M.; Heleno, S.; Melgar, B.; Ivanov, M.; Soković, M.; Carocho, M.; Ferreira, I.C.F.R.; Barros, L. Antimicrobial Activity of Aqueous Plant Extracts as Potential Natural Additives. *Proceedings* **2021**, *70*, 79. [CrossRef]
50. Moussa, H.; Dahmoune, F.; Mahdjoub, M.M.; Kadri, N.; Remini, H. Definitive Screening Design and I-Optimal Design for Optimization of Ultrasound-Assisted Extraction of Phenolic Content and Antioxidant Capacity from *Salvia officinalis* L. Leaves. *Sustain. Chem. Pharm.* **2022**, *29*, 100820. [CrossRef]
51. Medeiros, D.C.D.; Mizokami, S.S.; Sfeir, N.; Georgetti, S.R.; Urbano, A.; Casagrande, R.; Verri, W.A.; Baracat, M.M. Preclinical Evaluation of Rutin-Loaded Microparticles with an Enhanced Analgesic Effect. *ACS Omega* **2019**, *4*, 1221–1227. [CrossRef]
52. Guazelli, C.F.S.; Fattori, V.; Colombo, B.B.; Ludwig, I.S.; Vicente, L.G.; Martinez, R.M.; Georgetti, S.R.; Urbano, A.; Casagrande, R.; Baracat, M.M.; et al. Development of Trans-Chalcone Loaded Pectin/Casein Biodegradable Microcapsules: Efficacy Improvement in the Management of Experimental Colitis. *Int. J. Pharm.* **2023**, *642*, 123206. [CrossRef] [PubMed]
53. Mebarki, N.; Ziane, H.; Fazouane, F.; Nabiev, M. Microencapsulation of Thymus Fontanesii Extracts in Pectin/Casein: Characterization, Release Behavior and Storage Stability. *Iran. Polym. J. (Engl. Ed.)* **2022**, *31*, 301–316. [CrossRef]
54. Feyzioglu, G.C.; Tornuk, F. Development of Chitosan Nanoparticles Loaded with Summer Savory (*Satureja hortensis* L.) Essential Oil for Antimicrobial and Antioxidant Delivery Applications. *LWT* **2016**, *70*, 104–110. [CrossRef]
55. Hamid, S.; Oukil, N.F.; Moussa, H.; Mahdjoub, M.M.; Djihad, N.; Berrabah, I.; Bouhenna, M.M.; Chebrouk, F.; Hentabli, M. Enhancing Basil Essential Oil Microencapsulation Using Pectin/Casein Biopolymers: Optimization through D-Optimal Design, Controlled Release Modeling, and Characterization. *Int. J. Biol. Macromol.* **2024**, *265*, 130948. [CrossRef]
56. Benammar, S.; Haffas, M.; Hamitouche, A.; Boudjemaa, A.; Bachari, K. Relevance of Anethum Graveolens to Remove Rhodamine B in Aqueous Solution: Characterization, Kinetic and Isotherm Study. *React. Kinet. Mech. Catal.* **2023**, *136*, 465–490. [CrossRef]
57. Bakry, A.M.; Huang, J.; Zhai, Y.; Huang, Q. Myofibrillar Protein with κ- or λ-Carrageenans as Novel Shell Materials for Microencapsulation of Tuna Oil through Complex Coacervation. *Food Hydrocoll.* **2019**, *96*, 43–53. [CrossRef]
58. Chaib, S.; Benali, N.; Arhab, R.; Sadraoui Ajmi, I.; Bendaoued, H.; Romdhane, M. Preparation of *Thymus vulgaris* Essential Oil Microcapsules by Complex Coacervation and Direct Emulsion: Synthesis, Characterization and Controlled Release Properties. *Arab. J. Sci. Eng.* **2021**, *46*, 5429–5446. [CrossRef]
59. Hosseini, S.F.; Zandi, M.; Rezaei, M.; Farahmandghavi, F. Two-Step Method for Encapsulation of Oregano Essential Oil in Chitosan Nanoparticles: Preparation, Characterization and in Vitro Release Study. *Carbohydr. Polym.* **2013**, *95*, 50–56. [CrossRef]
60. Moussa, H.; Hamid, S.; Mameri, A.; Lekmine, S.; Tahraoui, H.; Kebir, M.; Touzout, N.; Dahmoune, F.; Ola, M.S.; Zhang, J.; et al. From Green Chemistry to Healthy Environments: Silver Nanoparticles as a Dual Antioxidant and Antibacterial Agents for Advancing Biomedicine and Sustainable Wastewater Treatment. *Bioengineering* **2024**, *11*, 1205. [CrossRef]
61. de Souza, V.B.; Thomazini, M.; Echalar Barrientos, M.A.; Nalin, C.M.; Ferro-Furtado, R.; Genovese, M.I.; Favaro-Trindade, C.S. Functional Properties and Encapsulation of a Proanthocyanidin-Rich Cinnamon Extract (*Cinnamomum zeylanicum*) by Complex Coacervation Using Gelatin and Different Polysaccharides. *Food Hydrocoll.* **2018**, *77*, 297–306. [CrossRef]
62. Teofilović, B.; Grujić-Letić, N.; Gligorić, E.; Rašković, A.; Igić, R.; Vastag, G.; Gadžurić, S. Experimental and Computational Evaluation of Extraction Procedure and Scavenging Capacity of Sweet Basil Extracts (*Ocimum basilicum* L.). *Plant Foods Hum. Nutr.* **2021**, *76*, 240–247. [CrossRef]
63. Djihad, N.; Fadloun, N.; Hamid, S.; Attia, A.; Petronilho, S. Microencapsulation of a Green Microalga (*Chlorella vulgaris*) by Complex Coacervation for Its Valuation as a Feasible Ingredient in Pear Snacks. *Algal Res.* **2024**, *83*, 103727. [CrossRef]
64. Trinh, T.H.; Shaari, K.Z.K.; Basit, A.; Azeem, B. Effect of Particle Size and Coating Thickness on the Release of Urea Using Multi-Diffusion Model. *Int. J. Chem. Eng. Appl.* **2014**, *5*, 58–63. [CrossRef]
65. Toledo Hijo, A.A.C.; Da Costa, J.M.G.; Silva, E.K.; Azevedo, V.M.; Yoshida, M.I.; Borges, S.V. Physical and Thermal Properties of Oregano (*Origanum vulgare* L.) Essential Oil Microparticles. *J. Food Process Eng.* **2015**, *38*, 1–10. [CrossRef]
66. Liu, X.; Renard, C.M.G.C.; Bureau, S.; Le Bourvellec, C. Revisiting the Contribution of ATR-FTIR Spectroscopy to Characterize Plant Cell Wall Polysaccharides. *Carbohydr. Polym.* **2021**, *262*, 117935. [CrossRef] [PubMed]
67. Kumar, M.; Mishra, R.K.; Banthia, A.K. Development of Pectin Based Hydrogel Membranes for Biomedical Applications. *Int. J. Plast. Technol.* **2010**, *14*, 213–223. [CrossRef]

68. Souza, K.S.; Moreira, L.S.; Silva, B.T.; Oliveira, B.P.M.; Carvalho, A.S.; Silva, P.S.; Verri, W.A.; Sá-Nakanishi, A.B.; Bracht, L.; Zanoni, J.N.; et al. Low Dose of Quercetin-Loaded Pectin/Casein Microparticles Reduces the Oxidative Stress in Arthritic Rats. *Life Sci.* **2021**, *284*, 119910. [CrossRef]
69. Radünz, M.; dos Santos Hackbart, H.C.; Camargo, T.M.; Nunes, C.F.P.; de Barros, F.A.P.; Dal Magro, J.; Filho, P.J.S.; Gandra, E.A.; Radünz, A.L.; da Rosa Zavareze, E. Antimicrobial Potential of Spray Drying Encapsulated Thyme (*Thymus vulgaris*) Essential Oil on the Conservation of Hamburger-like Meat Products. *Int. J. Food Microbiol.* **2020**, *330*, 108696. [CrossRef]
70. Pinilla, C.M.B.; Noreña, C.P.Z.; Brandelli, A. Development and Characterization of Phosphatidylcholine Nanovesicles, Containing Garlic Extract, with Antilisterial Activity in Milk. *Food Chem.* **2017**, *220*, 470–476. [CrossRef]
71. Okur, İ.; Baltacıoğlu, C.; Ağçam, E.; Baltacıoğlu, H.; Alpas, H. Evaluation of the Effect of Different Extraction Techniques on Sour Cherry Pomace Phenolic Content and Antioxidant Activity and Determination of Phenolic Compounds by FTIR and HPLC. *Waste Biomass Valorization* **2019**, *10*, 3545–3555. [CrossRef]
72. Baltacıoğlu, H.; Baltacıoğlu, C.; Okur, I.; Tanrıvermiş, A.; Yalıç, M. Optimization of Microwave-Assisted Extraction of Phenolic Compounds from Tomato: Characterization by FTIR and HPLC and Comparison with Conventional Solvent Extraction. *Vib. Spectrosc.* **2021**, *113*, 103204. [CrossRef]
73. He, J.; Rodriguez-Saona, L.E.; Giusti, M.M. Midinfrared Spectroscopy for Juice Authentication-Rapid Differentiation of Commercial Juices. *J. Agric. Food Chem.* **2007**, *55*, 4443–4452. [CrossRef] [PubMed]
74. Moussa, H.; Dahmoune, F.; Lekmine, S.; Mameri, A.; Tahraoui, H.; Hamid, S.; Benzitoune, N.; Moula, N.; Zhang, J.; Amrane, A. Optimization of Ultrasound-Assisted Extraction of Bioactive Compounds from *Carthamus caeruleus* L. Rhizome: Integrating Central Composite Design, Gaussian Process Regression, and Multi-Objective Grey Wolf Optimization Approaches. *Process Biochem.* **2024**, *147*, 476–488. [CrossRef]
75. Darrag, H.M.; Almuhanna, H.T.; Hakami, E.H. Secondary Metabolites in Basil, Bio-Insecticide, Inhibition Effect, and In Silico Molecular Docking against Proteolytic Enzymes of the Red Palm Weevil (*Rhynchophorus ferrugineus*). *Plants* **2022**, *11*, 1087. [CrossRef]
76. Fernandes, F.; Pereira, E.; Prieto, M.A.; Calhelha, R.C.; Ćirić, A.; Soković, M.; Simal-Gandara, J.; Barros, L.; Ferreira, I.C.F.R. Optimization of the Extraction Process to Obtain a Colorant Ingredient from Leaves of *Ocimum basilicum* Var. Purpurascens. *Molecules* **2019**, *24*, 686. [CrossRef]
77. Fayoumi, L.; Khalil, M.; Ghareeb, D.; Chokr, A.; Bouaziz, M.; El-Dakdouki, M.H. Phytochemical Constituents and Therapeutic Effects of the Essential Oil of Rose Geranium (*Pelargonium hybrid*) Cultivated in Lebanon. *S. Afr. J. Bot.* **2022**, *147*, 894–902. [CrossRef]
78. Rezzoug, M.; Bakchiche, B.; Gherib, A.; Roberta, A.; Guido, F.; Kilinçarslan, Ö.; Mammadov, R.; Bardaweel, S.K. Chemical Composition and Bioactivity of Essential Oils and Ethanolic Extracts of *Ocimum basilicum* L. and Thymus Algeriensis Boiss. & Reut. from the Algerian Saharan Atlas. *BMC Complement. Altern. Med.* **2019**, *19*, 146. [CrossRef]
79. García-Saldaña, J.S.; Campas-Baypoli, O.N.; López-Cervantes, J.; Sánchez-Machado, D.I.; Cantú-Soto, E.U.; Rodríguez-Ramírez, R. Microencapsulation of Sulforaphane from Broccoli Seed Extracts by Gelatin/Gum Arabic and Gelatin/Pectin Complexes. *Food Chem.* **2016**, *201*, 94–100. [CrossRef]
80. Calderón-Oliver, M.; Pedroza-Islas, R.; Escalona-Buendía, H.B.; Pedraza-Chaverri, J.; Ponce-Alquicira, E. Comparative Study of the Microencapsulation by Complex Coacervation of Nisin in Combination with an Avocado Antioxidant Extract. *Food Hydrocoll.* **2017**, *62*, 49–57. [CrossRef]
81. Campini, P.A.L.; de Oliveira, É.R.; Camani, P.H.; da Silva, C.G.; Yudice, E.D.C.; de Oliveira, S.A.; dos Santos Rosa, D. Assessing the Efficiency of Essential Oil and Active Compounds/Poly (Lactic Acid) Microcapsules against Common Foodborne Pathogens. *Int. J. Biol. Macromol.* **2021**, *186*, 702–713. [CrossRef] [PubMed]
82. Karimi Sani, I.; Alizadeh, M.; Pirsa, S.; Moghaddas Kia, E. Impact of Operating Parameters and Wall Material Components on the Characteristics of Microencapsulated Melissa Officinalis Essential Oil. *Flavour. Fragr. J.* **2019**, *34*, 104–112. [CrossRef]
83. Pozippe, M.; Favaro-trindade, C.S.; Matias, S.; Alencar, D.; Thomazini, M.; Camargo, J.C.D.; Contreras, C.J. LWT—Food Science and Technology Microencapsulation of Propolis Extract by Complex Coacervation. *LWT—Food Sci. Technol.* **2011**, *44*, 429–435. [CrossRef]
84. Rasteh, I.; Pirnia, M.; Miri, M.A.; Sarani, S. Encapsulation of Zataria Multiflora Essential Oil in Electrosprayed Zein Microcapsules: Characterization and Antimicrobial Properties. *Ind. Crops Prod.* **2024**, *208*, 117794. [CrossRef]
85. Silva, F.M.R.; Magalhaes, F.E.A.; Batista, F.L.A.; da Silva, L.M.R.; Ricardo, N.M.P.S.; Sabino, L.B.d.S.; de Figueiredo, R.W. Microencapsulation of Green Tea (*Camellia sinensis*) Phenolic Extract: Physical-Chemical Characterization, Antimicrobial and Toxicological Properties. *Food Chem. Adv.* **2023**, *3*, 100360. [CrossRef]
86. Silva, D.F.; Favaro-Trindade, C.S.; Rocha, G.A.; Thomazini, M. Microencapsulation of Lycopene by Gelatin-Pectin Complex Coacervation. *J. Food Process. Preserv.* **2012**, *36*, 185–190. [CrossRef]

87. Vonghirundecha, P.; Chusri, S.; Meunprasertdee, P.; Kaewmanee, T. Microencapsulated Functional Ingredients from a Moringa Oleifera Leaf Polyphenol-Rich Extract: Characterization, Antioxidant Properties, in Vitro Simulated Digestion, and Storage Stability. *LWT* **2022**, *154*, 112820. [CrossRef]
88. Wang, B.; Adhikari, B.; Barrow, C.J. Optimisation of the Microencapsulation of Tuna Oil in Gelatin-Sodium Hexametaphosphate Using Complex Coacervation. *Food Chem.* **2014**, *158*, 358–365. [CrossRef]
89. Tavares, L.; Zapata Noreña, C.P. Encapsulation of Garlic Extract Using Complex Coacervation with Whey Protein Isolate and Chitosan as Wall Materials Followed by Spray Drying. *Food Hydrocoll.* **2019**, *89*, 360–369. [CrossRef]
90. da Silva, F.T.; da Cunha, K.F.; Fonseca, L.M.; Antunes, M.D.; El Halal, S.L.M.; Fiorentini, Â.M.; da Rosa Zavareze, E.; Dias, A.R.G. Action of Ginger Essential Oil (*Zingiber officinale*) Encapsulated in Proteins Ultrafine Fibers on the Antimicrobial Control in Situ. *Int. J. Biol. Macromol.* **2018**, *118*, 107–115. [CrossRef]
91. Stojanovic, R.; Belscak-Cvitanovic, A.; Manojlovic, V.; Komes, D.; Nedovic, V.; Bugarski, B. Encapsulation of Thyme (*Thymus serpyllum* L.) Aqueous Extract in Calcium Alginate Beads. *J. Sci. Food Agric.* **2012**, *92*, 685–696. [CrossRef] [PubMed]
92. Su, H.; Huang, C.; Liu, Y.; Kong, S.; Wang, J.; Huang, H.; Zhang, B. Preparation and Characterization of Cinnamomum Essential Oil-Chitosan Nanocomposites: Physical, Structural, and Antioxidant Activities. *Processes* **2020**, *8*, 834. [CrossRef]
93. Jain, A.; Thakur, D.; Ghoshal, G.; Katare, O.P.; Shivhare, U.S. Characterization of Microcapsulated β-Carotene Formed by Complex Coacervation Using Casein and Gum Tragacanth. *Int. J. Biol. Macromol.* **2016**, *87*, 101–113. [CrossRef]
94. Grenha, A.; Guerreiro, F.; Lourenço, J.P.; Lopes, J.A.; Cámara-Martos, F. Microencapsulation of Selenium by Spray-Drying as a Tool to Improve Bioaccessibility in Food Matrix. *Food Chem.* **2023**, *402*, 134463. [CrossRef]

surfaces

MDPI

Article

Effects of Temperature on the Fracture Response of EMC-Si Interface Found in Multilayer Semiconductor Components

João Valdoleiros [1], Alireza Akhavan-Safar [2,*], Payam Maleki [2], Pedro F. C. Videira [1], Ricardo J. C. Carbas [2], Eduardo A. S. Marques [1], Bala Karunamurthy [3] and Lucas F. M. da Silva [1,*]

1 Faculty of Engineering, University of Porto, Rua Dr. Roberto Frias, 4200-465 Porto, Portugal
2 Institute of Science and Innovation in Mechanical and Industrial Engineering (INEGI), Rua Dr. Roberto Frias, 4200-465 Porto, Portugal
3 Infineon Technologies Austria AG, Siemensstrasse 2, 9500 Villach, Austria
* Correspondence: aakhavan-safar@inegi.up.pt (A.A.-S.); lucas@fe.up.pt (L.F.M.d.S.)

Abstract: Despite the fact that temperature is an important condition that affects the behavior of material interfaces used in integrated circuits (ICs), such as the case for epoxy molding compound (EMC) and silicon (Si), this has not been thoroughly studied. To fill this gap, the present work aims to examine the fracture of the bi-material interfaces in multilayered semiconductor components and, more specifically, the EMC-Si, through the experimental quasi-static mode I fracture experiments conducted at different temperatures. The experiments were followed by numerical simulations using cohesive zone modeling (CZM) implemented using Abaqus. Simulation results were aimed at matching experimental data using an inverse CZM approach to determine cohesive properties such as initial stiffness and maximum traction. Experimental results revealed temperature-dependent variations in fracture behavior, with low temperature (−20 °C) showing a decrease in stiffness with values around 650 MPa/mm and a maximum tensile strength of 48 MPa; high temperature (100 °C) revealed a maximum traction and stiffness of 120 MPa and 1200 MPa/mm, respectively. A possible explanation for the results obtained at high temperatures is that temperature changes cause a significant redistribution of residual stresses in the sample and at the interfaces, reducing the stiffness at lower temperatures.

Keywords: semiconductor; cohesive zone modeling; bi-material interface; quasi-static test; silicon

1. Introduction

Over the past decades, microelectronic devices and semiconductors have become an immeasurably important part of our daily lives. The industry responsible for manufacturing these devices is driven by three fundamental objectives: enhancing reliability, reducing costs, and minimizing the size of microprocessors.

Research in semiconductors began to intensify in the 20th century with the emergence of modern electronics. The discovery and development of semiconductors as essential materials for the manufacturing of electronic devices, such as transistors and ICs, spurred a series of studies to understand the physical and chemical properties of these materials [1]. Advances in nanotechnology and microelectronics have sparked a growing interest in the study of interfaces between different layers in semiconductor components. Commercial solutions incorporating these technologies are becoming increasingly prevalent across industries such as electronics, photonics, and energy. Companies like Henkel and Dow offer thermal interface materials (TIMs) that use bi-material interfaces to improve heat

https://doi.org/10.3390/surfaces8010002

dissipation in electronic devices. Solar cell manufacturers explore bi-material interfaces to improve light absorption and charge transport in photovoltaic devices. The precise control of the material interfaces significantly enhances device performance and functionality, paving the way for next-generation electronic technologies

Among different types of bi-material interfaces found in semiconductor devices, the analysis of the EMC-Si interface has been relatively limited. Schlottig et al. [2] investigated the interfacial fracture toughness of the epoxy molding compound (EMC) to silicon interface using the Mixed Mode Chisel setup (MMC). They identified thermal residual stresses, induced during the cooling process, as the most critical parameter impacting the energy release rate and mode mix angle when interpreting fracture data of the EMC/Si interface. Oh et al. [3] developed an adhesion shear test jig to measure the adhesion strength of the EMC/Si chip interface at high temperatures (200 °C). This temperature was selected due to the potential occurrence of interfacial failure in semiconductor packages during the reflow process, which exceeds 200 °C. Conversion et al. [4] employed ANSYS simulation software to undertake a thermomechanical analysis of adhesion forces at the interface between EMC (epoxy molding compound) and Si (silicon) chips, as well as the relationship between maximum shear stress and principal stress. The findings indicate that a mismatch in the coefficient of thermal expansion (CTE) between the EMC and Si chip results in a progressive increase in both shear stress and principal stress. This escalating stress ultimately surpasses the adhesion force of the EMC/Si chip interface, leading to inevitable interfacial delamination.

Xiao [5] investigated the delamination of the EMC and copper interface using a bi-material two-layer beam structure. In this study, EMC and copper were bonded to create a mixed-mode bending test, allowing for an assessment of the delamination toughness of the EMC-Cu interface. The study revealed that residual stresses had a considerable impact on the crack tip singularity and the interface toughness between EMC and copper. Additionally, Sadeghinia [6] examined the effects of moisture, temperature, and mode mixity on the interfacial fracture toughness of the EMC–copper interface. The study concluded that residual stresses play a significant role in promoting delamination at the EMC–copper interface in microelectronic components. Poshtan et al. [7], by performing a Miniaturized Sub-Critical Bending (MSCB) test, demonstrated that interfacial crack propagation is the dominant failure mode, with MC remaining contaminants observed on the Load Frame (LF) surface. They stated that the amount of these embedded particles is influenced by temperature, surface roughness, and mode mixity. Due to the significantly different testing conditions and sample types, a direct comparison of all the results is not feasible. However, a recent review paper has attempted to address this by normalizing the parameters, enabling a comparative analysis of the results obtained in various studies [8]. Fan et al. [9] utilized Atomic Force Microscopy (AFM) to investigate the cohesion between EMC and copper (EMC-Cu), while Khan [10] focused on the adhesion between transferred graphene and silicon. As the results show, adhesion varies significantly with changes in the graphene used. The values found for the interface of silicon and graphene are very similar to the Mode I fracture energy (G_{Ic}) found for the interface of EMC and copper, as Samet and Woodruff [11] showed. Wang et al. [12] assessed interfacial fracture toughness in flip-chip packages and bi-material systems. The results show a value of 0.02 N/mm for the fracture toughness under Mode II loading conditions. Krieger et al. [13] analyzed the mixed-mode fracture of EMC–copper using the CZM approach, and Raghavan et al. [14] developed a framework based on the cohesive zone modeling approach to investigate interfacial delamination in sub-micron-thick layers. However, none of these studies have examined the behavior of the EMC-Si interface across varying temperatures. Consequently, this paper seeks to address this specific research gap by investigating the temperature-dependent characteristics of the EMC-Si interface.

The properties of some materials, such as EMC, that are commonly employed to protect and shield other fragile components from exterior conditions like heat, moisture, shocks, etc., and silicon, one of the main materials used in microprocessors, have been thoroughly investigated due to their vast array of applications, yet there has not been much focus on studying the interfaces of these two materials when utilized in a semiconductor. The lack of sufficient knowledge about how these interfacial properties change with environmental factors, mainly temperature, poses significant challenges. This can lead to problems because any damage at the interface may simply result in the initiation and subsequent propagation of cracks through the component. Therefore, studying and analyzing the variation in interface properties of these two materials with different temperatures is an important barrier that needs to be taken in order to achieve further development of new products and ideas. Because, as products get more sophisticated, the interaction between some of the most common materials used in this sector turns into a pivotal aspect in making new ideas come to life.

This paper aims to analyze crack propagation at the EMC-Si bi-material interface under diverse temperatures and study interfacial damage evolution by determining the interfacial cohesive contact characteristics of the examined bi-material interface. These characteristics are essential for designers and engineers to forecast interface strength and durability, especially under Mode I (tensile or opening mode) loading conditions. Therefore, the cohesive zone model was essential to simulate the interfacial failure of the joint.

2. Materials and Methods

2.1. Design and Geometry

This study focused on the EMC-Si interface. EMC and silicon are typically joined through a process that involves the application of heat and pressure during the molding process [15,16]. The EMC, which consists of thermosetting epoxy polymers, is deposited on the silicon wafer, allowing it to cure and create a strong adhesive bond between the two materials. This bond is crucial for the reliability of electronic packages, helping to prevent void formation and interface delamination. Figure 1 is a technical illustration of the wafer's layers; the specimen also comprises an inbuilt gold layer that serves as a pre-crack to facilitate crack propagation along the bi-material interface during the DCB (double cantilever beam) quasi-static tests.

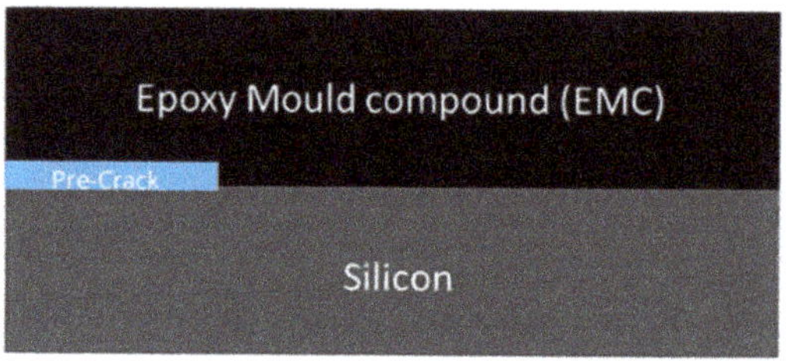

Figure 1. Schematic representation of the tested specimens.

In consideration of the DCB testing setup in universal testing machines, the width and length dimensions have been thoughtfully pre-selected to ensure compatibility, and the holes incorporated into the steel bars serve the purpose of applying load to the wafers via loading pins. The technical dimensions of the DCB and the loading pin reference point can be seen in Figure 2.

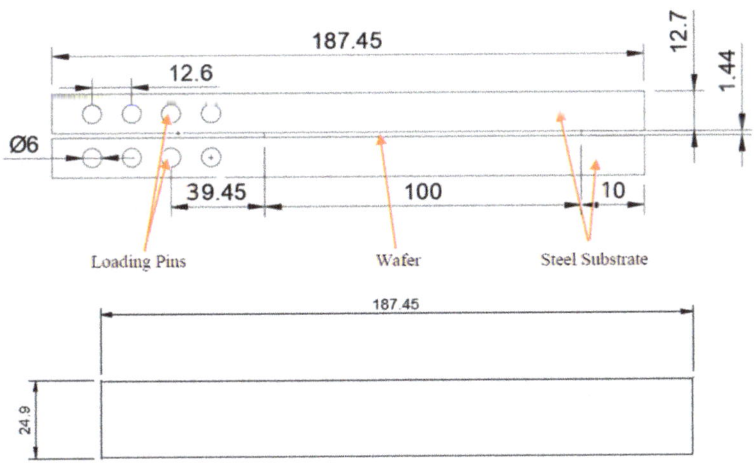

Figure 2. Illustration of the dimensions of the DCB (all dimensions are in mm).

2.2. Materials

The relevant mechanical properties of EMC and silicon are given in Table 1. It should be noted that silicon exhibits anisotropic behavior [17]; the possible values of Young's modulus (E) for silicon range from 130 to 188 GPa [18]. For silicon in the <100> crystallographic direction, the Young's modulus is typically around 130 GPa [19]. This value is commonly used in simulations and approximations, especially in contexts where a simplified model is appropriate to describe the mechanical behavior of the material. Among the materials comprising the wafers, silicon is the most common material to be found in semiconductors, with its ultra-high purity single crystal structure and very brittle characteristics, which play a critical role in the operation of computers, smartphones, and various other electronic gadgets. Another advantage of using a semiconductor, such as silicon, is that at low temperatures the electrons are held in place by covalent bonds, acting as insulators. However, at higher temperatures, these electrons gain enough energy to break free from their bonds and move about the crystal lattice, allowing conduction to occur.

On the opposite side, EMC is essentially a ductile glass-fiber-reinforced epoxy made of silica, hardener, epoxy resin, and, optionally, other fillers and additives that play a vital role in protecting circuits on semiconductor devices from moisture, heat, and shock. It can, not only enhance the reliability of the component by minimizing the impact of electromagnetic interference (EMI) from external sources but also play a vital role in packaging semiconductor devices, where their interaction with silicon significantly influences the structural integrity of these components [20,21].

Table 1. List of material properties [22,23].

Materials	Ultimate Tensile Strength (MPa)	Poisson's Ratio	Young's Modulus (GPa)
Silicon	165	0.28	130
EMC	90	0.38	24
Steel PM300	1020	0.33	210

Changes in temperature and humidity can influence the interfacial strength [24]. Moreover, adding a very thin layer of a different material may contribute to a better interface contact and, thus, a higher G_{Ic}. The reason behind choosing gold as a pre-crack material is because it is an inert material and has a low bonding strength with silicon and

EMC; this way, the chemical reactions that may occur between silicon or EMC and gold can be neglected.

It is important to note that when the wafers are produced, EMC and silicon are merged at high temperatures. During the curing and cooling processes, the materials may warp, resulting in residual stresses that must be considered. Introducing an initial defect such as a pre-crack minimizes the chance of peak loads prior to delamination.

Two different adhesives were used. One for testing at room and low temperature, Scotch Weld 163-2K (3M, Saint Paul, MN, USA), and another one for high temperature, Delo Monopox AD286 (DELO, Windach, Germany). This was necessary due to the loss of mechanical properties with high temperatures of the first adhesive. The parameters for both adhesives were given by the producer and can be seen in Table 2.

Table 2. Adhesive properties.

Adhesive	Tensile Strength [MPa]	Poisson's Ratio	Young's Modulus [GPa]
Delo Monopox AD286	64	-	3.8
Scotch Weld 163-2	48.3	0.34	1.1

According to the literature, silicon properties' variation with temperature can be ignored [25]. As for the EMC, the variation in its properties with temperature relies heavily on its composition. EMC is a composite made with epoxy and fillers, and if, for instance, its ratio changes from producer to producer, it is hard to tell how the variation in its properties with temperature may occur [21,26]. Although there is strong evidence that with higher temperatures, EMC's Young's modulus decreases, this variation did not significantly affect the numeric simulations.

2.3. Manufacturing Process

The manufacturing process starts with surface preparation of the steel bars and wafer surfaces. The first step is sandblasting the steel bars. This technique enhances surface roughness while deeply eliminating and cleaning contaminants and leftovers, guaranteeing this way superior adhesion. Then, spare sand and dust are then removed from the steel bars with the aid of compressed air and afterward cleaned with acetone.

The preparation of the wafer starts by applying 800-grit sandpaper at 45° angles. The surfaces are then cleaned with acetone and followed by a plasma treatment of 6 s on each face. This process enhances the surface energy of the EMC by approximately 200%, facilitating the application of the adhesive in them and, also, a better adhesion to the steel substrates [20]. After this, two layers of tape are applied to the sides of the wafer to prevent any excess adhesive from flowing out and to bind the interface at the sides.

Following the preparation of the bar and the wafer, it is necessary to use a mold to regulate the alignment of the DCBs and the applied pressure during the curing process (Figure 3). The adhesive is then manually applied. The curing process is carried out at a temperature of 120 °C for 90 min with the use of three 1 kg masses that are placed on the top aluminum plate along the length of the wafer.

The mold consists of two aluminum plates with small holes designed for inserting pins to secure the DCBs in position. Nevertheless, given the short length of the steel bars employed in the DCB assembly, it is imperative to incorporate spacers behind them to ensure a snug fit onto the mold. Not only that, but tapes also (with the wafer height) were positioned on the extremities of the wafer to ensure a more compact fit of the weights during the curing process.

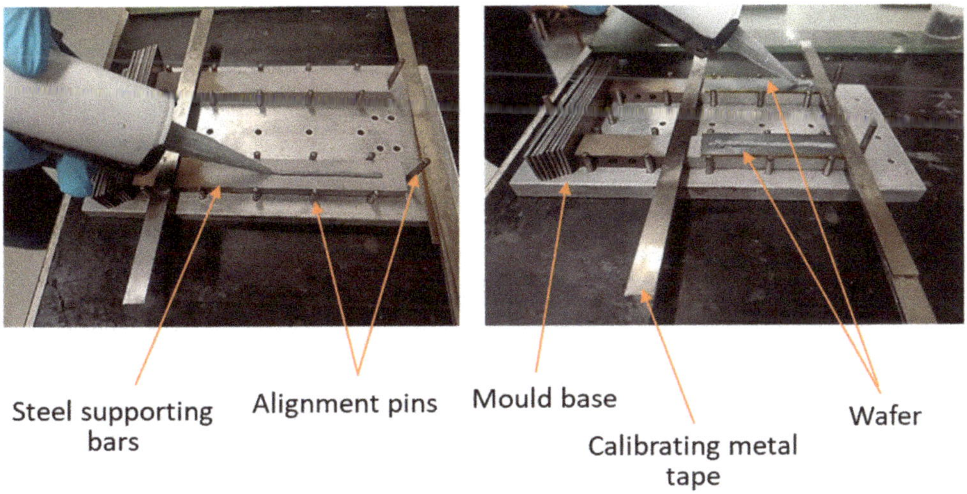

Figure 3. DBC joint in the mold and application of adhesive.

2.4. Test Procedure

After completing all the steps mentioned and removing the tape and eventual excess adhesive covering the edges of the wafer, the DCB joints will be tested under three different conditions: room temperature (around 23 °C), low temperature (−20 °C), and high temperature (100 °C). It should be noted that all tests are conducted at a relative humidity of 50% to 60%. This variation in the relative humidity has a negligible impact on the results.

During the room temperature and low-temperature testing procedures, the INSTRON 8801 (INSTRON, Norwood, MA, USA) machine was used, and the displacement rate for the tests was set at 0.2 mm/min. The load was applied using a Mode I loading technique throughout the test. For both low- and high-temperature testing, a thermocouple (see Figure 4) was used to measure the temperature accurately during the test. By placing the thermocouple on the specimen or within the testing apparatus, it is possible to monitor temperature changes accurately and adjust as needed to maintain the desired testing conditions.

Figure 4. Testing procedure under low and high temperatures.

The machine's retrieved data were treated using the Compliance-Based Beam Method (CBBM). This method assists in discerning the energy release rate of a crack without requiring direct measurement of its size during testing.

Instead, the compliance (*C*) of the specimen is used to calculate an equivalent crack length (a_{eq}) using Equation (1). This equivalent crack length is then utilized in Equation (2) to determine the critical fracture energy. Equation (2) includes various parameters such as the modulus of the specimen (*G*), the flexural modulus (E_f), which is obtained from Equation (3), the load (*P*), the thickness of the substrate (*h*), and the width of the specimen (*B*) [20,22].

This method relies exclusively on the joint's compliance, which is ascertained by measuring the load (*P*) and displacement (δ) values during the experiment. By utilizing this compliance, it becomes possible to estimate both the equivalent crack length (a_{eq}) and the critical fracture energy (G_{Ic}) with a high degree of accuracy. CBBM's superiority to classical methods stems from its consideration of the fracture process zone (FPZ) (located ahead of the crack tip), which can greatly impact the fracture behavior of adhesive joints. The FPZ is a critical area at the crack tip where complex interactions, such as micro-cracking and plastic deformation, significantly affect the material's response to loading. The energy dissipated within this zone must be accounted for when evaluating fracture toughness, particularly for ductile adhesives that exhibit considerable deformation during crack propagation. The FPZ influences the overall fracture toughness by absorbing energy. This absorption is paramount in preventing catastrophic failure and can lead to an increase in the effective fracture energy measured by tests using the CBBM. The method's flexibility allows for adjustments to the calculated elastic modulus to account for the differing energy dissipation patterns in varying adhesives and joint configurations, thereby enhancing accuracy [27].

$$C = \frac{\delta}{P} = \frac{8{a_{eq}}^3}{bh^3E_f} + \frac{12a_{eq}}{5Gbh}. \tag{1}$$

$$G_{Ic} = \frac{6P}{B^2h}\left(\frac{2{a_{eq}}^2}{h^2E_f} + \frac{1}{5G}\right). \tag{2}$$

$$E_f = \left(C_0 - \frac{12(a_0 + |\Delta|)}{5Gbh} + \frac{1}{5G}\right)^{-1}\frac{8(a_0 + |\Delta|)^3}{bh^3}. \tag{3}$$

The flexural modulus represents the stress concentration around the crack tip and is affected by the substrate dimensions (*h* and *b*), the initial crack length (a_0), the adherent shear modulus (*G*), and the initial compliance (C_0). *C* in Equation (1) is calculated by the ratio of displacement (δ) to the applied load (*P*). A correction factor for the crack length (Δ) is applied and determined by Equation (4)

$$\Delta = h\sqrt{\frac{E\left(3 - 2\frac{\tau}{1+\tau}\right)^2}{11G}}. \tag{4}$$

$$\tau = \frac{1.18E}{G}. \tag{5}$$

To accurately interpret the experimental values, it is important to have a numerical model that defines how the crack propagates. The interaction between silicon and EMC is influenced by the stiffness coefficients, maximum traction, and fracture energy of the interfaces. Since the value of G_{Ic} is already known, an inverse approach was used to determine the stiffness and maximum traction. This involved testing different values for the two parameters and analyzing the resulting load–displacement curve. The final values for maximum traction and initial stiffness were determined when the experimental and numerical load–displacement curves matched.

All these parameters are important for a CZM analysis, which has been widely used for modeling crack propagation through interfaces. This technique works on the basis that there is a unified surface where damage can occur and spread.

2.5. Numerical Simulation

From the experimental values retrieved from the quasi-static tests, an iterative process using an inverse CZM approach was employed. The simulation tried to match the load–displacement curves obtained experimentally by manipulating the values for initial stiffness and maximum traction.

2.5.1. Cohesive Behavior

Cohesive elements are designed to simulate crack growth and cohesive behavior within a homogeneous material. However, they may not capture the intricacies of interface failure in the EMC-Si bi-material system. Accordingly, cohesive surface or cohesive contact is the best choice for our specific application, since it provided a more robust solution for the observed failure mechanisms. This approach assumes the existence of a cohesive surface through which the damage will initiate and propagate through.

CZM is a model that uses the cohesive law to predict the behavior of materials under fracture while regarding fracture formation as a gradual phenomenon (see Figure 5a), where the crack propagation is resisted by cohesive tractions. The interface exhibits varying degrees of damage along its length, with the crack tip showing the most severe damage, as indicated by higher values in Figure 5a. The damage decreases for points farther from the crack tip, eventually reaching zero. The damage index ranges from 0 (no damage) far from the crack tip to 1 (complete failure) at the crack tip. Figure 5a also demonstrates that traction on the crack surfaces is lowest near the crack tip and increases with distance from the crack tip. The traction-separation curve in cohesive contact represents the cohesive forces and energy dissipation at the bi-material interface between two adjacent surfaces that are modeled as a distinct region with its own material properties and behavior.

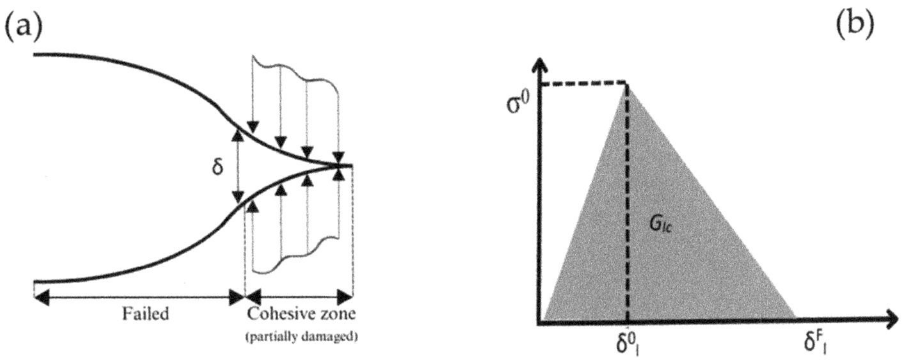

Figure 5. (**a**) CZM visual approach and (**b**) CZM traction separation curve.

Three parameters are required to conduct a triangular CZM analysis (Figure 5b): fracture energy, initial stiffness, and maximum traction. The G_{Ic} was obtained experimentally; as for the other two, an inverse cohesive contact method was utilized, where values for maximum nominal stress and initial stiffness were applied until the numerical and experimental load–displacement curves matched.

The slope of the initial segment ($\delta < \delta^0{}_I$) indicates the typical cohesive stiffness, while σ_0 represents the maximum traction in this model. When $\delta^0{}_I$ is reached, damage begins at

0 and progresses until it reaches 1 at $\delta^F{}_I$. At this point, tension decreases to zero, indicating the absence of interactions between the contacting surfaces.

The damage gradually spreads along the bi-material interface, affecting each point to varying degrees. Notably, the damage at the crack tip is more severe, with higher values observed. However, as one moves away from this location, the level of damage decreases and may even reach zero. The damage index ranges from 0 (indicating no damage) for points far from the crack tip to 1 (indicating complete element failure) for the crack tip itself.

2.5.2. Numerical Model

The dimensions used for both the wafer and the steel bars are the ones mentioned in the previous section, with the exception for the adhesive thickness, which was added to the simulation with a value of 0.15 mm (on both sides). The modeled bonded joint, as well as the different zones of the part, can be seen in Figure 6.

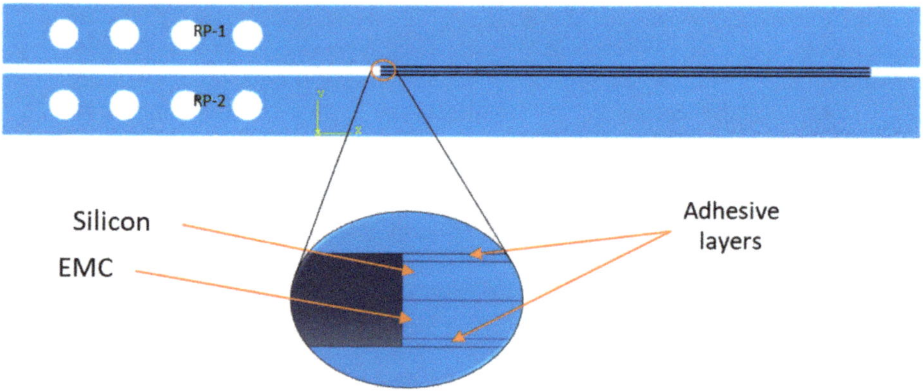

Figure 6. DCB assembly and different parts (wafer).

The selected simulation method was quasi-static analysis, and considering the geometric configuration, it is imperative to establish interactions among the manufactured components to ensure a precise simulation of DCB's behavior on Abaqus software (2021). Empirical tests have revealed crack propagation occurring between the silicon and EMC interface. Hence, a cohesive zone model was used to define the damage path along these interfaces.

Three constraints, more specifically two couplings (RP-1 and RP-2) and the fixed tip, had to be defined. The two couplings in the center of the holes of the DCB, the second ones from right to left, simulate the place where the vertical load was applied during the experimental tests. Moreover, all the interfaces, except the EMC-Si surface, were tied together, as can be seen in Figure 7.

To simulate the displacement applied through the loading pins, the y-axis displacement was set to 1.5 mm, while the other two axes were not constrained. The simple support only had the y-axis displacement constrained. These conditions restrict the model's y-directional movement and bottom bar rotations. A visual representation of each boundary condition in the DCB can be seen in Figure 7.

To better simulate the pre-crack, a frictionless contact was added for the first 2 cm (pre-crack length) of the wafer geometry, as shown in Figure 7. The CZM contact properties were then defined for the EMC-Si interface, more specifically the G_{Ic} and the viscosity coefficient, and the other two parameters, the elasticity module and the shear tension, were estimated to begin the inverse cohesive approach.

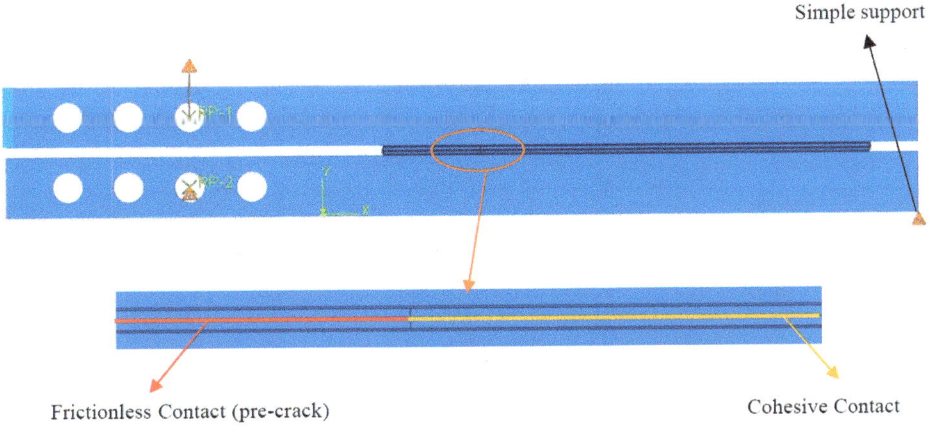

Figure 7. Loading and boundary conditions (the load was applied to RP-1 while RP-2 was simply supported.

In order to obtain a simple but refined mesh, a global mesh with a medium density of 0.1 mm was applied in the entire wafer, and a local mesh with a maximum element size of 3 mm and a minimum element size of 0.1 mm was used along the steel bars. The DCB's mesh geometry can be observed in Figure 8. All the mesh element types are 4-node bilinear plane stress.

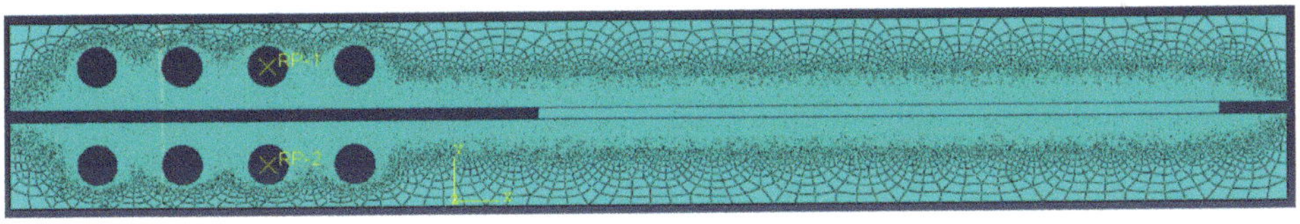

Figure 8. Mesh configuration.

As mentioned before, it is needed to develop a cohesive law that can accurately predict damage propagation; an iterative approach was utilized. This involved determining the values of the cohesive zone properties that would yield a similar curve to the experimental one. The final values were determined once the curves matched each other. The same procedure was used for the other temperatures.

3. Results and Discussion

3.1. Failure Mode

Considering the experimental results, it was discovered that the crack did not always propagate exclusively at the bi-material interface but instead partially or entirely through another layer of the tested wafer. Consequently, the results may vary according to their corresponding failure mechanisms, which include interfacial failure (when the crack propagated through the bi-material interface), silicon failure (when the crack extended into the silicon layer of the wafer), and mixed failure (when the crack simultaneously propagated through the bi-material interface and the silicon layer); see Figure 9. It is worth noting that the weakest part of the joint is the one to which the crack will extend.

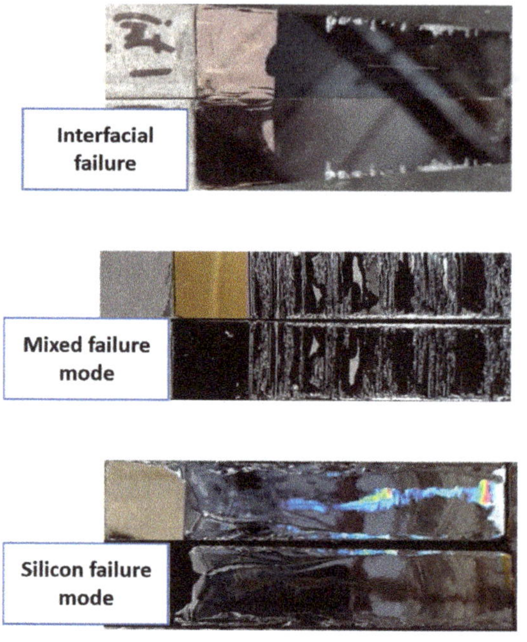

Figure 9. Different types of failure modes in wafers.

According to Oh et al. [3], interfacial failure is due to low EMC/Si interfacial adhesion strength. They concluded that plasma treatment could significantly enhance the adhesion strength of the EMC/Si chip interface by removing the contamination and forming the functional groups like carboxylic acids on the surface of the Si chip. These carboxylic acid functional groups can improve the interfacial adhesion by reacting chemically with the epoxy and thus providing a mechanism of interfacial bridging through the formation of covalent bonds, leading to better adhesion strength.

Xiao et al. [28] confirmed that the failure path of delaminated EMC/Cu samples depends on the loading mode mixture, environmental conditions, and surface roughness. Their analysis of various delaminated specimens revealed three scenarios. First, chunks of molding compound remained on the leadframe surface, with heights up to several tens of micrometers. These chunks contained less filler than the bulk molding compound, and their distribution depended on the load mixture and temperature conditions, leading to holes in the EMC counterparts. In the second scenario, cracking occurred at texture features on the leadframe surface, where deep trenches were filled with sheared-off molding compound. The third scenario involved clean detachment of the EMC from the leadframe surface.

According to Tian et al. [16], the entry of H_2O and O_2 through polymer-based epoxy plastic sealers or polyimide (PI) layers can lead to corrosion and oxidation reactions. These reactions, combined with pre-existing thermal stresses, result in hybrid delamination failures at the interface. They suggest that addressing delamination failures can be achieved through two primary methods: enhancing production process technology or altering the structure or material of the device package.

Schlottig et al. [2] observed that kinking cracks in the silicon material were primarily caused by flaws in the silicon strip flanks. They emphasized the importance of careful handling of this part of the sample during the preparation steps to avoid pre-damage. The flaws were predominantly found at the flanks of prototype samples, resulting in cracks in every experiment.

3.2. Experimental Results

All three testing temperatures revealed an interfacial failure mode during the initial moments of testing. However, under low temperature, crack propagation across the wafer displayed a mixed mode failure after 8 mm of interfacial failure. This last phase, characterized by the unpredictability of the crack show, has not been subject to study. The quasi-static Mode I experimental results for the three temperatures are shown in Figure 10.

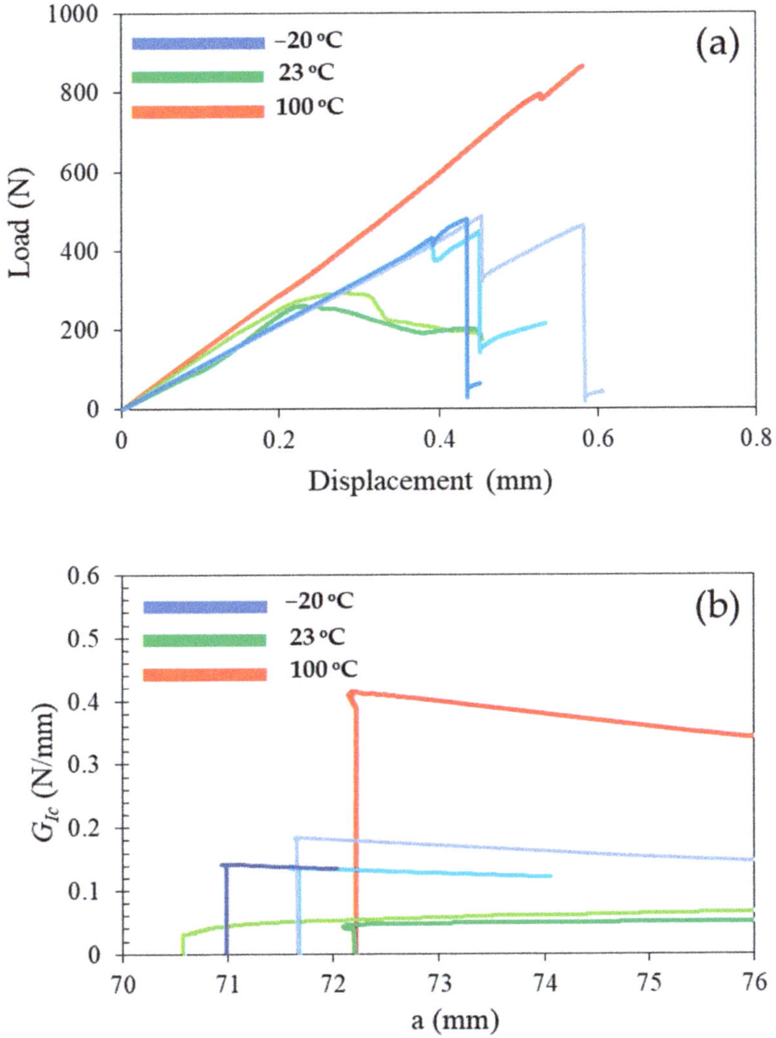

Figure 10. Comparison of (**a**) the experimental load–displacement curves and (**b**) the experimental R-curves at the respective temperatures.

The experimental results shown in Figure 10a,b are load–displacement and R-curves for six specimens at three different temperatures. As for Table 3, the results presented for the room temperature are a mean of two tests carried out during the experimental procedures; the low temperature values shown were obtained from a mean of three experimental tests conducted, and for high temperature, only one result was obtained.

Table 3. Comparison of the interfacial properties at different temperatures.

	G_{Ic} (N/mm)	Failure Mode
Room Temperature (23 °C)	0.05 ± 0.001	Interfacial
Low Temperature (−20 °C)	0.13 ± 0.005	Interfacial
High Temperature (100 °C)	0.37	Interfacial

It is important to notice that the value of the parameters for the room temperature curve is lower than that of the other curves. Theoretically, there must be a correlation between the variation in values from low to high temperatures. This anomaly can be explained by the residual stresses that are significantly changed by changing the testing temperature. This factor affects the behavior of the wafer and the tested joint. Defining loading conditions is often not straightforward because stresses can be introduced throughout the processing history. Typically, the constituents are stress-free at the processing temperature, but chemical stressors like migration, oxidation, or polymerization can still occur at this stage [29]. These residual stresses are usually compounded by stresses induced by thermal mismatches during cooling from the processing temperature or environmental temperature fluctuations, as well as time-dependent properties such as Young's modulus (because of the viscoelastic behavior of EMC) [29]. At low temperatures, materials typically exhibit increased stiffness and reduced ductility. The differences in thermal contraction rates between the polymer matrix (EMC) and the silicon layer lead to higher locked-in residual stresses [5,30,31]. In addition to this, the residual stress induced in the previous steps of wafer production and DCB manufacturing is also re-distributed by changing the testing temperature. At 23 °C, there is some degree of molecular mobility within the polymer matrix, allowing for partial relaxation of the previously induced residual stresses. However, this relaxation may still be insufficient to significantly improve energy absorption during fracture due to the material's inherent properties at this temperature. Room temperature behavior can be influenced by competing effects; while some residual stresses may be relieved, the low toughness of the polymer at this operational range may not facilitate effective energy dissipation, resulting in the lowest measured G_{Ic} observed in the experiments. The stress distribution might still manifest as tension at the crack tip, which could make the material more susceptible to crack propagation at this temperature, further lowering the fracture energy. At elevated temperatures, the polymer matrix experiences significant thermal activation. The molecular mobility increases, allowing for the redistribution and relaxation of residual stresses. This thermally driven process helps mitigate the tensile stresses that may have been present at lower temperatures. As a result of stress relaxation, the material can undergo larger deformations before fracturing, leading to an increase in the energy absorption capability. The fracture energy observed at this temperature stands at G_{Ic} = 0.37 N/mm, reflecting a significant enhancement in toughness.

3.3. Numerical vs. Experimental Results

- Room Temperature

The numerical simulation for the room temperature curve gave us a good approximation, with a maximum experimental load of 260 N and a fracture energy of 0.051 N/mm, as can be seen in Figure 11. The CZM parameters obtained were 720 MPa/mm for the initial stiffness and a maximum tensile strength of 31 MPa.

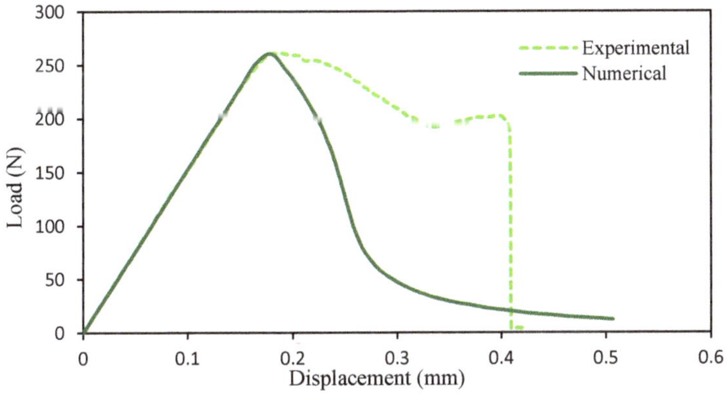

Figure 11. Comparison of numerical and experimental load–displacement curves at room temperature.

As observed in Figure 12, the decline in value for the numerical G_{Ic} happens after its peak. The explanation for this phenomenon resides in the numerical load–displacement curve format. In contrast with the experimental load–displacement curve, the numerical curve presents the steepest decline in maximum load after its peak. This can possibly be explained because mixed failure in crack propagation occurs after an initial interfacial failure. The CZM model describes only crack propagation between a bi-material interface (EMC-Si).

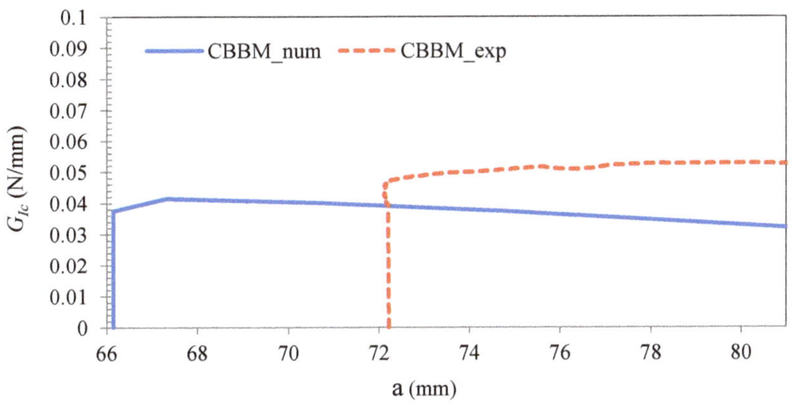

Figure 12. Comparison of numerical and experimental R-curves for the EMC-Si interface (Room temperature).

- Low Temperature

For the tested configurations at low temperature (−20 °C), after damage initiation, the initial crack propagation, after peak loads, shows a relatively abrupt reduction in load. The first peak of maximum load sits around 420 N (Figure 13), and the CZM parameters obtained were 650 MPa/mm for the initial stiffness and a maximum tensile strength of around 48 MPa. The R-curves are shown in Figure 14.

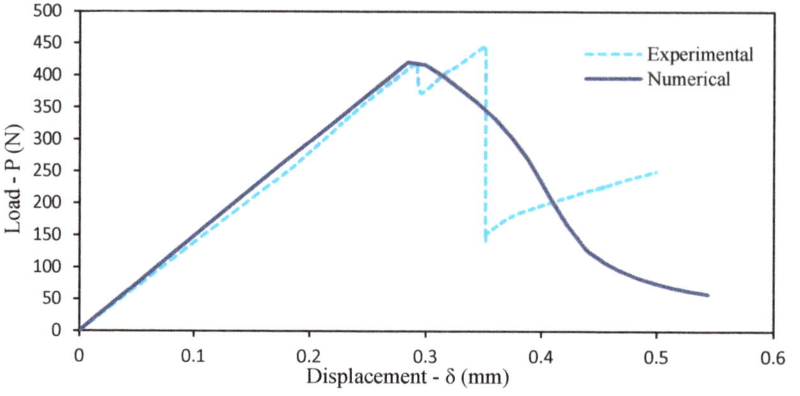

Figure 13. Comparison of numerical and experimental load–displacement curves at low temperature.

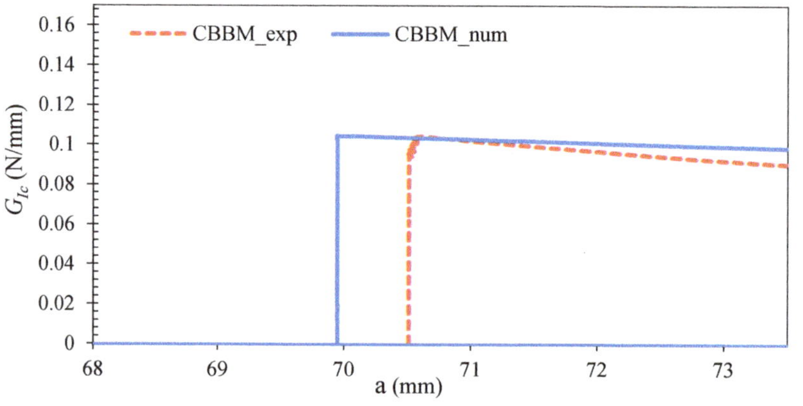

Figure 14. Comparison of numerical and experimental R-curves for the EMC-Si interface (low temperature).

- High Temperature

At high temperature (100 °C), a maximum load of around 850 N with a fracture energy of 0.37 N/mm can be observed (Figure 15). For the CZM parameters, an initial stiffness of 1200 MPa/mm and a maximum tensile strength of 120 MPa.

Given that wafers are assembled at high temperatures and are composed of a different range of materials, each with different coefficients of thermal expansion, there are stresses distributed through these specimens, which can alter the properties of the interface. This can impact the results obtained; for example, it is possible that at lower temperatures these stresses contribute to an easier delamination, allowing one to achieve lower mechanical properties at low temperatures than at higher temperatures.

It should also be considered that since wafer manufacturing is carried out at high temperatures, the redistribution of thermal stresses can play a great influence on the variability of results of low vs. high temperature testing. It is important to highlight that T_g of the epoxy can vary significantly within a range based on different curing conditions and formulations, typically measured between 120 °C and 190 °C for various epoxy systems [2,8,28]. The high-temperature conditions tested are close but still below the T_g of the EMC. This also may affect the results at high temperature. In EMC interfaces, interfacial strength only exhibits a significant decrease after reaching a testing temperature of around 150 °C [8]. This can be attributed to the glass transition temperature (T_g) of the

EMC, which causes the EMC to shift from a hard and glassy state to a soft and rubbery state, accompanied by a significant decrease in Young's modulus (up to 90% decrease), leading to a considerable loss of interfacial adhesion strength.

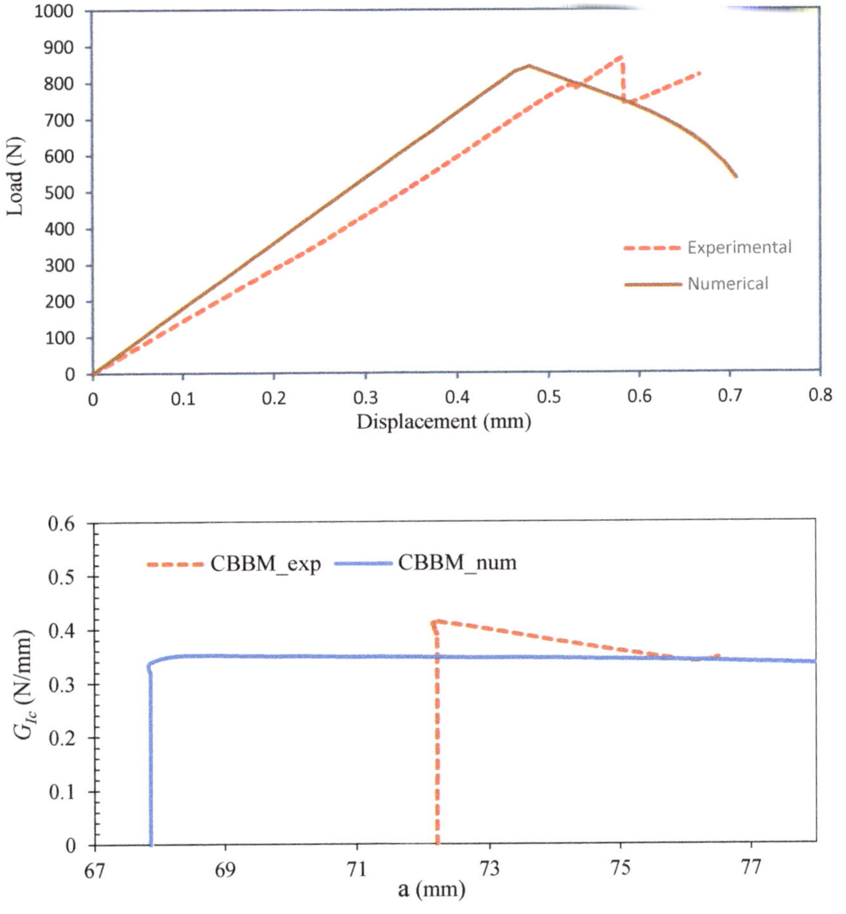

Figure 15. Comparison of numerical and experimental load–displacement and R-curves at high temperature (100 °C).

The numerical results were, as mentioned before, obtained from the iterative process that employed the inverse CZM approach, and the values presented in Table 4 are representative of the last iteration.

Table 4. Numerical simulation parameters.

	Initial Stiffness (MPa/mm)	G_{Ic} (N/mm)	Max. Tensile Strength (MPa)
Room Temperature (23 °C)	720	0.051	31
Low Temperature (−20 °C)	650	0.13	48
High Temperature (100 °C)	1200	0.37	120

4. Conclusions

In this study, crack propagation and interfacial failure at the EMC-Si bi-material interface under varying temperatures were analyzed using experimental and numerical approaches. The study was driven by the imperative need to enhance understanding of material interfaces at different service temperatures crucial to the semiconductor industry's development. Through experimental mode I fracture tests, valuable data on fracture energy and crack propagation behavior under different temperature conditions was obtained. The results showed significant temperature-dependent variations in fracture behavior, highlighting the complex interactions of thermal and mechanical stresses at the bi-material interface.

Additionally, numerical simulations using CZM allowed for a deeper understanding of damage propagation and interface behavior. By employing an inverse contact CZM approach, numerical simulations were matched with experimental data, determining this way the cohesive properties of the interface under study. For room temperature testing, the results showed an initial cohesive stiffness of 720 MPa/mm and a maximum traction of 31 MPa. For low temperature, these parameters are 650 MPa/mm and 48 MPa for the stiffness and traction, respectively. And last, for high-temperature testing, the findings were 1200 MPa/mm for the initial stiffness and 120 MPa for the maximum traction. As the results show, a linear relationship of fracture energy with temperature was not observed, as the fracture energy at room temperature was lower than at low and high temperatures. It should be noted that thermal stresses in the tested wafers may significantly contribute to the failure mechanism and mechanical properties.

Understanding the behavior of bi-material interfaces is crucial for optimizing the design and durability of microelectronic devices, which play pivotal roles in various sectors, including health, communication, security, and education. Still further analysis is required to better understand the interaction effects of thermal and mechanical stresses on the fracture response of such bi-material interfaces.

Author Contributions: Conceptualization, L.F.M.d.S. and A.A.-S.; methodology, A.A.-S. and B.K.; software, J.V. and P.F.C.V.; validation, R.J.C.C. and E.A.S.M.; formal analysis, R.J.C.C. and E.A.S.M.; investigation, J.V. and P.M.; resources, B.K. and R.J.C.C.; data curation, J.V.; writing—original draft preparation, J.V. and P.F.C.V.; writing—review and editing, A.A.-S., P.M. and L.F.M.d.S.; supervision, A.A.-S. and L.F.M.d.S.; project administration, L.F.M.d.S. All authors have read and agreed to the published version of the manuscript.

Funding: This research received no external funding.

Institutional Review Board Statement: Not applicable.

Informed Consent Statement: Not applicable.

Data Availability Statement: Data is contained within the article.

Conflicts of Interest: Bala Karunamurthy was employed by Infineon Technologies Austria AG. The remaining authors declare that the research was conducted in the absence of any commercial or financial relationships that could be construed as a potential conflict of interest.

References

1. Alferov, Z.I. The semiconductor revolution in the 20th century. *Russ. Chem. Rev.* **2013**, *82*, 587. [CrossRef]
2. Schlottig, G.; Maus, I.; Walter, H.; Jansen, K.M.B.; Pape, H.; Wunderle, B. Interfacial fracture parameters of silicon-to-molding compound. In Proceedings of the 2010 60th Electronic Components and Technology Conference (ECTC), Las Vegas, NV, USA, 1–4 June 2010; pp. 1939–1945.
3. Oh, G.-H.; Joo, S.-J.; Jeong, J.-W.; Kim, H.-S. Effect of plasma treatment on adhesion strength and moisture absorption characteristics between epoxy molding compound/silicon chip (EMC/chip) interface. *Microelectron. Reliab.* **2019**, *92*, 63–72. [CrossRef]

4. Conversion, A.; Ubando, A.T.; Gonzaga, J. Interfacial Delamination Validation on Fan-Out Wafer-Level Package Using Finite Element Method. *Solid State Phenom.* **2023**, *343*, 73–78. [CrossRef]
5. Xiao, A. Interface characterization and failure modeling for semiconductor application. In *Precision and Microsystems Engineering, Citeseer*; Delft University of Technology: Delft, The Netherlands, 2012.
6. Sadeghinia, M. Failure and Delamination in Microelectronic Packages. Ph.D. Thesis, Delft University of Technology, Delft, The Netherlands, 2013.
7. Poshtan, E.A.; Rzepka, S.; Silber, C.; Wunderle, B. An in-situ numerical–experimental approach for fatigue delamination characterization in microelectronic packages. *Microelectron. Reliab.* **2016**, *62*, 18–25. [CrossRef]
8. Ferreira, R.A.; Akhavan-Safar, A.; Carbas, R.J.C.; Marques, E.A.S.; Karunamurthy, B.; da Silva, L.F.M. Advancements in mechanical characterization techniques and environmental effects on bi-material interfaces in microelectronics: A literature review. *J. Adhes.* **2024**, *101*, 413–445. [CrossRef]
9. Fan, H.B.; Wong, C.K.Y.; Yuen, M.M.F. A new method to predict delamination in electronic packages. In Proceedings of the Electronic Components and Technology, 2005, ECTC'05, Lake Buena Vista, FL, USA, 31 May–3 June 2005; pp. 145–150.
10. Khan, Z.H. Investigation of Mechanical Properties of Graphene on Silicon Wafers. Ph.D. Thesis, University of Technology Sydney, Ultimo, Australia, 2017.
11. Samet, D.S. Development of a Fatigue-Compatible Cohesive Zone Method for a Copper-Epoxy Molding Compound Bimaterial Interface. Ph.D. Thesis, Georgia Institute of Technology, Atlanta, GA, USA, 2018.
12. Wang, J.; Zou, D.; Lu, M.; Ren, W.; Liu, S. Evaluation of interfacial fracture toughness of a flip-chip package and a bimaterial system by a combined experimental and numerical method. *Eng. Fract. Mech.* **1999**, *64*, 781–797. [CrossRef]
13. Krieger, W.E.R.; Raghavan, S.; Kwatra, A.; Sitaraman, S.K. Cohesive zone experiments for copper/mold compound delamination. In Proceedings of the 2014 IEEE 64th Electronic Components and Technology Conference (ECTC), Orlando, FL, USA, 27–30 May 2014; pp. 983–989.
14. Raghavan, S.; Schmadlak, I.; Leal, G.; Sitaraman, S.K. Mixed-mode cohesive zone parameters for sub-micron scale stacked layers to predict microelectronic device reliability. *Eng. Fract. Mech.* **2016**, *153*, 259–277. [CrossRef]
15. Liu, S.L.; Chen, G.; Yong, M.S. EMC characterization and process study for electronics packaging. *Thin Solid Films* **2004**, *462*, 454–458. [CrossRef]
16. Tian, W.; Chen, X.; Zhang, G.; Chen, Y.; Luo, J. Delamination of Plasticized Devices in Dynamic Service Environments. *Micromachines* **2024**, *15*, 376. [CrossRef]
17. Van Driel, W.D.; Zhang, G.Q.; Fan, X.J. Thermo-mechanics of integrated circuits and packages. In *Mechanics of Microelectronics*; Springer: Berlin/Heidelberg, Germany, 2006; pp. 169–279.
18. Hopcroft, M.A.; Nix, W.D.; Kenny, T.W. What is the Young's Modulus of Silicon? *J. Microelectromechanical Syst.* **2010**, *19*, 229–238. [CrossRef]
19. Boyd, E.J.; Uttamchandani, D. Measurement of the anisotropy of young's modulus in single-crystal silicon. *J. Microelectromechanical Syst.* **2011**, *21*, 243–249. [CrossRef]
20. Adler, C.; Morais, P.; Akhavan-Safar, A.; Carbas, R.J.; Marques, E.A.; Karunamurthy, B.; Silva, L.F.D. Cohesive Properties of Bimaterial Interfaces in Semiconductors: Experimental Study and Numerical Simulation Using an Inverse Cohesive Contact Approach. *Materials* **2024**, *17*, 289. [CrossRef] [PubMed]
21. Al Rashid, J.; Koohestani, M.; Saintis, L.; Barreau, M. A State-of-the-Art Review on IC EMC Reliability. In Proceedings of the 31st European Safety and Reliability Conference, Research Publishing Services, Angers, France, 19–23 September 2021; pp. 1850–1857.
22. Morais, P.; Akhavan-Safar, A.; Carbas, R.J.C.; Marques, E.A.S.; Karunamurthy, B.; da Silva, L.F.M. Mode I Fatigue and Fracture Assessment of Polyimide–Epoxy and Silicon–Epoxy Interfaces in Chip-Package Components. *Polymers* **2024**, *16*, 463. [CrossRef] [PubMed]
23. A Background to Silicon and Its Applications. Available online: https://www.azom.com/properties.aspx?ArticleID=599 (accessed on 5 December 2023).
24. Brillson, L.J. The structure and properties of metal-semiconductor interfaces. *Surf. Sci. Rep.* **1982**, *2*, 123–326. [CrossRef]
25. Liu, Z. Temperature-dependent elastic constants and Young's modulus of silicon single crystal. *Young* **2021**, *4*, 23.
26. Kim, J.; Song, M.; Gu, C.Y.; Ma, S.; Lee, J.H.; Lee, W.S.; Kim, T.S. Enhancing predictability of thermal warpage by applying temperature-dependent Poisson's ratio of epoxy molding compound. *Polym. Test.* **2023**, *125*, 108140. [CrossRef]
27. Correia, D.S.; Costa, I.D.; Marques, E.A.S.; Carbas, R.J.C.; da Silva, L.F.M. Development of a Unified Specimen for Adhesive Characterization—Part 2: Experimental Study on the Mode I (mDCB) and II (ELS) Fracture Components. *Materials* **2024**, *17*, 1049. [CrossRef] [PubMed]
28. Xiao, A.; Schlottig, G.; Pape, H.; Wunderle, B.; Jansen, K.M.B.; Ernst, L.J. Delamination and combined compound cracking of EMC-copper interfaces. In Proceedings of the 2010 60th Electronic Components and Technology Conference (ECTC), Las Vegas, NV, USA, 1–4 June 2010; pp. 114–120.

29. Zhang, G.-Q.; Van Driel, W.D.; Fan, X.J. *Mechanics of Microelectronics*; Springer Science & Business Media: Berlin/Heidelberg, Germany, 2006; Volume 141.
30. Guo, S.; Dillard, D.A.; Nairn, J.A. Effect of residual stress on the energy release rate of wedge and DCB test specimens. *Int. J. Adhes. Adhes.* **2006**, *26*, 285–294. [CrossRef]
31. Sicot, O.; Gong, X.-L.; Cherouat, A.; Lu, J. Influence of residual stresses on the mechanical behavior of composite laminate materials. *Adv. Compos. Mater.* **2005**, *14*, 319–342. [CrossRef]

surfaces

Article

Improving Cotton Fabric Dyeability by Oxygen Plasma Surface Activation

Víctor M. Serrano-Martínez *, Carlos Ruzafa-Silvestre, Carlota Hernández-Fernández, Elena Bañón-Gil, Francisca Arán-Ais and Elena Orgilés-Calpena

Footwear Technology Centre, Campo Alto Campground, 03600 Alicante, Spain
* Correspondence: vmserrano@inescop.es

Abstract: This paper focuses on investigating the use of low-pressure oxygen plasma as a surface treatment aimed at enhancing the wettability and dyeability of cotton fabrics for use in textiles and footwear materials. Plasma activation modified the cotton fabric surfaces, increasing their affinity for polar liquids. The research thoroughly characterised the treated fabrics through a combination of analytical methods and physical testing. Plasma treatment was performed using a 13.56 MHz RF generator at 90 W power, with an oxygen flow rate of 500 sccm and a pressure of 0.30 mbar, for treatment durations of 30, 60, and 120 s. Changes in surface chemistry were analysed with XPS, while SEM was used to observe morphological changes. Static water contact angle measurements confirmed a reduction from 128.5° in untreated cotton to 25.6° in samples treated for 30 s, indicating a significant increase in hydrophilicity. Water absorption tests showed a maximum absorption capacity of 119.6% after 60 min for plasma-treated samples, compared to 65.7% for untreated cotton. Contact angle measurements verified that surface hydrophilicity increased following the treatment. Furthermore, physical tests, such as rub fastness, colourimetry, and water absorption, were carried out to evaluate improvements in wettability, dyeability, and overall performance. The results showed notable enhancements in the wetting properties of cotton textiles, enabling better absorption of water and dyes, along with improved fixation. The comprehensive characterisations provided insights into the mechanisms behind these improvements. This research offers a sustainable approach for the textile industry, as plasma technology is a dry process that enhances dyeing efficiency while maintaining fabric performance and lowering environmental impact.

Keywords: oxygen plasma activation; wettability; dyeability; cotton fabric; textile; footwear; surface treatment

1. Introduction

Textiles, comprising materials such as cotton, silk, wool, and various other fabric constituents, play a pivotal role in diverse applications, including clothing, footwear, home furnishings, medical products, automotive interiors, and industrial gear. Additionally, they serve purposes in civil engineering, sports equipment, aerospace and defence, agriculture, and environmental applications, showcasing their wide-ranging utility [1]. Among these, cotton is globally valued for its biodegradability, comfort, and versatility across industries, making it a cornerstone of textile production and a key material in sustainable practices [2,3]. Within the clothing and footwear industry, the differentiation between hydrophobic and hydrophilic properties holds utmost importance [4,5]. Depending on the specific product type and its intended use, a diverse range of textile characteristics is needed. In the realm of both sectors, there is a perpetual quest for textiles to inherently exhibit hydrophobic properties, thereby repelling water. However, in certain scenarios, it becomes imperative for textiles to possess innate hydrophilicity, fostering an affinity for water. This hydrophilic trait facilitates enhanced dye absorption during the textile dyeing

weft threads, weighing 480 g/m^2 and a thickness of 0.93 ± 0.05 mm. Carburos Metálicos S.A. (Barcelona, Spain) supplied oxygen (O_2, 99.995% purity) as the gas precursor for the activation process. For dyeing both untreated and treated samples, Zaneline (Madrid, Spain) provided the reactive dye, Panreac Quimica S.L. (Barcelona, Spain) supplied sodium hydroxide (NaOH, 98.0% purity), and sodium chloride (NaCl, 99.5% purity) came from Merck Life Science S.L.U. (Madrid, Spain).

Secondly, A low-pressure plasma activation setup was developed for treating raw cotton samples using a low-pressure plasma system (Diener Electronic Vertriebs GmbH, Ebhausen, Germany) with a 24 L stainless-steel chamber and a dedicated sample holder. The reactor is equipped with a 13.56 MHz RF generator, delivering up to 300 W of power, and includes two gas inlet channels and a microdosing pump. It supports various plasma surface modification processes such as cleaning, activation, coating, and etching, and it is suitable for laboratory-scale applications. The chamber's temperature can be controlled during treatment. Sample designations and corresponding process parameters are shown in Table 1 [47].

Table 1. Sample nomenclature according to the operating conditions of the plasma activation treatments studied.

Cotton Samples	Gas Precursor	Time (s)	Power (W)
AC-0	-	-	-
AC-0.5	O_2	30	90
AC-1	O_2	60	90
AC-2	O_2	120	90

Figure 1 visually presents the low-pressure oxygen plasma activation treatment of cotton textiles, showing the transition from untreated to plasma-activated cotton. During this procedure, raw cotton samples measuring 110 × 110 mm are placed inside the chamber of the low-pressure plasma system. The chamber is first evacuated to a base pressure of 0.10 mbar. Oxygen is then introduced via a mass flow controller at a flow rate of 500 sccm, allowing the pressure to stabilise at a 0.30 mbar, ensuring uniform distribution of oxygen molecules throughout the chamber. Once the gas supply phase is complete, the RF generator is activated at 90 W of power input, ionising the oxygen molecules and generating oxygen plasma between 30 and 120 s of activation time. This plasma uniformly spreads across the chamber, modifying the surface of the exposed cotton samples. The process ends when the preset activation time is reached, at which point the plasma is turned off. The chamber is then brought back to atmospheric pressure by venting air, after which the treated samples are removed [47,48].

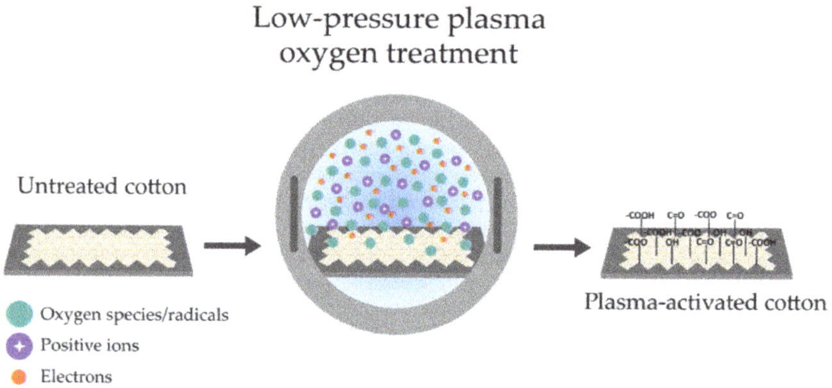

Figure 1. Sketch of the low-pressure oxygen plasma activation treatment of cotton textiles.

On the other hand, to study the improvement of the dyeing capacity of the treated samples, the samples must be dyed after plasma treatment. For this purpose, a well-established dye bath method by the supplier of the reactive dye has been carried out [49], where the following reagents were used. For every unit of the fabric's weight, a 1.5% proportion of reactive dye is employed. Additionally, a solution of sodium hydroxide (NaOH) is prepared, consisting of 3.6 g of NaOH per litre of water. Sodium chloride (NaCl) is introduced into the distilled water at a concentration of 50 g per litre.

The methodology for dyeing samples begins with immersing the untreated cotton fabric in distilled water for 10 min to remove impurities. For treated samples, fabric is placed in a 2 L beaker with a magnetic stirrer (700–1000 rpm). Sodium chloride is added, followed by reactive dye, adjusted to the fabric's weight. The fabric is mixed for 5 min, after which sodium hydroxide is gradually introduced for dye fixation, following a specific formula. The temperature is raised to 60 °C and maintained for 40–60 min. The fabric is then cooled to room temperature in a controlled environment and finally rinsed with 1.5 litres of water to remove excess dye, ensuring colourfastness. Samples are drained and dried at room temperature. For improved dye fixation, some samples were ironed at 90 °C, with 3 bar pressure for 30 s. The ironing was conducted using a 'cotton mode' for optimal heat and pressure. The process follows standard textile practices to ensure reproducibility and aligns with industry norms. Figure 2 clearly outlines the dyeing process performed on the cotton samples, detailing each step from fabric preparation to the final dyed product. The diagram highlights key stages, including the addition of NaCl and NaOH, reactive dyeing under controlled conditions, cooling, washing, drying, and ironing of the cotton fabric.

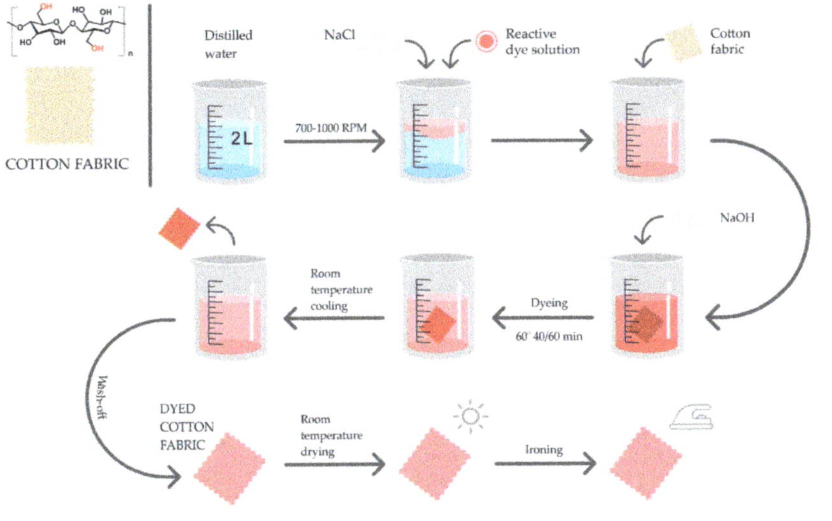

Figure 2. Sketch diagram about the dyeing process carried out for the different cotton samples.

Various characterisation techniques were employed, providing comprehensive insights into both the chemical interactions and physical attributes of the treated fabrics. The techniques used are as follows:

X-ray Photoelectron Spectroscopy (XPS): The chemical composition of both treated and untreated samples was examined using X-Ray Photoelectron Spectroscopy (XPS) with a Thermo Scientific K-ALPHA instrument in Waltham, MA, USA. XPS determined the film's chemical properties with a 3 mA current, 12 kV voltage, Al-K radiation, and Avantage software analysis, version 5.9925. No pre-treatment was performed on the samples prior to analysis to maintain their native state. A flood gun prevented charging. Technical Research Services (SSTTI) at the University of Alicante carried out the XPS analysis.

Scanning Electron Microscopy (SEM): Surface modifications and morphological analyses of treated and untreated cotton samples were conducted using a JEOL IT500HR/LA high-resolution scanning electron microscope with EDS capabilities. This system, featuring a field emission gun, provides resolutions of 1.5 nm at 30 kV and 4.0 nm at 1 kV, operating from 0.5 to 30 kV. The microscope is also integrated with a Raith Elphy Quantum EBL system. Samples were analysed in their native state without any metal coating to preserve the original surface characteristics. The analysis was performed by the Technical Research Services (SSTTI) at the University of Alicante [47,50].

Static Contact Angles: A Theta Flow optical tensiometer (Biolin Scientific Oy, Espoo, Finland) was utilised to assess the wettability of both treated and untreated surfaces. Contact angle measurements were conducted using three 4 μL drops of a polar liquid (distilled water, H_2O) [51] in accordance with the EN 828 standard [52]. The measurements were performed in a controlled environment of 23 ± 2 °C and 50 ± 5% relative humidity.

Colour Difference: Plasma-treated and untreated textile colour variation was assessed using a portable CM600d spectrophotometer (Konica Minolta, Tokyo, Japan), in line with ISO 105-J03:2009 [53]. This device accommodates samples of diverse sizes and shapes, employing an 8 mm fixed aperture. At three central points on the sample, measurements were taken with the lighting/observer combination D65/10°, with the specular component (SPEX) taken out. The mean value of the colour difference CMC (ΔEcmc (l:c)) was determined.

Water absorption: Due to the lack of a standardised test method for water absorption in textiles, ISO 22649:2016 has been used (method A) [54] to determine the water absorption of the textile under static conditions. Prior to testing, the fabrics were oven-dried at 23 °C for 24 h and 50% humidity to ensure consistent initial moisture content. This standard, specific to footwear regardless of the material, involves immersing a test specimen of textile mass in water for a specified period, followed by the measurement of the percentage of water absorbed based on the mass. The oven used was sourced from Memmert GmbH + Co. KG, Büchenbach, Germany.

Colour fastness to rubbing and fade: The colour fastness of textiles can be determined by colour fastness to rubbing and colour fastness to water tests. Colour fastness to rubbing of textiles according to ISO 17700:2019 [55] may be performed in three different test methods (method A, method B, and method C) for evaluating the degree of colour transfer from the surfaces of materials during a dry or wet rubbing process. The methods apply to uppers, linings, and insoles, regardless of the material. Method D also applies to sewing threads and laces. The methods are method A, a reciprocating abrasion tester with a square head; method B, a rotary abrasion tester; and method C, a reciprocating abrasion tester with a circular head. The methods apply to footwear uppers, linings, and insoles, regardless of the material. Method B was used in the investigation. According to the standard, the samples were tested both dry and wet using circular pieces of washed pure wool felt, rotating with a force of 24.5 N at 15.6 rad/s for 200 revolutions. Colour fastness to water according to ISO 105-E01:2013 [56] was performed to determine the probability of sample colour bleeding using a standardised multi-fibre textile. The change in colour of the sample and the staining of the felt or the multi-fibre textile have been measured according to the ISO 105-A03:2020 standard [57], where the samples are observed using D65 light and tilted at 45 degrees. The Grey Scale Rate for each determination was established.

3. Results and Discussion

This Section presents a detailed account of the experimental results, followed by their interpretation and the corresponding conclusions derived from the experiments.

In the following Section, a detailed characterisation of raw cotton samples subjected to low-pressure oxygen plasma treatment using X-ray Photoelectron Spectroscopy (XPS) is being conducted. Figure 3 presents the XPS survey of raw cotton and plasma-oxidised cotton samples, showing the binding energy spectra. The Figure highlights the presence of oxygen (O 1s) and carbon (C 1s) peaks, indicating surface modifications induced by plasma treatment. The untreated sample (AC-0) exhibits prominent peaks for oxygen (O 1s) and carbon (C 1s) at binding energies of approximately 532 and 285 eV, respectively [58]. In contrast, the plasma-activated samples revealed a notable decrease in carbon peak intensity and a corresponding increase in oxygen intensity, indicating the incorporation of oxygen-functional groups, such as hydroxyl (–OH), ethers (–O–), and carbonyl (C=O), onto the cotton surface. These surface modifications occur due to the oxidation of carbon-carbon (C–C) bonds and the introduction of new chemical functionalities by the plasma treatment, such as the formation of carboxyl groups (–COOH). These changes at the molecular level have a significant impact on the chemical and physical properties of the treated cotton surface, as seen below [59].

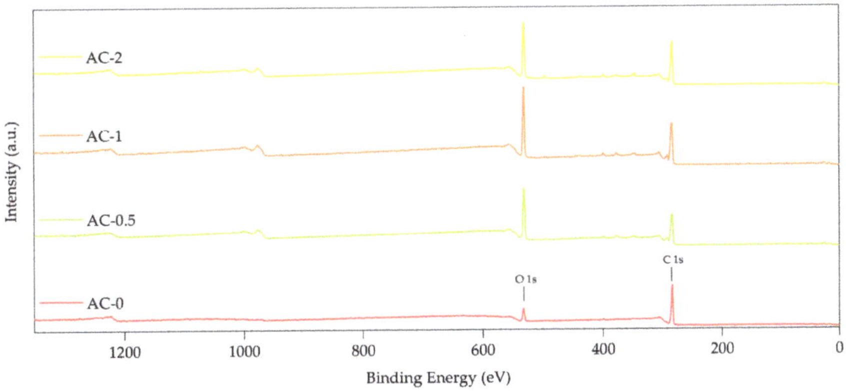

Figure 3. XPS survey of raw cotton and plasma-oxidised cotton samples.

As summarised in Table 2, the elemental composition data clearly show that carbon content decreased significantly, from 86.64% in the untreated sample (AC-0) to 62.78% in AC-0.5 and 62.14% in AC-1, followed by a slight increase to 65.49% in AC-2. This decline in carbon is coupled with a significant increase in oxygen content, reaching 33.87% in AC-1. Such changes confirm the effective removal of carbon-based contaminants and the oxygen enrichment of the cotton surface. The O/C ratio increased notably, from 0.14 in AC-0 to 0.55 in AC-1, further highlighting the oxygen-induced surface modification. However, a slight decline at extended plasma exposure (AC-2) suggests a saturation effect. Overall, oxygen plasma treatment reduces carbon and increases oxygen on the cotton surface, enhancing its properties for scientific and technological applications [59,60].

Table 2. Elemental composition of untreated and plasma-activated cotton samples.

Element	Atomic Percentages (% at.)			
	AC-0	AC-0.5	AC-1	AC-2
C	86.64	62.78	62.14	65.49
N	0.97	1.85	1.85	1.71
O	12.39	33.42	33.87	31.40
O/C	0.14	0.53	0.55	0.48

To gain further insights into the chemical species formed, the C 1s and O 1s peaks were deconvoluted, and the resulting atomic concentrations are presented in Table 3. The aliphatic carbon (C1s) drops from 54.36% in the control to lower levels in treated samples (23.37% for 30 s, 24.55% for 1 min, 30.25% for 2 min), indicating C–C and C–H bond breakdown due to oxidation. C–O species at 285.6 eV initially increase after 30 s but decrease with longer treatment, suggesting dynamic surface modifications. Conversely, the carbonyl group (C=O) at 286.4 eV decreases after 30 s but significantly increases with extended exposure, reflecting the formation of stable oxygen-functional groups due to plasma exposure. Carboxyl content (287.9 eV) rises in all treated samples, indicating the formation of these groups through oxidation processes [60].

Table 3. The atomic percentages (at%) of chemical species identified by XPS at the C1s and O1s peaks for both untreated and plasma-oxidised cotton samples.

Element	Species	Binding Energy (eV)	Atomic Percentages (% at.)			
			AC-0	AC-0.5	AC-1	AC-2
C1s	C–C/C–H	284.6	54.36	23.37	24.55	30.25
	C–O	285.6	11.99	14.46	0.36	3.86
	C=O	286.4	15.30	10.30	22.70	18.22
	O–C=O	287.9	4.99	14.65	14.53	13.16
O1s	C=O/O–C=O	531.2	3.55	7.68	8.98	8.46
	O–C/OH	532.6	5.73	18.86	23.25	20.12
	O=C–O/O=C=O	533.8	2.22	6.29	1.64	2.82
	O=C=O	536.2	0.89	0.59	0	0

For the O 1s spectrum, Figure 3 illustrates notable changes in the chemical environment of oxygen. As indicated in Table 3, the contributions of ketone groups (531.2 eV) and ether/hydroxyl groups (532.6 eV) increase significantly in all treated samples, reflecting their incorporation and oxidation produced on the cotton surface. Ester groups (533.8 eV) exhibit dynamic behaviour, increasing after 30 s, then decreasing, followed by another rise in longer treatments. Additionally, the peak at 536.2 eV, assigned to weakly adsorbed CO_2 or volatile oxygen-containing species, completely diminishes with increased plasma exposure [61]. This disappearance reflects a surface cleansing effect, where plasma treatment removes loosely adsorbed contaminants. Detailed deconvolution results, including spectra for all samples, are provided in the Supplementary Material (Table S1), offering further insights into the chemical composition and bonding configurations at different activation levels.

In summary, XPS analysis highlights substantial chemical transformations induced by low-pressure oxygen plasma treatment. The reduction in aliphatic carbon content and the incorporation of oxygen-functional groups, such as hydroxyl, carbonyl, and ketone species, are key contributors to the improved surface properties observed in plasma-treated cotton. These changes are directly linked to the enhanced hydrophilicity and dyeing performance discussed in subsequent sections.

Furthermore, Scanning Electron Microscopy (SEM) analyses were conducted on samples of textile cotton subjected to low-pressure oxygen plasma treatments. Figure 4 presents Scanning Electron Microscopy (SEM) images of the cotton samples at × 1500 magnification, showing the progression in surface morphology from untreated cotton (AC-0) to plasma-activated samples (AC-0.5, AC-1, and AC-2). The images clearly highlight the increasing surface modifications as a function of plasma exposure time.

As illustrated in Figure 4, the untreated sample (AC-0) exhibits smooth fibres with minimal texture, which is characteristic of untreated cotton. After 30 s of plasma treatment (AC-0.5), slight surface changes are observed, indicating the onset of etching due to plasma exposure. The 1 min treated sample (AC-1) shows a more pronounced increase in surface roughness, enhancing surface area and wettability. Finally, the 2 min treated sample (AC-2) displays the most significant morphological changes, with a substantial increase in surface texture and roughness, suggesting a more advanced etching effect as a result of prolonged exposure to the oxygen plasma. This increased modification improves properties such as dye uptake, wettability, and fibre-to-fibre friction [59]. The progression from (a) to (d) presents that low-pressure oxygen plasma treatment has a cumulative effect on the surface of cotton fibres, with more extended treatment times leading to greater surface roughness. This can be associated with increased hydrophilicity and changes in other surface properties due to the plasma treatment's ability to remove surface impurities and oxidise the fibre surface [62].

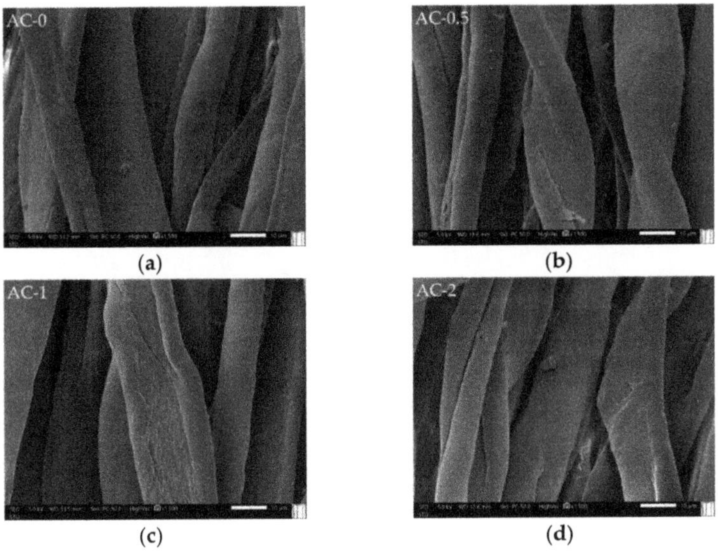

Figure 4. Scanning Electron Microscopy (SEM) images obtained at × 1500 magnification for (**a**) AC-0, untreated cotton sample; (**b**) AC-0.5; (**c**) AC-1; and (**d**) AC-2, plasma-activated cotton samples.

The static water contact angle analysis investigated the impact of low-pressure oxygen plasma on cotton fibre wettability. Table 4 presents the contact angle measurements for untreated and plasma-activated cotton samples, while Figure 5 shows the corresponding contact angle images for each sample, illustrating the effect of plasma treatment on the cotton surface. Cotton naturally exhibits slight hydrophobicity due to its structure and chemical composition, with high contact angles indicating hydrophobic surfaces. As shown in Table 4, untreated cotton (AC-0) had a contact angle of 128.50°, confirming its hydrophobic nature. This is due to natural oils and waxes in the fibres, causing water to form droplets rather than spread, indicating poor initial water absorption. These substances are usually removed during textile finishing processes [63].

Table 4. Contact angle measurements for the untreated and plasma-activated cotton samples.

Sample	Contact Angle (°)
AC-0	128.50 ± 3.40
AC-0.5	25.60 ± 5.30
AC-1	36.00 ± 2.80
AC-2	31.20 ± 1.30

Conversely, cotton samples subjected to plasma treatment demonstrated significant reductions in their contact angles. Specifically, fibres treated for durations of 30 s, 1 min, and 2 min exhibited an average water contact angle reduction of approximately 76% compared to the untreated sample. This transformation is consistent with findings in the literature, where plasma-induced surface oxidation introduces polar oxygen-functional groups such as hydroxyl (–OH), carbonyl (C=O), and carboxyl (–COOH), enhancing the fibre's hydrophilicity [64].

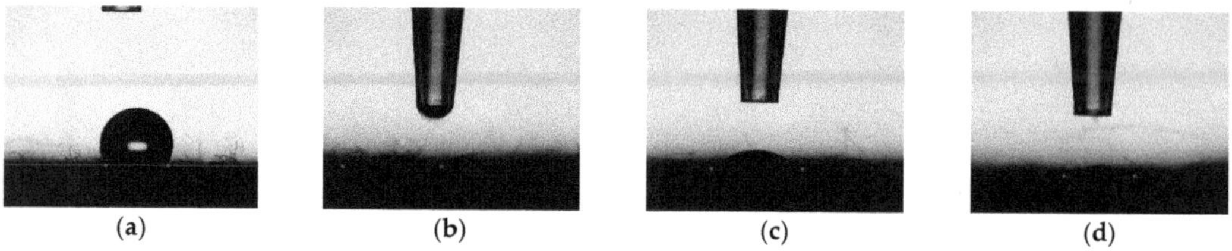

Figure 5. Contact angle images were taken for (**a**) AC-0, an untreated cotton sample; (**b**) AC-0.5, 30 s; (**c**) AC-1, 1 min; and (**d**) AC-2, 2 min of plasma activation treatment.

This considerable decrease underscores a transition to a markedly hydrophilic nature. For instance, Zemljic et al. reported a similar reduction in contact angles for cellulosic fabrics treated with oxygen plasma, achieving values as low as 15–20° [42]. This change can be attributed to the action of oxygen plasma, which introduces oxygen-functional groups on the fibre surface and potentially modifies its surface morphology by a surface oxidation process [43,65,66]. This transition from hydrophobic to hydrophilic character enhances cotton's interaction and affinity with water molecules [67].

Furthermore, the hydrophilic transformation achieved here aligns with practical applications requiring improved moisture management and dye uptake in textiles, confirming the effectiveness of plasma treatment for industrial purposes.

It is noteworthy that, regardless of the plasma exposure time, the treated samples exhibited similar hydrophilic behaviour, which underscores the efficiency of plasma treatment in inducing significant changes in cotton's surface properties within a short timeframe.

In this study, colour evaluations were carried out to assess potential colour changes in cotton samples subjected to low-pressure plasma treatments with oxygen in comparison to an untreated cotton control sample (AC-0). These colour changes were analysed in triplicate using the CIELAB colour space, characterised by L* (luminance), a* (red/green), and b* (yellow/blue) values, following the ISO 22700:2019 standard [68]. Table 5 shows the variations in colour parameters after plasma treatment compared to the untreated sample.

Table 5. CIELAB component values untreated and plasma-activated cotton samples in accordance with ISO 22700 standards.

CIELAB Components	AC-0	AC-0.5	AC-1	AC-2
L*	86.55	85.97	86.24	86.22
a*	1.29	1.41	1.36	1.34
b*	10.61	9.97	9.83	9.88
ΔE*	0.00	0.66	0.78	0.70

As shown in Table 5, the results showed no notable colour differences between the treated samples and the untreated control. The ΔE* (total colour difference) value was measured following the criteria established by the upper materials lab at INESCOP, with the threshold set at ≤ 2.5. The values obtained for the AC-0.5, AC-1, and AC-2 samples were 0.66, 0.78, and 0.70, respectively. It is important to note that these ΔE* values demonstrated a slight but non-significant increase, which is attributed to the exposure time to the low-pressure plasma treatment with oxygen [66,69].

The limited colour change can be attributed to the shallow penetration depth of the plasma treatment, which primarily modifies the surface layer of the fibres without affecting their bulk properties. This shallow penetration is consistent with findings in the literature regarding plasma technologies, where the treatment selectively enhances surface properties (e.g., hydrophilicity) while preserving the original aesthetic qualities of the material [70].

Such colour preservation is advantageous for practical applications, as it ensures that the visual appearance and colour consistency of the cotton fabric remain unchanged during plasma surface modification. This is particularly relevant for industrial applications where maintaining the fabric's aesthetic properties is essential, alongside enhancing its functional performance.

In addition, in the study of water absorption in cotton samples, the influence of low-pressure oxygen plasma treatment on fibre absorption capacity was investigated [71,72]. Cotton samples were categorised based on treatment exposure time: untreated (AC-0), 30 s (AC-0.5), 1 min (AC-1), and 2 min (AC-2). Each sample underwent absorption tests at 15, 30, and 60 min intervals.

Table 6 presents the results on the water absorption capabilities of cotton fibres treated with low-pressure oxygen plasma. The untreated sample (AC-0) displayed a gradual increase in absorption over time, moving from 35.62% at 15 min to 65.72% at 60 min. This progression illustrates the natural moisture-absorbing properties of cotton.

Table 6. Water absorption results obtained for untreated and plasma-activated cotton samples.

Sample	Water Absorption (%)		
	15 min	30 min	60 min
AC-0	35.62	55.47	65.72
AC-0.5	112.29	108.75	98.19
AC-1	118.45	115.51	119.64
AC-2	113.37	117.57	118.63

The plasma-treated samples, however, exhibited significantly enhanced absorption from the onset. For instance, AC-0.5 absorbed 112.29% of water at 15 min, which gradually decreased to 98.19% by the 60 min mark. While this could partially be attributed to saturation, the relatively large difference (13%) suggests additional factors. One plausible explanation is the redistribution of absorbed water within the fibre matrix, where water initially adheres to surface-modified regions and subsequently diffuses into less accessible areas of the fibre over time. Another potential factor involves structural rearrangements within the fibre matrix or partial relaxation of the plasma-modified surface, which could alter the water retention dynamics.

The samples treated for longer durations, AC-1 and AC-2, demonstrated even more remarkable absorption capacities, maintaining high levels of water retention throughout the 60 min. This indicates that extending the duration of plasma treatment not only enhances but also stabilises the hydrophilicity of the cotton fibres, allowing for sustained high water absorption rates over time [63]. Conversely, plasma-treated samples showed a marked improvement in absorption from the outset. Figure 6 shows images of untreated and plasma-activated cotton samples after performing the water absorption characterisation technique. The Figure compares the visual appearance of the samples, labelled as AC-0 (a), AC-0.5 (b), AC-1 (c), and AC-2 (d), highlighting the effects of plasma treatment on water absorption behaviour.

The significant increase in water absorption in plasma-treated samples is due to physical and chemical modifications in the fibres. Oxygen plasma adds hydrophilic functional groups to the cotton surface, improving water adhesion while also altering surface roughness and internal morphology, enhancing water retention [73].

Regarding the plasma exposure time, it shows an increase in absorption, but not proportionally. The sample treated for 30 s had similar absorption to those treated for longer, indicating that surface and morphological changes occur quickly. Longer exposure may intensify these changes without significantly improving water absorption. These findings highlight plasma treatment's potential for enhancing rapid liquid absorption in textiles.

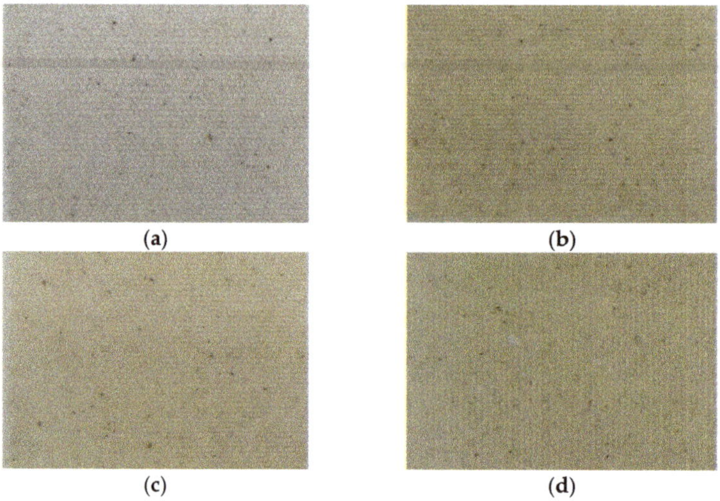

Figure 6. Untreated and plasma-activated samples after performing the water absorption characterisation technique. Being AC-0 (**a**), AC-0.5 (**b**), AC-1 (**c**), and AC-2 (**d**) samples, respectively.

Lastly, the data presented in Table 7 and the images in Figure 7 show the effects of surface activation treatments on the colour fastness to rubbing of cotton samples tested under dry and wet conditions. The samples were treated using low-pressure oxygen plasma for varying durations: 30 s for AC-0.5, 1 min for AC-1, and 2 min for AC-2. This treatment was aimed at enhancing the cotton's surface characteristics to improve dye fixation without the use of chemicals. Additionally, some samples were ironed after dyeing to assess the impact of ironing on dye fixation [74,75].

Table 7. Colour fastness to rubbing obtained results for the different treated and untreated samples.

Sample	Fixing Method	Test Method	Transfer	Deterioration	Aspect
AC-0	Unironed	Dry	5	3	Bleaching
		Wet	1/2	2/3	Bleaching
	Ironed	Dry	5	4	Bleaching
		Wet	2/3	2	Bleaching
AC-0.5	Unironed	Dry	5	3	Bleaching
		Wet	2/3	2	Bleaching
	Ironed	Dry	5	4	Bleaching
		Wet	3/4	2	Bleaching
AC-1	Unironed	Dry	5	4	Light bleaching
		Wet	4	3/4	Light bleaching
	Ironed	Dry	5	4	Light bleaching
		Wet	3/4	3/4	Light bleaching
AC-2	Unironed	Dry	5	4/5	Very light bleaching
		Wet	3/4	4/5	Very light bleaching
	Ironed	Dry	5	4/5	Light bleaching
		Wet	3/4	4	Light bleaching

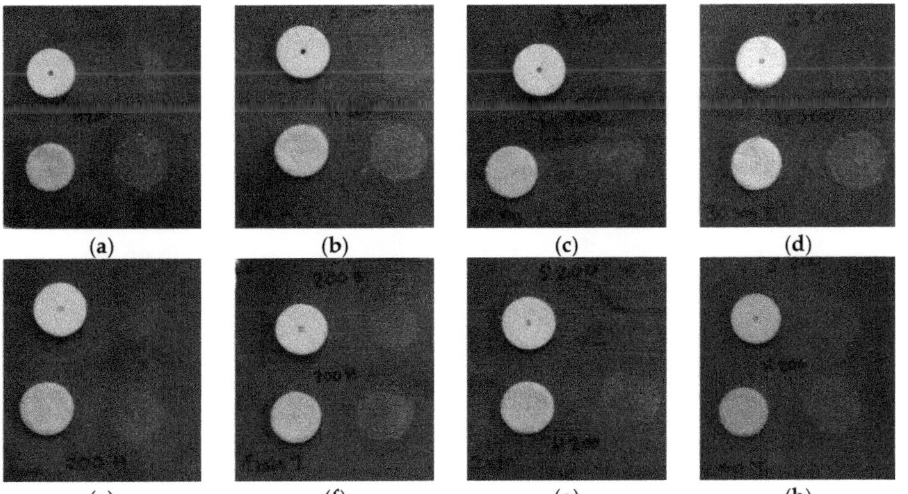

Figure 7. Colour fastness to rubbing characterisation of the different samples studied. (**a**) AC-0 Unironed, (**b**) AC-0 Ironed, (**c**) AC-0.5 Unironed, (**d**) AC-0.5 Ironed, (**e**) AC-1 Unironed, (**f**) AC-1 Ironed, (**g**) AC-2 Unironed, and (**h**) AC-2 Ironed. In each image, the top circular wool is dry, and the bottom circular wool is wet.

The results show consistent colour fastness under dry conditions, with all samples achieving a rating of 5, indicating strong dye fixation, unaffected by plasma treatment duration or ironing. However, under wet conditions, the untreated sample (AC-0) had poor fastness (ratings of 1/2 unironed, 2/3 ironed), with significant colour transfer and a whitish appearance. Plasma treatment improved wet fastness, with AC-0.5 rated 2/3 unironed and 3/4 ironed and AC-2 achieving 3/4 for both. The improved colour fastness in plasma-treated samples, such as AC-0.5 and AC-2, can be attributed to the surface modifications induced by the plasma, which increase surface roughness and introduce oxygen-functional groups like hydroxyl (–OH) and carbonyl (C=O), enhancing dye adhesion to the fibre. This result aligns with previous studies reporting that plasma surface treatments enhance dye fixation by improving fibre hydrophilicity and creating a more receptive surface for dye molecules [76–78]. Ironing also enhanced fastness, suggesting it helps set the dye. Thus, combining plasma treatment with thermal fixation may further stabilise the dye molecules on the fibre surface, reducing dye transfer in wet conditions [79]. In conclusion, both plasma treatment duration and ironing improve colour fastness, especially in wet conditions, enhancing dyed textile durability [60,62,80].

4. Conclusions

This study has comprehensively investigated the effects of low-pressure oxygen plasma treatment on cotton's surface chemistry, morphology, wettability, water absorption, and rubbing fastness. Employing a range of analytical techniques, including XPS and SEM, significant chemical and physical modifications in the cotton samples were observed.

On the one hand, the XPS analyses were particularly revealing, showing changes in the chemical composition, notably in the oxidation of cellulose, and a decrease in surface carbon content. These changes have enhanced the surface properties of cotton, expanding its practical applications in textiles requiring high hydrophilicity, such as improved dye fixation for the fashion and textile industries. Additionally, this treatment holds potential for medical textiles and absorbent products, where high water retention is essential.

On the other hand, SEM analysis, although more subtle, confirmed morphological alterations like increased surface roughness, which, in combination with the chemical modifications, contribute to the overall improved properties of cotton. This is particularly evident in the enhanced hydrophilicity and dyeing capacity of the treated fabric.

According to contact angle measurements and water absorption tests, the plasma-treated cotton functionally demonstrated an increase in hydrophilicity and water absorption capacity. Compared to conventional chemical treatments, this surface modification was achieved efficiently without introducing harmful chemicals, thereby reducing environmental impact and water consumption. These improvements are crucial for applications in sustainable textile manufacturing, addressing industry challenges such as reducing water pollution and resource consumption.

Furthermore, the rubbing fastness tests showed enhanced resistance to rubbing post-treatment in wet conditions as plasma exposure time increases, while colour difference analysis confirmed that these functional improvements do not compromise the cotton's visual appeal, with only minimal variations in ΔE^* values observed. The shallow penetration depth of the plasma treatment ensures that the surface properties are modified without affecting the bulk characteristics of the fabric, preserving its original aesthetic qualities.

In comparison to traditional chemical treatments, low-pressure oxygen plasma treatment emerges as a sustainable alternative, offering reduced chemical usage, energy efficiency, and improved material properties. This research highlights the practical and environmental advantages of plasma technology, positioning it as a promising solution for industrial applications in the fashion, textile, and technical sectors.

Supplementary Materials: The following supporting information can be downloaded at: https://www.mdpi.com/article/10.3390/surfaces7040071/s1, Table S1: Deconvoluted C1s and O1s peaks during the oxygen plasma activation process on cotton samples: (a) C1s spectrum of AC-0; (b) C1s spectrum of AC-0.5; (c) C1s spectrum of AC-1; (d) C1s spectrum of AC-2; (e) O1s spectrum of AC-0; (f) O1s spectrum of AC-0.5; (g) O1s spectrum of AC-1; (h) O1s spectrum of AC-2. Each spectrum provides insights into the chemical composition and bonding configurations at different activation levels.

Author Contributions: Conceptualization and investigation, V.M.S.-M. and C.R.-S.; methodology, V.M.S.-M. and C.R.-S.; validation and formal analysis, V.M.S.-M., C.R.-S. and C.H.-F.; writing—original draft preparation, V.M.S.-M. and C.R.-S.; writing—review and editing, V.M.S.-M., C.R.-S. and C.H.-F.; supervision, E.B.-G., F.A.-A. and E.O.-C. All authors have read and agreed to the published version of the manuscript.

Funding: This research was funded by the Valencian Institute for Business Competitiveness (IVACE) of the Generalitat Valenciana and the European Regional Development Fund (ERDF) under grant number IMDEEA/2024/42, SurfIAtech.

Data Availability Statement: The data presented in this study are available upon request from the corresponding author.

Conflicts of Interest: The authors declare no conflicts of interest.

References

1. Anis, M.; AlTaher, G.; Sarhan, W.; Elsemary, M. Textile Applications. In *Nanovate*; Springer International Publishing: Cham, Switzerland, 2017; pp. 163–181.
2. Sawant, J.; Guru, R.; Grewal, D.; Talekar, S.C.; Kulkarni, S.P. Sustainability in textiles: A critical review of eco-friendly practices and materials. *ShodhKosh J. Vis. Perform. Arts* **2024**, *5*, 85–97. [CrossRef]
3. Vitale, G.S.; Scavo, A.; Zingale, S.; Tuttolomondo, T.; Santonoceto, C.; Pandino, G.; Lombardo, S.; Anastasi, U.; Guarnaccia, P. Agronomic Strategies for Sustainable Cotton Production: A Systematic Literature Review. *Agriculture* **2024**, *14*, 1597. [CrossRef]
4. Peran, J.; Razic, S.E. Application of Atmospheric Pressure Plasma Technology for Textile Surface Modification. *Text. Res. J.* **2019**, *90*, 1174–1197. [CrossRef]
5. Förch, R.; Zhang, Z.; Knoll, W. Soft Plasma Treated Surfaces: Tailoring of Structure and Properties for Biomaterial Applications. *Plasma Process. Polym.* **2005**, *2*, 351–372. [CrossRef]
6. Zhang, C.; Zhao, M.; Wang, L.; Qu, L.; Men, Y. Surface Modification of Polyester Fabrics by Atmospheric-Pressure Air/He Plasma for Color Strength and Adhesion Enhancement. *Appl. Surf. Sci.* **2017**, *400*, 304–311. [CrossRef]
7. Shang, S.M. Process Control in Dyeing of Textiles. In *Process Control in Textile Manufacturing*; Elsevier Ltd.: Amsterdam, The Netherlands, 2012; pp. 300–338; ISBN 9780857090270.

8. Giolando, S.T.; Rapaport, R.A.; Larson, R.J.; Federle, T.W.; Stalmans, M.; Masscheleyn, P. Environmental Fate and Effects of DEEDMAC: A New Rapidly Biodegradable Cationic Surfactant for Use in Fabric Softeners. *Chemosphere* **1995**, *30*, 1067–1083. [CrossRef]
9. Johnson, P.; Trybala, A.; Starov, V.; Pinfield, V.J. Effect of Synthetic Surfactants on the Environment and the Potential for Substitution by Biosurfactants. *Adv. Colloid. Interface Sci.* **2021**, *288*, 102340. [CrossRef]
10. Huber, L.H. Ecological Behavior of Cationic Surfactants from Fabric Softeners in the Aquatic Environment. *J. Am. Oil Chem. Soc.* **1984**, *61*, 377–382. [CrossRef]
11. Weber, E.J.; Colón, D.; Baughman, G.L. Sediment-Associated Reactions of Aromatic Amines. 1. Elucidation of Sorption Mechanisms. *Environ. Sci. Technol.* **2001**, *35*, 2470–2475. [CrossRef] [PubMed]
12. Chen, X.; Ning, X.; Lai, X.; Wang, Y.; Zhang, Y.; He, Y. Chlorophenols in Textile Dyeing Sludge: Pollution Characteristics and Environmental Risk Control. *J. Hazard. Mater.* **2021**, *416*, 125721. [CrossRef]
13. Lara, L.; Cabral, I.; Cunha, J. Ecological Approaches to Textile Dyeing: A Review. *Sustainability* **2022**, *14*, 8353. [CrossRef]
14. Murugesh Babu, K. Overview on Significant Approaches to Waterless Dyeing Technology. In *Emerging Technologies for Textile Coloration*; CRC Press: Boca Raton, FL, USA, 2022; pp. 171–188; ISBN 9781003140467.
15. Zaman, M.; Liu, H.; Xiao, H.; Chibante, F.; Ni, Y. Hydrophilic Modification of Polyester Fabric by Applying Nanocrystalline Cellulose Containing Surface Finish. *Carbohydr. Polym.* **2013**, *91*, 560–567. [CrossRef] [PubMed]
16. GAJDZICKI, B. Softeners and Their Impact on the Functional Properties of Textile Products. *Text. Rev.-Fiber Cloth. Leather* **2019**, *1*, 32–39. [CrossRef]
17. Jhanji Dhir, Y. Natural Fibers: The Sustainable Alternatives for Textile and Non-Textile Applications. In *Natural Fiber*; IntechOpen: London, UK, 2022.
18. Brueckner, T.; Eberl, A.; Heumann, S.; Rabe, M.; Guebitz, G.M. Enzymatic and Chemical Hydrolysis of Poly(Ethylene Terephthalate) Fabrics. *J. Polym. Sci. A Polym. Chem.* **2008**, *46*, 6435–6443. [CrossRef]
19. Hsieh, Y.-L.; Cram, L.A. Enzymatic Hydrolysis to Improve Wetting and Absorbency of Polyester Fabrics. *Text. Res. J.* **1998**, *68*, 311–319. [CrossRef]
20. Daria, M.; Krzysztof, L.; Jakub, M. Characteristics of Biodegradable Textiles Used in Environmental Engineering: A Comprehensive Review. *J. Clean. Prod.* **2020**, *268*, 122129. [CrossRef]
21. Radetic, M.; Jovancic, P.; Puac, N.; Petrovic, Z.L. Environmental Impact of Plasma Application to Textiles. *J. Phys. Conf. Ser.* **2007**, *71*, 012017. [CrossRef]
22. Ruzafa-Silvestre, C.; Carbonell-Blasco, P.; Orgiles-Calpena, E.; Aran Ais, F. Low-Pressure Plasma Treatment Applied to Polymeric Materials for a Sustainable Footwear Industry. In *Proceedings of the 8th International Conference on Advanced Materials and Systems*; INCDTP—Leather and Footwear Research Institute (ICPI): Bucharest, Romania, 2020; pp. 455–460.
23. Caschera, D.; Cortese, B.; Mezzi, A.; Brucale, M.; Ingo, G.M.; Gigli, G.; Padeletti, G. Ultra Hydrophobic/Superhydrophilic Modified Cotton Textiles through Functionalized Diamond-Like Carbon Coatings for Self-Cleaning Applications. *Langmuir* **2013**, *29*, 2775–2783. [CrossRef]
24. Yacoub, A.A.; Patel, B.H. Atmospheric Plasma Technology for Sustainable Wet Processing of Textiles. *Mater. Today Proc.* **2024**. [CrossRef]
25. Domonkos, M.; Tichá, P. Low-Temperature Atmospheric Pressure Plasma Treatment in the Polymer and Textile Industry. *IEEE Trans. Plasma Sci.* **2023**, *51*, 1671–1681. [CrossRef]
26. Kakad, G.S.; Rathod, A.R.; Suman, B. Plasma Treatment for Textiles. *Man-Made Text. India* **2006**, *3*, 85–89.
27. Shahidi, S.; Ahmadi, M.; Rashidi, A.; Ghoranneviss, M. Effect of Plasma Treatment on Self-cleaning of Textile Fabric Using Titanium Dioxide. *Micro Nano Lett.* **2015**, *10*, 408–413. [CrossRef]
28. Petraconi, A.; Miranda, F.; Prado, E.; Braite, B.; Gasi, F.; Bittencourt, E.; Valadares, G.; Massi, M.; Petraconi, G.; da Silva Sobrinho, A. Hybrid Corona–Dielectric Barrier Discharge for Permethrin Polymerisation on Polyamide Fabric at Atmospheric Pressure. *Fibers Polym.* **2023**, *24*, 373–382. [CrossRef]
29. Cools, P.; Morent, R.; De Geyter, N. Plasma Modified Textiles for Biomedical Applications. In *Advances in Bioengineering*; InTech: Rijeka, Croatia, 2015.
30. Morais, D.; Guedes, R.; Lopes, M. Antimicrobial Approaches for Textiles: From Research to Market. *Materials* **2016**, *9*, 498. [CrossRef] [PubMed]
31. Shintani, H.; Sakudo, A.; Burke, P.; McDonnell, G. Gas Plasma Sterilization of Microorganisms and Mechanisms of Action. *Exp. Ther. Med.* **2010**, *1*, 731–738. [CrossRef] [PubMed]
32. Vohrer, U. *Plasma Technol. Text*; Woodhead Publishing, Limited: Cambridge, UK, 2007.
33. Gotmare, V.D.; Samanta Kartick, K.; Patil, V.; Basak, S.; Chattopdhayay, S.K. Surface Modification of Cotton Textile Using Low-Temperature Plasma. *Int. J. Bioresour. Sci.* **2015**, *2*, 37–45.
34. Teli, M.D.; Pandit, P.; Samanta, K.K. Application of Atmospheric Pressure Plasma Technology on Textile. *J. Text. Assoc.* **2015**, *75*, 422.
35. Legein, F.; Loulidi, S. Plasma Nanocoating of Nonwovens. *Vak. Forsch. Prax.* **2022**, *34*, 36–39. [CrossRef]
36. Yilma, B.B.; Luebben, J.F.; Nalankilli, G. Cold Plasma Treatment in Wet Chemical Textile Processing. *Fibres Text. East. Eur.* **2020**, *28*, 118–126. [CrossRef]

37. Ullah, M.H.; Akther, H.; Rahman, M.M.; Foisal, A.B.M.; Hasan, M.M.; Amir-Al Zumahi, S.M.; Amri, A. Surface Modification and Improvements of Wicking Properties and Dyeability of Grey Jute-Cotton Blended Fabrics Using Low-Pressure Glow Discharge Air Plasma. *Heliyon* **2021**, *7*, e07893. [CrossRef]
38. El-Khatib, E.M.; Raslan, W.M.; El-Halwagy, A.A.; Galab, S. Effect of Low Temperature Plasma Treatment on the Properties of Wool/Polyester Blend. *Res. J. Text. Appar.* **2013**, *17*, 124–132. [CrossRef]
39. Karahan, H.A.; Özdoğan, E. Improvements of Surface Functionality of Cotton Fibers by Atmospheric Plasma Treatment. *Fibers Polym.* **2008**, *9*, 21–26. [CrossRef]
40. Tudoran, C.; Roşu, M.; Coroş, M. A Concise Overview on Plasma Treatment for Application on Textile and Leather Materials. *Plasma Process. Polym.* **2020**, *17*, 2000046. [CrossRef]
41. Bhat, N.V.; Bharati, R.N.; Gore, A.V.; Patil, A.J. Effect of Atmospheric Pressure Air Plasma Treatment on Desizing and Wettability of Cotton Fabrics. *Indian J. Fibre Text. Res.* **2011**, *36*, 42–46.
42. Fras Zemljič, L.; Peršin, Z.; Stenius, P. Improvement of Chitosan Adsorption onto Cellulosic Fabrics by Plasma Treatment. *Biomacromolecules* **2009**, *10*, 1181–1187. [CrossRef] [PubMed]
43. Cai, Z.; Qiu, Y. Dyeing Properties of Wool Fabrics Treated with Atmospheric Pressure Plasmas. *J. Appl. Polym. Sci.* **2008**, *109*, 1257–1261. [CrossRef]
44. Peršin, Z.; Vesel, A.; Kleinschek, K.S.; Mozetič, M. Characterisation of Surface Properties of Chemical and Plasma Treated Regenerated Cellulose Fabric. *Text. Res. J.* **2012**, *82*, 2078–2089. [CrossRef]
45. Hwang, Y.J.; Mccord, M.G.; An, J.S.; Kang, B.C.; Park, S.W. Effects of Helium Atmospheric Pressure Plasma Treatment on Low-Stress Mechanical Properties of Polypropylene Nonwoven Fabrics. *Text. Res. J.* **2005**, *75*, 771–778. [CrossRef]
46. Jelil, R.A. A Review of Low-Temperature Plasma Treatment of Textile Materials. *J. Mater. Sci.* **2015**, *50*, 5913–5943. [CrossRef]
47. Silvestre, C.R.; Blasco, M.P.C.; López, S.R.; Aguilar, H.P.; Limiñana, M.Á.P.; Gil, E.B.; Calpena, E.O.; Ais, F.A. Hydrophobic Leather Coating for Footwear Applications by a Low-Pressure Plasma Polymerisation Process. *Polymers* **2021**, *13*, 3549. [CrossRef] [PubMed]
48. Randhawa, H. Review of Plasma-Assisted Deposition Processes. *Thin Solid Film.* **1991**, *196*, 329–349. [CrossRef]
49. Zaneline Instructions for Dyeing with Reactive Dyes. Available online: https://www.zaneline.com. (accessed on 23 October 2024).
50. Ruzafa-Silvestre, C.; Juan-Fernández, B.; Carbonell-Blasco, M.P.; Bañón-Gil, E.; Orgilés-Calpena, E.; Arán-Ais, F. Organosilicon-Based Plasma Nanocoating on Crust Leather for Water-Repellent Footwear. *Materials* **2022**, *15*, 7255. [CrossRef] [PubMed]
51. Carbonell-Blasco, M.P.; Ruzafa-Silvestre, C.; Mateu-Romero, B.; Orgilés-Calpena, E.; Arán-Ais, F. Cleaner Technologies to Minimise the Environmental Footprint of the Footwear Bonding Process. *Proc. Inst. Mech. Eng. C J. Mech. Eng. Sci.* **2024**, *238*, 2568–2582. [CrossRef]
52. *EN 828:2013*; Adhesives Wettability Determination by Measurement of Contact Angle and Surface Free Energy of Solid Surface. European Committee for Standardization (CEN): Brussels, Belgium, 2013.
53. *UNE-EN ISO 105-J03:2009*; Tests for Colour Fastness—Part J03: Calculation of Colour Differences. UNE: Madrid, Spain; ISO: Geneva, Switzerland, 2009.
54. *UNE-EN ISO 22649:2016*; Test Methods for Insoles and Insocks—Water Absorption and Desorption. UNE: Madrid, Spain; ISO: Geneva, Switzerland, 2016.
55. *ISO 17700:2019*; Test Methods for Upper Components and Insocks. Colour Fastness to Rubbing and Bleeding. UNE: Madrid, Spain; ISO: Geneva, Switzerland, 2019.
56. *UNE-EN ISO 105-E01:2013*; Tests for Colour Fastness—Part E01: Colour Fastness to Water. UNE: Madrid, Spain; ISO: Geneva, Switzerland, 2013.
57. *UNE-EN ISO 105-A03:2020*; Tests for Colour Fastness—Part A03: Grey Scale for Assessing Staining. UNE: Madrid, Spain; ISO: Geneva, Switzerland, 2020.
58. Caschera, D.; Mezzi, A.; Cerri, L.; de Caro, T.; Riccucci, C.; Ingo, G.M.; Padeletti, G.; Biasiucci, M.; Gigli, G.; Cortese, B. Effects of Plasma Treatments for Improving Extreme Wettability Behavior of Cotton Fabrics. *Cellulose* **2014**, *21*, 741–756. [CrossRef]
59. Kan, C.-W.; Man, W.-S. Surface Characterisation of Atmospheric Pressure Plasma Treated Cotton Fabric—Effect of Operation Parameters. *Polymers* **2018**, *10*, 250. [CrossRef] [PubMed]
60. McCoustra, M.R.S.; Mather, R.R. Plasma Modification of Textiles: Understanding the Mechanisms Involved. *Text. Prog.* **2018**, *50*, 185–229. [CrossRef]
61. Udomsap, P.; Meesiri, S.; Chollacoop, N.; Eiad-Ua, A. Biomass Nanoporous Carbon-Supported Pd Catalysts for Partial Hydrogenation of Biodiesel: Effects of Surface Chemistry on Pd Particle Size and Catalytic Performance. *Nanomaterials* **2021**, *11*, 1431. [CrossRef] [PubMed]
62. Wang, X.; Zhao, H.; Chen, F.; Ning, X.; Chen, S.; Guan, Q.; Jiang, S.; Miao, D. The Application of Atmospheric Plasma for Cotton Fabric Desizing. *Fibers Polym.* **2019**, *20*, 2334–2341. [CrossRef]
63. Liu, Y. Chemical Composition and Characterization of Cotton Fibers. In *Cotton Fiber: Physics, Chemistry and Biology*; Springer International Publishing: Cham, Switzerland, 2018; pp. 75–94.
64. Dufour, T. From Basics to Frontiers: A Comprehensive Review of Plasma-Modified and Plasma-Synthesized Polymer Films. *Polymers* **2023**, *15*, 3607. [CrossRef]

65. Urgilés, I.; Fabián, D. Activación Superficial de Polietileno, Mediante Plasma Frío de Baja Presión Como Pre-Tratamiento Para Mejorar Adhesión de Tintes en Impresiones. Bachelor's Thesis, Universidad Politécnica Salesiana, Cuenca, Spain, 2014.
66. Rombaldoni, F.; Montarsolo, A.; Mossotti, R.; Innocenti, R.; Mazzuchetti, G. Oxygen Plasma Treatment to Reduce the Dyeing Temperature of Wool Fabrics. *J. Appl. Polym. Sci.* **2010**, *118*, 1173–1183. [CrossRef]
67. Oliveira, F.R.; Steffens, F.; de Holanda, F.S.B.; do Nascimento, J.H.O.; Matsui, K.N.; Souto, A.P. Physical, Chemical and Morphological Characterization of Polyamide Fabrics Treated with Plasma Discharge. *Mater. Res.* **2017**, *20*, 60–68. [CrossRef]
68. *ISO 22700:2019*; Measuring the Colour and Colour Difference of Finished Leather. ISO: Geneva, Switzerland, 2009.
69. Kan, C.; Cheung, H.; Chan, Q. A Study of Plasma-Induced Ozone Treatment on the Colour Fading of Dyed Cotton. *J. Clean. Prod.* **2016**, *112*, 3514–3524. [CrossRef]
70. Žigon, J.; Pavlič, M.; Kibleur, P.; Van den Bulcke, J.; Petrič, M.; Van Acker, J.; Dahle, S. Treatment of Wood with Atmospheric Plasma Discharge: Study of the Treatment Process, Dynamic Wettability and Interactions with a Waterborne Coating. *Holzforschung* **2021**, *75*, 603–613. [CrossRef]
71. Pérez-Limiñana, M.A.; Pérez-Aguilar, H.; Ruzafa-Silvestre, C.; Orgilés-Calpena, E.; Arán-Ais, F. Effect of Processing Time of Steam-Explosion for the Extraction of Cellulose Fibers from Phoenix Canariensis Palm Leaves as Potential Renewable Feedstock for Materials. *Polymers* **2022**, *14*, 5206. [CrossRef] [PubMed]
72. Deshmukh, R.R.; Bhat, N.V. Pretreatments of Textiles Prior to Dyeing: Plasma Processing. In *Textile Dyeing*; InTech: London, UK, 2011.
73. Ghimire, B.; Subedi, D.P.; Khanal, R. Improvement of Wettability and Absorbancy of Textile Using Atmospheric Pressure Dielectric Barrier Discharge. *AIP Adv.* **2017**, *7*, 085213. [CrossRef]
74. Iftikhar, M.; Jamil, N.A.; Shahbaz, B. Rubbing, Ironing and Dry Cleaning Fastness of Reactive Dyed Cotton Knitted Fabric as Influenced by Salt, Alkali and Dye. *Int. J. Agric. Biol.* **2001**, *1*, 109–112.
75. Yasukawa, R.; Higashitani, H.; Yasunaga, H.; Urakawa, H. Dye Fixation Process in Ink-Jet Printing of Cotton Fabric by Reactive Dye. *Sen'i Gakkaishi* **2008**, *64*, 113–117. [CrossRef]
76. Haji, A.; Kan, C.-W. Plasma Treatment for Sustainable Functionalization of Textiles. In *Green Chemistry for Sustainable Textiles*; Elsevier: Amsterdam, The Netherlands, 2021; pp. 265–277.
77. Vankar, P.S.; Gangwar, A. Natural Dyeing Mediated by Atmospheric Air Pressure Plasma Treatment of Polyester. *Pigment Resin Technol.* **2024**, *53*, 569–575. [CrossRef]
78. Yari, M.R.; Zakerhamidi, M.S.; Ghomi, H. Plasma Immobilization of Azobenzene Dye on Polyamide 6 Polymer. *Sci. Rep.* **2023**, *13*, 983. [CrossRef] [PubMed]
79. Broadbent, A.D.; Thérien, N.; Zhao, Y. Comparison of the Thermal Fixation of Reactive Dyes on Cotton Using Infrared Radiation or Hot Air. *Ind. Eng. Chem. Res.* **1998**, *37*, 1781–1785. [CrossRef]
80. Kan, C.W.; Yuen, C.W.M. Effect of Atmospheric Pressure Plasma Treatment on the Desizing and Subsequent Colour Fading Process of Cotton Denim Fabric. *Color. Technol.* **2012**, *128*, 356–363. [CrossRef]

MDPI AG
Grosspeteranlage 5
4052 Basel
Switzerland
Tel.: +41 61 683 77 34

Surfaces Editorial Office
E-mail: surfaces@mdpi.com
www.mdpi.com/journal/surfaces

www.ingramcontent.com/pod-product-compliance
Lightning Source LLC
Chambersburg PA
CBHW061555210926
51850CB00009B/2114
9783725860173